SOLAR, STELLAR AND GALACTIC CONNECTIONS BETWEEN PARTICLE PHYSICS AND ASTROPHYSICS

ASTROPHYSICS AND SPACE SCIENCE PROCEEDINGS

SOLAR, STELLAR AND GALACTIC CONNECTIONS BETWEEN PARTICLE PHYSICS AND ASTROPHYSICS

Edited by

ALBERTO CARRAMIÑANA

Instituto Nacional de Astofísica,
Óptica y Electrónica, Tonantzintla,
México

FRANCISCO SIDDHARTHA GUZMÁN

Instituto de Física y Matemáticas, Universidad
Michoacana de San Nicolás de Hidalgo,
México

and

TONATIUH MATOS

Centro de Investigación y Estudios Avanzados del IPN,
México DF, México

Springer

A C.I.P. Catalogue record for this book is available from the Library of Congress.

ISBN-13 978-90-481-7399-0
ISBN-10 1-4020-5575-7 (e-book)
ISBN-13 978-1-4020-5575-1 (e-book)

Published by Springer,
P.O. Box 17, 3300 AA Dordrecht, The Netherlands.

www.springer.com

Printed on acid-free paper

Contents

Preface vii

List of Contributors ix

Part I Extended Topics

Nuclear Astrophysics: Evolution of Stars from Hydrogen Burning to Supernova Explosion
K. Langanke 3

Pulsars as Probes of Relativistic Gravity, Nuclear Matter, and Astrophysical Plasmas
James M. Cordes 43

Theory of Gamma-Ray Burst Sources
Enrico Ramirez-Ruiz 77

Understanding Galaxy Formation and Evolution
Vladimir Avila-Reese 115

Ultra-high Energy Cosmic Rays: From GeV to ZeV
Gustavo Medina Tanco 165

Part II Astronomical Technical Reviews

Radio Astronomy: The Achievements and the Challenges
Luis F. Rodríguez 199

Gamma-ray Astrophysics - Before GLAST
Alberto Carramiñana 215

Gravitational Wave Detectors: A New Window to the Universe
Gabriela González, for the LIGO Scientific Collaboration 231

Part III Research Short Contributions

Hybrid Extensive Air Shower Detector Array at the University of Puebla to Study Cosmic Rays
O. Martínez, E. Pérez, H. Salazar, L. Villaseñor 243

Search for Gamma Ray Bursts at Sierra Negra, México
H. Salazar, L. Villaseñor, C. Alvarez, O. Martínez 253

Are There Strangelets Trapped by the Geomagnetic Field?
J.E. Horvath, G.A. Medina Tanco, L. Paulucci 263

Late Time Behavior of Non Spherical Collapse of Scalar Field Dark Matter
Argelia Bernal, F. Siddhartha Guzmán 271

Inhomogeneous Dark Matter in Non-trivial Interaction with Dark Energy
Roberto A. Sussman, Israel Quiros and Osmel Martín González 279

Mini-review on Scalar Field Dark Matter
L. Arturo Ureña-López 295

Index 303

Preface

The very small and the very large are intimately connected in Nature. Particle physics and astrophysics meet in fundamental questions: the structure and evolution of stars; their end and how this is manifested; how we think galaxies are created from matter we have yet to discover and why we believe the most energetic particles cannot come from the most distant universe.

During the IV Escuela Mexicana de Astrofísica (EMA-2005), held in the beautiful colonial city of Morelia between 18 and 23 July 2005, we reviewed and explored the numerous connections between astrophysics and particle physics. The core of the school program, aimed to advanced postgraduated students and young researchers in physics and astrophysics, was formed by half a dozen extended lecture courses delivered by recognized experts in their fields. The written versions of these courses became the main substance of this book. Three review talks were devoted to the techniques and results of novel astronomical windows of the XX and XXI centuries: radioastronomy, gamma-ray astronomy and gravitational wave astronomy. This volume includes also six short contributions, presented as single talks during the EMA-2005, examples of experimental and theoretical research work presently conducted in México and Latin-America.

This book is the final product of a two year process centered on the EMA-2005. We believe it will serve as a guide not just to the participants but also to the communities of all interrelated fields.

As editors and organizers of the EMA-2005 we are grateful to the sponsors institutions:
- Centro de Investigación y Estudios Avanzados (CINVESTAV) of the Instituto Politécnico Nacional;
- Consejo Estatal de Ciencia y Tecnología (COECyT) del Estado de Michoacán;
- Instituto de Físico de la Universidad de Guanajuato (IFUG);
- Instituto Nacional de Astrofísica, Óptica y Electrónica (INAOE);
- la Universidad Michoacana de San Nicolás de Hidalgo (UMSNH);
- la Universidad Nacional Autónoma de México (UNAM).

The Organizing Committee, chaired by Tonatiuh Matos (CINVESTAV), included Vladimir Ávila-Reese (UNAM), Ricardo Becerril (UMSNH), Alberto Carramiñana (INAOE), José García (IFUG), Efraín Chávez (UNAM), Jorge Hirsch (UNAM), Lukas Nellen (UNAM), Dany Page (UNAM), Luis Felipe Rodríguez (UNAM), José Valdés Galicia (UNAM) and Arnulfo Zepeda (CINVESTAV). The government of the State of Michoacán was very supportive of this event and is specially thanked for taking charge of the splendid Conference Dinner.

Morelia, Michoacán,
June 2006

Alberto Carramiñana
Francisco Siddhartha Guzmám
Tonatiuh Matos

List of Contributors

Karlheinz Langanke
Technische Universität Darmstadt, D-64291 Darmstadt, Germany
k.langanke@gsi.de

James M. Cordes
Cornell University, Ithaca NY 14853, USA
cordes@astro.cornell.edu

Enrico Ramirez-Ruiz
Institute for Advanced Study, Einstein Drive, Princeton NJ 08540, USA
enrico@ias.edu

Vladimir Ávila-Reese
Instituto de Astronomía, Universidad Nacional Autónoma de México, AP 70-264, México DF 04510
avila@astroscu.unam.mx

Gustavo Medina Tanco
Instituto Astronômico e Geofísico, USP, Brazil
Instituto de Ciencias Nucleares, UNAM, México
gmtanco@gmail.com

Luis Felipe Rodríguez
Centro de Radioastronomía y Astrofísica, UNAM,
Morelia, Michoacán 58089, México
l.rodriguez@astrosmo.unam.mx

Alberto Carramiñana
Instituto Nacional de Astrofísica, Óptica y Electrónica,
Tonantzintla, Puebla 72840, México
alberto@inaoep.mx

Gabriela González - for the LIGO Scientific Collaboration
Department of Physics and Astronomy, Louisiana State University
202 Nicholson Hall, Tower Drive, Baton Rouge, LA 70803, USA
gonzalez@lsu.edu

Oscar Martínez
Facultad de Ciencias Físico-Matemáticas, Benemérita Universidad Autónoma de Puebla, Puebla, Puebla 72570, México
omartin@fcfm.buap.mx

E. Perez
Facultad de Ciencias Físico-Matemáticas, Benemérita Universidad

Autónoma de Puebla, Puebla,
Puebla 72570, México

Humberto Salazar
Facultad de Ciencias Físico-Matemáticas, Benemérita Universidad
Autónoma de Puebla, Puebla,
Puebla 72570, México
hsalazar@fcfm.buap.mx

Luis Villaseñor
Instituto de Física y Matemáticas, Universidad
Michoacana de San Nicolás de Hidalgo.
Edificio C3, Cd. Universitaria.
Morelia Michoacán, 58040 México
villasen@ifm.umich.mx

Cesar Álvarez
Facultad de Ciencias Físico-Matemáticas, Benemérita Universidad
Autónoma de Puebla, Puebla,
Puebla 72570, México
calvarez@fcfm.buap.mx

Jorge Horvath
Instituto de Astronomia, Geofísica e Ciências
Atmosféricas IAG/USP, Rua do Matão, 1226, 05508-900 São Paulo SP, Brazil
foton@astro.iag.usp.br

L. Paulucci
Instituto de
Física, Universidade de São Paulo, Rua do Matão, Travessa
R, 187. CEP 05508-090 Ciudade Universitária, São Paulo - Brazil
paulucci@fma.if.usp.br

Argelia Bernal
Departmento de Física, Centro De Investigación y De
Estudios Avanzados Del IPN, AP 14-740,07000 México D.F., México
abernalresca@fis.cinvestav.mx

Francisco Siddhartha Guzmán
Instituto de Física y Matemáticas, Universidad
Michoacana de San Nicolás de Hidalgo.
Edificio C3, Cd. Universitaria.
Morelia Michoacán, 58040 México
guzman@ifm.umich.mx

Roberto Sussman
Instituto de Ciencias Nucleares, Apartado Postal
70543, UNAM, México DF, 04510, México
sussman@nuclecu.unam.mx

Israel Quiros
Departamento de Física, Universidad Central de las Villas, Santa Clara, Cuba
israel@uclv.edu.cu

Osmel Martín
Departamento de Física, Universidad Central de las Villas, Santa Clara, Cuba
osmel@uclv.edu.cu

Luis Arturo Ureña-López
Instituto de Física de la Universidad de Guanajuato, A.P. E-143,
C.P. 37150, León, Guanajuato, México.
lurena@fisica.ugto.mx

Part I

Extended Topics

Nuclear Astrophysics: Evolution of Stars from Hydrogen Burning to Supernova Explosion

K. Langanke

Gesellschaft für Schwerionenforschung and
Technische Universität Darmstadt,
D-64291 Darmstadt, Germany

1 Introduction

Nuclear astrophysics has developed in the last twenty years into one of the most important subfields of 'applied' nuclear physics. It is a truly interdisciplinary field, concentrating on primordial and stellar nucleosynthesis, stellar evolution, and the interpretation of cataclysmic stellar events like novae and supernovae.

The field has been tremendously stimulated by recent developments in laboratory and observational techniques. In the laboratory the development of radioactive ion beam facilities as well as low-energy underground facilities have allowed to remove some of the most crucial ambiguities in nuclear astrophysics arising from nuclear physics input parameters. This work has been accompanied by significant progress in nuclear theory which makes it now possible to derive some of the input at stellar conditions based on microscopic models. Nevertheless, much of the required nuclear input is still insufficiently known. Here, decisive progress is expected once radioactive ion beam facilities of the next generation, like the one at GSI, are operational. The nuclear progress goes hand-in-hand with tremendous advances in observational data arising from satellite observations of intense galactic gamma-sources, from observation and analysis of isotopic and elemental abundances in deep convective Red Giant and Asymptotic Giant Branch stars, and abundance and dynamical studies of nova ejecta and supernova remnants. Recent breakthroughs have also been obtained for measuring the solar neutrino flux, giving clear evidence for neutrino oscillations and confirming the solar models. Also, the latest developments in modeling stars, novae, x-ray bursts, type I supernovae, and the identification of the neutrino wind driven shock in type II supernovae as a possible site for the r-process allow now much better predictions from nucleosynthesis calculations to be compared with the observational data.

A. Carramiñana et al. (eds.), Solar, Stellar and Galactic Connections between Particle Physics and Astrophysics, 3–41.

It is impossible to present all these exciting developments in a set of four hour-long lectures. We will rather focus on a classical aspect, the evolution of massive stars in hydrostatic equilibrium from the initial hydrogen burning phase up to their cataclysmic final fate as a core-collapse supernova. This means, however, that other important aspects of nuclear astrophysics have to be omitted. These include evolution of binary systems and their related nucleosynthesis (novae, x-ray bursters, type Ia supernovae), nucleosynthesis beyond iron (s-process, r-process, p-process) or big-bang nucleosynthesis. For the interested reader I will at least point to some excellent recent reviews which discuss aspects of nuclear astrophysics. We mention here a few:

- General Nucleosynthesis: G. Wallerstein *et al.*, Rev. Mod. Phys. 69 (1997) 795; M. Arnould and K. Takahashi, Rep. Progr. Phys. 62 (1999) 395; F. Käppeler, F.-K. Thielemann and M. Wiescher, Annu. Rev. Nucl. Part. Sci. 48 (1998) 175
- Core-collapse supernovae: H.Th. Janka, K. Kifonidis and M. Rampp, in *Physics of Neutron Star Interiors*; eds. D. Blaschke, N.K. Glendenning and A. Sedrakian, Lecture Notes in Physics 578 (Springer, Berlin) 333; A. Burrows, Prog. Part. Nucl. Phys. 46 (2001) 59; H.A. Bethe, Rev. Mod. Phys. 62 (1990) 801
- Type-Ia supernovae: W. Hillebrandt and J.C. Niemeyer, Annu. Rev. Astron. Astrophys. 38 (2000) 191
- S-process: F. Käppeler, Prog. Part. Nucl. Phys. 43 (1999) 419; M. Busso, R. Gallino and G.J. Wasserburg, Annu. Rev. Astron. Astrophys. 37 (1999) 239
- R-process: J.J. Cowan, F.-K. Thielemann and J.W. Truran, Phys. Rep. 208 (1991) 267; Y.-Z. Qian, Prog. Part. Nucl. Phys. 50 (2003) 153

Of course, it is still very much recommended to read the two pioneering papers: E.M. Burbidge, G.R. Burbidge, W.A. Fowler and F. Hoyle, Rev. Mod. Phys. 29 (1957) 547 and A.G.W. Cameron, *Stellar Evolution, Nuclear Astrophysics, and Nucleogenesis*, Report CRL-41, Chalk River, Ontario.

2 The Nuclear Physics Input

2.1 Rate Equations and Reaction Rates

Nuclear reactions play an essential role in the evolution of a star and in many other astrophysical scenarios. Obviously, they change the chemical composition of the environment in a manner that can be described by a set of rate equations,

$$\frac{\delta Y_i}{\delta t} = \sum_j C^i_j Y_j + \sum_{j,k} C^i_{jk} Y_j Y_k - \sum_{j,k} C^k_{ij} Y_i Y_j \tag{1}$$

where Y_i is the relative abundance, by number, of the nuclide i. Alternatively, the rate equation can be expressed in terms of the mass fraction X_i of a nuclide, which is related to the relative abundance via $X_i = A_i Y_i$, where A_i is the number of nucleons in the nuclide i. For a complete description of the astrophysical scenarios with which we are concerned in the chapter, the rate equations have to be supplemented by equations that, in the case of a star, describe energy and momentum conservation, energy transport, the state of matter, etc., or in the early universe, the time evolution of the temperature.

The coefficients C in Eq. (1) are the rate constants. In the case of the destruction of the nuclide j, as in photodissociation ($\gamma + j \rightarrow i + y$), the nuclide i will be generated and the coefficient C^i_j is positive. Similarly, the nuclide i can either be generated ($e^- + j \rightarrow i + \nu$) or destroyed ($e^- + i \rightarrow j' + \nu$) by electron capture. Correspondingly, the coefficients C^i_j would be positive or negative. In two-body reactions, the nuclide i can be produced ($j + k \rightarrow i + ...$) or destroyed ($i + j \rightarrow k + ...$). The (positive) rate coefficients are then given by

$$
\begin{aligned}
C^i_{jk} &= \frac{\rho(1+\delta_{jk})}{N_j N_k m_u} R_{jk} = \frac{\rho}{m_u} \langle \sigma v \rangle_{jk} \\
C^k_{ij} &= \frac{\rho(1+\delta_{ij})}{N_i N_j m_u} R_{ij} = \frac{\rho}{m_u} \langle \sigma v \rangle_{ij}
\end{aligned}
\tag{2}
$$

where ρ is the (local) mass density, $m_u = 931.502$ MeV is the atomic mass unit, and N_i is the number density of nuclide i. To derive an expression for the nuclear reaction rates R_{ij}, consider a process in which a projectile nucleus X reacts with a target nucleus $Y (X + Y \rightarrow ...)$. The cross section for this reaction depends on the relative velocity v of the two nuclei and is given by $\sigma(v)$. The number densities of the two species in the environment are N_x and N_y. Then, the nuclear reaction rate R_{xy} is simply the product of the effective reaction area $(\sigma(v) \cdot N_y)$ spanned by the target nuclei and the flux of projectile nuclei $(N_x \cdot v)$. Thus

$$
R_{xy} = \frac{1}{1+\delta_{xy}} N_x N_y \langle \sigma(v) v \rangle \tag{3}
$$

where we have taken account of the distribution of velocities of target and projectile nuclei in the astrophysical environment. Thus, the product $\sigma(v)v$ has to be averaged over the distribution of target and projectile velocities, as indicated by the $\langle\rangle$ brackets in Eq. (3). The Kronecker-symbol avoids double-counting for identical projectile and target nuclei. Sometimes, three-body reactions, like the fusion of 3α-particles to ^{12}C (see Section 3), play a role in the nuclear network requiring the rate equations (1) to be modified appropriately.

In all applications with which we are concerned below, the velocity distribution of the nuclei is well described by a Maxwell-Boltzmann distribution characterized by some temperature T. Then one has ($E = \frac{\mu}{2} v^2$) [1]

$$\langle \sigma(v)v \rangle = \left(\frac{8}{\pi\mu}\right)^{1/2} \left(\frac{1}{kT}\right)^{3/2} \int_0^\infty \sigma(E)E \quad \exp \quad \left(-\frac{E}{kT}\right) dE \tag{4}$$

The mean lifetime $\tau_y(X)$ of a nucleus X against destruction by the nucleus Y in a given environment is then defined as [1]

$$\tau_y(X) = \frac{1}{N_y \langle \sigma v \rangle} \tag{5}$$

2.2 Neutron-Induced Reactions

The interstellar medium (ISM) from which a star forms by gravitational condensation contains only ($Z \geq 1$) nuclei. Because the neutron half-life is about 10 minutes, which is short on most astrophysical time scales, the ISM does not contain *free* neutrons. However, neutrons are produced in stellar evolution stages by (α, n) reactions like $^{13}C(\alpha, n)^{16}O$ and $^{22}Ne(\alpha, n)^{25}Mg$ (Section 3). These neutrons thermalize very quickly in a star and can therefore also be described by Maxwell-Boltzmann distributions.

At low energies, nonresonant neutron-induced reactions are dominated by *s*-wave capture and the cross section σ_n approximately follows a $1/v$ law [2]. Thus, $\langle \sigma_n v \rangle \approx$ constant. At somewhat higher energies, partial waves with $l > 0$ may contribute. To account for these contributions, the product $\sigma_n v$ may conveniently be expanded in a MacLaurin series in powers of $E^{1/2}$,

$$\sigma_n v = S(0) + \dot{S}(0)E^{1/2} + \frac{1}{2}\ddot{S}(0)E \tag{6}$$

resulting in

$$\langle \sigma_n v \rangle = S(0) + \sqrt{\frac{4}{\pi}}\dot{S}(0)(kT)^{1/2} + \frac{3}{4}\ddot{S}(0)kT \tag{7}$$

where the parameters $S(0)$, $\dot{S}(0)$, $\ddot{S}(0)$ (the dots indicate derivatives with respect to $E^{1/2}$) have to be determined from experiment (or theory).

2.3 Nonresonant Charged-Particle Reactions

During the hydrostatic burning stages of a star, charged-particle reactions most frequently occur at energies far below the Coulomb barrier, and are possible only via the *tunnel effect*, the quantum mechanical possibility of penetrating through a barrier at a classically forbidden energy. At these low energies, the cross section $\sigma(E)$ is dominated by the penetration factor,

$$P(E) = \frac{|\ \psi(R_n)\ |^2}{|\ \psi(R_c)\ |^2} \tag{8}$$

the ratio of the squares of the nuclear wave functions at the sum of the nuclear radii, R_n (several fermis), and at the classical turning point R_c. By solving the Schrödinger equation for s-wave ($l = 0$) particles interacting via the Coulomb potential of two point-like charges

$$V(r) = \frac{Z_1 Z_2 e^2}{r} \tag{9}$$

one obtains [3]

$$P = \exp\left\{-2KR_c\left[\frac{\arctan\left(\frac{R_c}{R_n}-1\right)^{1/2}}{\left(\frac{R_c}{R_n}-1\right)^{1/2}} - \frac{R_n}{R_c}\right]\right\} \tag{10}$$

with

$$K = \sqrt{\frac{2\mu}{\hbar^2}\left[V(R_n) - E\right]} \tag{11}$$

Expression (10) simplifies significantly in most astrophysical applications, for which $E \ll V(R_n)$ or, relatedly, $R_c \gg R_n$. In these limits one obtains the well-known expression

$$P(E) = \exp\left(-\frac{2\pi Z_1 Z_2 e^2}{\hbar v}\right) \equiv \exp\left[-2\pi\eta(E)\right] \tag{12}$$

where $\eta(E)$ is often called the Sommerfeld parameter. In numerical units,

$$2\pi\eta(E) = 31.29 Z_1 Z_2 \sqrt{\frac{\mu}{E}} \tag{13}$$

where the energy E is defined in keV.

For the following discussion it is convenient and customary to redefine the cross section in terms of the astrophysical S-factor by factoring out the known energy dependences of the penetration factor (12) and the de Broglie factor, in the *model-independent* way,

$$S(E) = \sigma(E)(E)\exp\left[2\pi\eta(E)\right] \tag{14}$$

For low-energy, nonresonant reactions, the astrophysical S-factor should have only a weak energy dependence that reflects effects arising from the strong interaction, as from antisymmetrization, and from small contributions from partial waves with $l > 0$ and for the finite size of the nuclei. The validity of this approach has been justified in numerous (nonresonant) nuclear reactions for which the experimentally determined S-factors show only weak E-dependences at low energies. For heavier nuclei, the S-factor becomes somewhat more energy-dependent because of the finite-size effects, especially as E is increased.

Equation (4) may be rewritten in terms of the astrophysical S-factor

$$\langle \sigma v \rangle = \left(\frac{8}{\pi \mu} \right)^{1/2} \left(\frac{1}{kT} \right)^{3/2} \int_0^\infty S(E) \left[-\frac{E}{kT} - 2\pi\eta(E) \right] dE \tag{15}$$

For typical applications in hydrostatic stellar burning, the product of the two exponentials forms a peak ("Gamow-peak") which may be well approximated by a Gaussian,

$$\exp\left\{ -\frac{E}{kT} - 2\pi\eta(E) \right\} \cong I_{\mathrm{max}} \exp\left\{ -\left(\frac{E - E_0}{\Delta/2} \right)^2 \right\} \tag{16}$$

with [1]

$$E_0 = 1.22(Z_1^2 Z_2^2 \mu T_6^2)^{1/3} [\mathrm{keV}] \tag{17}$$

$$\Delta = \frac{4}{\sqrt{3}} \sqrt{E_0 kT}$$

$$= 0.749(Z_1^2 Z_2^2 \mu T_6^5)^{1/6} [\mathrm{keV}] \tag{18}$$

$$I_{\mathrm{max}} = \exp\left\{ -\frac{3E_0}{kT} \right\} \tag{19}$$

T_6 measures the temperature in units of 10^6 K.

Examples of E_0, Δ and I_{max}, evaluated for some nuclear reactions at the solar core temperature ($T_6 \approx 15.6$), are summarized in Table 1.

We conclude from Table 1 that the reactions operate in relatively narrow energy windows around the astrophysically most effective energy E_0. Furthermore, it becomes clear from inspecting the different I_{max} values that reactions of nuclei with larger charge numbers effectively cannot occur in the sun as, for these reactions, even the solar core is far too cold.

However, it usually turns out that the astrophysically most effective energy E_0 is smaller than the energies at which the reaction cross section can be measured directly in the laboratory. Thus for astrophysical applications, an extrapolation of the measured cross section to stellar energies is usually necessary, often over many orders of magnitude.

In the case of nonresonant reactions, the extrapolation can be safely performed in terms of the astrophysical S-factor, because of its rather weak energy dependence. One can then expand the S-factor in terms of a MacLaurin expansion in powers of E,

$$S(E) = S(0) + S(0)E + \frac{1}{2}\ddot{S}(0)E^2 + ... \tag{20}$$

Using Eq. (20) and correcting for slight asymmetries from the Gaussian approximation (16) one finds

$$\langle\sigma v\rangle = \left(\frac{2}{\mu}\right)^{1/2} \frac{\Delta}{(kT)^{3/2}} S_{\text{eff}}(E_0) \exp\left(-\frac{3E_0}{kT}\right) \tag{21}$$

with [4, 5]

$$S_{\text{eff}}(E_0) = S(0)\left[1 + \frac{5kT}{36E_0} + \frac{\dot{S}(0)}{S(0)}\left(E_0 + \frac{35}{36}kT\right) + \frac{1}{2}\frac{\ddot{S}(0)}{S(0)}\left(E_0^2 + \frac{89}{36}E_0 kT\right)\right] \tag{22}$$

From Eqs. (17), (18), (21) and (22), $\langle\sigma v\rangle$ can be written in terms of temperature alone:

$$\langle\sigma v\rangle = AT^{-2/3} \exp\left[-BT^{-1/3}\right] \sum_{n=0}^{5} \alpha_n T^{n/3} \tag{23}$$

where the parameters A, B, and α_n for most astrophysically important reactions are presented in the compilations of Fowler and collaborators [4, 6, 7, 8].

Table 1. Values for E_0, Δ, and I_{max} at solar core temperature ($T_6 = 15.6$)

Reaction	E_0 [keV]	$\Delta/2$ [keV]	I_{max}
$p+p$	5.9	3.2	1.1×10^{-6}
$^3\text{He} + {}^3\text{He}$	22.0	6.3	4.5×10^{-23}
$^3\text{He} + {}^4\text{He}$	23.0	6.4	5.5×10^{-23}
$p+{}^7\text{Be}$	18.4	5.8	1.6×10^{-18}
$p+{}^{14}\text{N}$	26.5	6.8	1.8×10^{-27}
$\alpha+{}^{12}\text{C}$	56.0	9.8	3.0×10^{-57}
$^{16}\text{O} + {}^{16}\text{O}$	237.0	20.2	6.2×10^{-239}

2.4 Resonant Reactions of Charged Particles

For resonant reactions, the assumption of an astrophysical S-factor that is only weakly dependent on energy is no longer valid. In fact, the cross section shows a strong variation over the energy range of the resonance that can usually be approximated by a Breit-Wigner single-resonance formula,

$$\sigma_{\text{BW}}(E) = \pi\lambda^2\omega \frac{\Gamma_a\Gamma_b}{(E - E_R)^2 + \Gamma^2/4} \tag{24}$$

where the Γ_i are the partial widths that define the decay (or formation) probabilities of the resonance in the channels i. (A nuclear resonance can in principle decay into all possible partitions of the nucleons that are allowed by the various conservation laws, e.g. energy, angular momentum, etc. Such a partition of nucleons is often called a channel. As an example, a resonance just above the ^{6}Li + p threshold can decay only into the ^{6}Li + p, ^{3}He + ^{4}He and ^{7}Be + γ "channels.") The total width Γ is the sum of the partial widths. The statistical factor ω is given by

$$\omega = \frac{(2J+1)}{(2J_P+1)(2J_T+1)}(1+\delta_{PT}) \tag{25}$$

where J is the total angular momentum of the resonance, while J_P, J_T are the spins of the projectile and target nuclei, respectively.

For further discussion, it is convenient to distinguish between narrow and broad resonances. By a *narrow resonance* we will understand a resonance for which the total width is much smaller than its resonance energy E_R, $\Gamma \ll E_R$. Then one can assume that the Maxwell-Boltzmann function and the E-factor in the integral (4) are nearly constant over the energy range of the resonance and obtain

$$\begin{aligned} \langle \sigma v \rangle &\sim \int_0^\infty \sigma_{\mathrm{BW}}(E)(E) \exp\left(-\frac{E}{kT}\right) dE \\ &\cong E_R \exp\left(-\frac{E_R}{kT}\right) \int_0^\infty \sigma_{BW}(E) dE \\ &= E_R \exp\left(-\frac{E_R}{kT}\right) 2\pi^2 \lambda^2 \omega \frac{\Gamma_a \Gamma_b}{\Gamma} \end{aligned} \tag{26}$$

where the product $\omega\Gamma_a\Gamma_b/\Gamma$ is often called the "resonance strength". If possible, the parameters $\Gamma_a, \Gamma_b, \Gamma, J$ and E_R should be determined experimentally by using either direct or indirect techniques. Note that the reaction rate depends sensitively on the resonance energy E_R because of its appearance in an exponent.

For broad resonances ($\Gamma \sim E_R$), the cross section is still given by a Breit-Wigner formula (24). However, now one has to remember that the partial width corresponding to the entrance channel, Γ_a, and possibly also the partial widths of the outgoing channels may be strongly energy dependent over the energy range of the resonance. Because this energy dependence stems mainly from the E dependence of the probability of penetration through the Coulomb barrier, one can approximate $\Gamma(E)$ for charged-particle reactions by the expression

$$\Gamma(E) = \frac{P_l(E, R_n)}{P_l(E_R, R_n)} \times \Gamma(E_R) \tag{27}$$

where $\Gamma(E_R)$ is the width at the resonance energy. The penetration factors in the partial wave l can be expressed in terms of regular and irregular Coulomb functions [9]

$$P_l(E, R_n) = \frac{1}{F_l^2(kR_n) + G_l^2(kR_n)} \tag{28}$$

where k is the wave number. Even for radiative capture reactions, the energy dependence of the width in the exit channel has to be considered. Here one finds

$$\Gamma_\gamma(E) = \left(\frac{E - E_f}{E_R - E_f}\right)^{2L+1} \Gamma_\gamma(E_R) \tag{29}$$

where L is the multipolarity of the electromagnetic transition, E_f is the energy of the final state in the transition, and $\Gamma_\gamma(E_R)$ is the radiative width at the resonance energy. For a reliable description of broad-resonance contributions to the nuclear reaction rate, quantities like $\Gamma(E_R)$ in Eq. (27) and $\Gamma_\gamma(E_R)$, L, and E_f in Eq. (29) should be determined experimentally.

The evaluation of $\langle\sigma v\rangle$ may be simplified by the fact that broad resonances frequently occur at energies E_R that are large compared with the most effective energy E_0. Thus, at astrophysical energies, only the slowly-varying tail of the resonance contributes. This tail can usually be expanded adequately in terms of a MacLaurin series for the resulting S-factor (20), so that the formalism developed for nonresonant reactions in Section 2.3 can then be applied to describe the reaction rates for broad resonances when E_R is far above E_0.

2.5 The General Case

In the most general situation the astrophysical S-factor might have contributions, in the relevant energy range near E_0, arising from narrow resonances, the low-energy tails of higher-energy, broad resonances, nonresonant reaction, and the high-energy tails of subthreshold states [for an important example, see the discussion of the $^{12}C(\alpha,\gamma)^{16}O$ reaction in Section 3]. For a subthreshold resonance, the cross section may also be described by a Breit-Wigner formula above the threshold. The energy dependences of the widths have to be taken into account, of course. While an analytical expression exists for the evolution of $\langle\sigma v\rangle$ for subthreshold resonances, the resulting integral in Eq. (15) is most simply calculated numerically.

Note that the contributions arising from the various sources listed above can interfere if they have the same J-value and parity. In many cases, the interference terms in the cross sections are the most important part of the extrapolation procedure. However, even the experimental determination of the sign of the interference term is sometimes not possible, and it is then only possible to put upper and lower limits on the extrapolated astrophysical cross sections.

For astrophysically important reactions, a series of regularly updated compilations gives conveniently parameterized presentations of the reaction rates as functions of temperature [4, 6, 7, 8, 10]. For a much more detailed discussion of the topics presented in this section, the reader is referred to the excellent textbook by Rolfs and Rodney [1] and references listed therein.

2.6 Plasma Screening

Up to now we have evaluated the reaction rates for the case of bare nuclei, in which the repulsive Coulomb barrier extends to infinity. In the astrophysical environment with which we are concerned here, the nuclei are surrounded by other nuclei and free electrons ("plasma"). The electrons tend to cluster around the nuclei, partially shielding the nuclear charges from one another. Consequently, in a plasma, two colliding nuclei have to penetrate an effective barrier that, at a given energy, is thinner than for two bare nuclei. As a result, nuclear reactions proceed faster in a plasma than would be deduced from the cross section for bare nuclei. This relation is usually defined by introducing an enhancement factor $f(E)$ [11]:

$$\langle \sigma v \rangle_{\text{plasma}} = f(E) \langle \sigma v \rangle_{\text{bare nuclei}} \tag{30}$$

where $\langle \sigma v \rangle_{\text{bare nuclei}}$ corresponds to the expressions derived in Sections 2.3 and 2.4.

In the plasmas of the stellar hydrostatic burning stages the average kinetic energy $\overline{kT}$ is much larger than the average Coulomb energy between the constituents $\overline{E}_{\text{cou}}$. In this "weak screening limit" ($\overline{E}_{\text{cou}} \ll \overline{kT}$), the Debye-Hückel theory is applicable and the Coulomb potential can be replaced by an effective shielded potential [11]:

$$V_{\text{eff}}(R) = \frac{Z_1 Z_2 e^2}{R} e^{-R/R_D} \tag{31}$$

where the Debye radius R_D is a characteristic parameter of the plasma with

$$R_D = \sqrt{\frac{kT}{4\pi e^2 \rho N_A \zeta}} \tag{32}$$

and

$$\zeta = \sum (Z_i^2 + Z_i) \frac{X_i}{A_i} \tag{33}$$

The sum in Eq. (33) runs over all different constituents of the plasma. As an example, $R_D = 0.218$ Å in the solar core.

During the barrier penetration process, the separation of the two colliding nuclei is usually much smaller than the Debye radius ($R \ll R_D$); accordingly, $V_{\text{eff}}(R)$ can be conveniently expanded as

$$V_{\text{eff}}(R) \cong \frac{Z_1 Z_2 e^2}{R} - \frac{Z_1 Z_2 e^2}{R_D}$$
$$= \frac{Z_1 Z_2 E^2}{R} - U_e \tag{34}$$

indicating that the effect of the plasma on the nuclear collision is approximately equivalent to providing a *constant* energy increment for the colliding particles of U_e. With Eqs. (4) and (34) it is simple to derive an expression for the enhancement factor $f(E)$:

$$\begin{aligned}\langle\sigma v\rangle_{\text{plasma}} &= \left(\frac{8}{\pi\mu}\right)^{1/2}\left(\frac{1}{kT}\right)^{3/2}\int_0^\infty \sigma(E+U_e)E\exp\left(-\frac{E}{kT}\right)dE \\ &\cong \exp\left(\frac{U_e}{kT}\right)\left(\frac{8}{\pi\mu}\right)^{1/2}\left(\frac{1}{kT}\right)^{3/2} \\ &\quad\times\int_0^\infty \sigma(E')E'\exp\left(-\frac{E'}{kT}\right)dE' \\ &= \exp\left(\frac{U_e}{kT}\right)\langle\sigma v\rangle_{\text{bare nuclei}}\end{aligned} \tag{35}$$

Applying the Debye-Hückel approach to the solar core (with $R_D = 0.218$ Å, $kT = 1.3$ keV), one finds that the plasma enhances reactions like ^{3}He(^{3}He, $2p$)^{4}He and ^{7}Be$(p,\gamma)^8$ by about 20%. We note that stellar and in particular solar models use screening descriptions which go beyond the simple Debye-Hückel treatment.

2.7 Electron Screening in Laboratory Nuclear Reactions

Electron screening effects also become important at the lowest energies currently feasible in laboratory measurements of light nuclear reactions [12]. Here, the electrons inevitably present in the target (and sometimes also bound to the projectile) partly shield the Coulomb barrier of the bare nuclei. Consequently, the measured cross section is larger than that of bare nuclei would be. Again, this can be expressed by introducing an enhancement factor defined as

$$S_{\text{meas}}(E) = f_{\text{lab}}(E)S(E)_{\text{bare nuclei}} \tag{36}$$

Note that the enhancement factor f_{lab} is not the same as defined in Eq. (30) as the physics behind the screening is quite different. In the plasma, the electrons are in continuum states, while the target electrons are bound to the nuclei. Thus, for astrophysical applications, it is important to deduce first the cross sections for bare nuclei from the measured data, by using a relation like Eq. (36). Then, in a second step, the resulting reaction rates have to be modified for plasma screening effects, e.g. using Eq. (35). It is also important to recognize that the general strategy in nuclear astrophysics of reducing the uncertainty in the cross section at the most effective energy E_0, by extending the measurements to increasingly lower energies, introduces a new risk, as it requires a precise knowledge of the enhancement factor $f_{\rm lab}(E)$. At this time, there is a significant, unexplained discrepancy between the experimentally extracted enhancement factors and the current theoretical predictions.

The ^{3}He$(d,p)^4$He reaction is probably the best studied example of laboratory electron screening effects, both experimentally and theoretically. As in the plasma case, the nuclear separation during the penetration process ($R \lesssim 0.02$ Å at $E = 6$ keV) is far inside the electron cloud of the atomic ^{3}He target, and the calculated screening effect of the electrons in the nuclear collision is to effectively provide a constant energy increase ΔE [12]. As $\Delta E \ll E$, one finds

$$
\begin{aligned}
S(E)_{\rm meas} &\cong S(E+\Delta E)_{\rm bare\,nuclei} \\
&\cong \exp\left[\pi\eta(E)\frac{\Delta E}{E}\right] S(E)_{\rm bare\,nuclei}
\end{aligned}
\qquad (37)
$$

At very low energies, the collision can be described in the adiabatic limit where the electrons remain in the lowest state of the combined projectile and target molecular system. It has been argued [12] that the adiabatic limit can already be applied at the lower energies at which ^{3}He$(d,p)^4$He data have been taken. This assumption was in fact approximately justified in a study of this reaction in which the electron wave functions were treated dynamically within the TDHF approach [13]. In the adiabatic limit, $\Delta E = 119$ eV for the d+^{3}He system, which corresponds to the difference in atomic binding energies between atomic He and the Li$^+$-ion. Using this value in Eq. (37), one underestimates the enhancement shown by the experimental data, which suggests $U_e \sim 220$ eV [14].

3 Hydrostatic Burning Stages

When the temperature and density in a star's interior rises as a result of gravitational contraction, it will be the lightest (lowest Z) species (protons) that can react first and supply the energy and pressure to stop the gravitational collapse of the gaseous cloud. Thus it is hydrogen burning (the fusion of four protons into a ^{4}He nucleus) in the stellar core that stabilizes the star first (and for the longest) time. However, because of the larger charge ($Z = 2$), helium, the ashes of hydrogen burning, cannot effectively react at the temperature and pressure present during hydrogen burning in the stellar core. After exhaustion of the core hydrogen, the resulting helium core will gravitationally contract, thereby raising the temperature and density in the core until the temperature and density are sufficient to ignite helium burning, starting with the triple-alpha reaction, the fusion of three ^{4}He nuclei to ^{12}C. In massive stars, this sequence of contraction of the core nuclear ashes until ignition of these nuclei in the next burning stage repeats itself several times. After helium burning, the massive star goes through periods of carbon, neon, oxygen, and silicon burning in its central core. As the binding energy per nucleon is a maximum near iron (the end-product of silicon burning), freeing nucleons from nuclei in and above the iron peak, to build still heavier nuclei, requires more energy than is released when these nucleons are captured by the nuclei present. Therefore, the procession of nuclear burning stages ceases. This results in a collapse of the stellar core and an explosion of the star as a type II supernova. As an example, Table 2 shows the timescales and conditions for the various hydrostatic burning stages of a 25 $M_\odot$ star. One observes that stars spend most of their lifetime ($\sim$ 90%) during hydrogen burning (then the stars will be found on the main sequence in the Hertzsprung-Russell diagram). The rest is basically spent during core helium burning. During this evolutionary stage, the star expands dramatically and becomes a Red Giant.

Table 2. Evolutionary stages of a 25 $M_\odot$ star (from [15])

Evolutionary stage	Temperature [keV]	Density [g/cm^3]	Time scale
Hydrogen burning	5	5	7×10^6 y
Helium burning	20	700	5×10^5 y
Carbon burning	80	2×10^5	600 y
Neon burning	150	4×10^6	1 y
Oxygen burning	200	10^7	6 mo
Silicon burning	350	3×10^7	1 d
Core collapse	700	3×10^9	seconds
Core bounce	15000	4×10^{14}	10 ms
Explosion	100-500	10^5-10^9	0.01 to 0.1 s

Stellar evolution depends very strongly on the mass of the star. On general grounds, the more massive a star the higher the temperatures in the core at which the various burning stages are ignited. Moreover, as the nuclear reactions depend very sensitively on temperature, the nuclear fuel is faster exhausted the larger the mass of the star (or the core temperature). This is quantitatively demonstrated in Table 3 which shows the timescales for core hydrogen burning as a function of the main-sequence of the star. One observes that stars with masses less than $\sim 0.5 M_\odot$ burn hydrogen for times which are significantly longer than the age of the Universe. Thus such low-mass stars have not completed one lifecycle and did also not contribute to the elemental abundances in the Universe. A star like our Sun has a life-expectation due to core hydrogen burning of about 10^{10} y, which is about double its current age ($\sim 4.55 \times 10^9$ y).

As the temperatures and densities required for the higher burning stages increase successively, stars need a minimum mass to ignite such burning phases. For example, a core mass slightly larger than $1M_\odot$ is required to ignite carbon burning. One also has to consider that stars, mainly during core helium burning, have mass losses due to flashes or stellar winds. In summary, as a rule-of-thumb, stars with main-sequence masses $\leq 8M_\odot$ end their lifes as White Dwarfs. These are stars which are dense enough so that their electrons are highly degenerate and are stabilized by the electron degeneracy pressure. There exists an upper limit for the mass of a star which can be stabilized by electron degeneracy. This is the Chandrasekhar mass of $\sim 1.4M_\odot$. Stars with masses of $\geq 13M_\odot$ go through the full cycle of hydrostatic burning stages and end with a collapse of their internal iron core. If the mass of the star is less than a certain limit, $M \sim 30M_\odot$, the star becomes a supernova leaving a compact remnant behind after the explosion; this is a neutron star. More massive stars might collapse directly into black holes.

Table 3. Hydrogen burning timescales τ_H as function of stellar mass (from [15])

M [$M_\odot$]	τ_H [y]
0.40	2×10^{11}
0.80	1.4×10^{10}
1.00	1×10^{10}
1.10	9×10^9
1.70	2.7×10^9
2.25	5×10^8
3.00	2.2×10^8
5.00	6×10^7
9.00	2×10^7
16.00	1×10^7
25.00	7×10^6
40.00	1×10^6

3.1 Hydrogen Burning

In low-mass stars, like our Sun, hydrogen burning proceeds mainly via the *pp* chains, with small contributions from the CNO cycle. The later becomes the dominant energy source in hydrogen-burning stars, if the temperature in the stellar core exceeds about 20 million degrees (the temperature in the solar core is 15.6 10^6 K).

The pp chains start with the fusion of two proton nuclei to the only bound state of the two-nucleon system, the deuteron. This reaction is mediated by the weak interaction, as a proton has to be converted into a neutron. Correspondingly, the cross section for this reaction is very small and no experimental data at low proton energies exist. Although the reaction rate at solar energies is based purely on theory, the calculations are generally considered to be under control and the uncertainty of the solar p+p rate is estimated to be better than a few percent [16]. The next reaction ($p + d \rightarrow {}^3$He) is mediated by the electromagnetic interaction. It is therefore much faster than the p+p fusion with the consequence that deuterons in the solar core are immediately transformed to ^{3}He nuclides and no significant abundance of deuterons is present in the core (there is about 1 deuteron per 10^{18} protons in equilibrium under solar core conditions). As no deuterons are available and ^{4}Li (the endproduct of p+^{3}He) is not stable, the reaction chain has to continue with the fusion of two ^{3}He nuclei via ^{3}He+^{3}He$\rightarrow$ 2p+^{4}He. This reaction terminates the ppI chain where in summary four protons are fused to one ^{4}He nucleus with an energy gain of 26.2 MeV; the rest of the mass difference is spent to produce two positrons and is radiated away by the neutrinos produced in the initial p+p fusion reaction. Once ^{4}He is produced in sufficient abundance, the pp chain can be completed by two other routes. At first ^{3}He and ^{4}He fuse to ^{7}Be which then either captures a proton (producing ^{8}B) or, more likely, an electron (producing ^{7}Li). The two chains are then terminated by the weak decay of ^{8}B to an excited state in ^{8}Be, which subsequently decays into two ^{4}He nuclei, and by the ^{7}Li(p,^{4}He)^{4}He reaction. In summary, both routes fuse 4 protons into one ^{4}He nucleus. The energy gain of these two branches of the pp chains is slightly smaller than in the dominating ppI chain, as neutrinos with somewhat larger energies are produced en route [17].

All of the pp chain reactions are non-resonant at low energies such that the extrapolation of data taken in the laboratory to stellar (solar) energies is quite reliable [16]. For two reactions (^{3}He+^{3}He$\rightarrow$ ^{4}He+2p and p+d $\rightarrow$^{3}He) the cross sections have been measured at the solar Gamow energies, thus making extrapolations of data unnecessary (see Fig. 1). These important measurements have been performed by the LUNA group at the Gran Sasso Laboratory [18] far underground to effectively remove background events due to the shielding by the rocks above the experimental hall. For many years the p+^{7}Be fusion reaction has been considered the most uncertain nuclear cross sections in solar models. However, impressive progress has been achieved in determining this reaction rate in recent years and it appears that the goal of knowing the solar reaction rate for this reaction better than the limit of 5%, as desired for solar models, has been reached [19].

Also in the CNO cycle four protons are fused to one ^{4}He nucleus. However, this reaction chain requires the presence of ^{12}C as catalyst. It then proceeds through the following sequence of reactions:

$$^{12}\mathrm{C}(\mathrm{p},\gamma)^{13}\mathrm{N}(\beta)^{13}\mathrm{C}(\mathrm{p},\gamma)^{14}\mathrm{N}(\mathrm{p},\gamma)^{15}\mathrm{O}(\beta)^{15}\mathrm{N}(\mathrm{p},\alpha)^{12}\mathrm{C}.$$

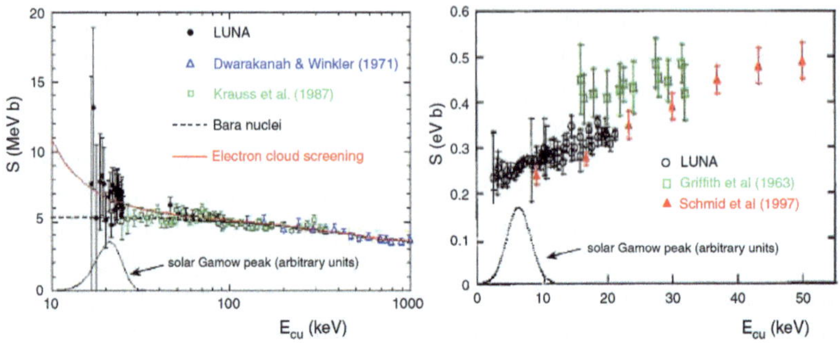

Fig. 1. S-factors for the low-energy ^{3}He+^{3}He and $d+p$ fusion reactions. The data have been taken at the Gran Sasso Underground Laboratory by the LUNA collaboration. For the first time, it has been possible to measure nuclear cross sections at the astrophysically most effective energies covering the regime of the Gamow peak. The ^{3}He+^{3}He data include electron screening effects which have been removed to obtain the cross section for bare nuclei.

The slowest step is the p+^{14}N reaction. There has been decisive experimental progress in determining this reaction rate in the last two years by the LUNA and LENA collaborations which both showed that the reaction rate is actually smaller by a factor 2 than previously believed [20, 21]. As a consequence the CNO cycle contributes less than 1% of the total energy generation in the Sun. The new p+^{14}N reaction rate has also interesting consequences for the age determination of globular clusters [22].

For many years the solar neutrino problem has been one of the outstanding puzzles in astrophysics [17]. It states that the flux of solar neutrinos measured by earthbound detectors was noticeably less than predicted by the solar models. The solution to the problem are neutrino oscillations. Within the solar pp chains only ν_e neutrinos are produced. Enhanced by matter effects some of these neutrinos are transformed into ν_μ or ν_τ neutrinos on their way out of the Sun. As the original solar neutrino detectors can only observe ν_e neutrinos, the observed neutrino flux in these detectors is smaller than the total neutrino flux generated in the Sun. This picture was confirmed by the SNO collaboration in the last two years [23]. The SNO detector has also the capability to observe neutral current events induced by neutrinos (the neutrino dissociation of deuterons into protons and neutrons in heavy water). As the neutral current events can be induced by all neutrino families such signal determines the total solar neutrino flux. It is found that it is larger than the ν_e flux and agrees nicely with the predictions of the solar models.

3.2 Helium Burning

No stable nuclei with mass numbers $A = 5$ and $A = 8$ exist. Thus, fusion reactions of p+^{4}He and ^{4}He+^{4}He lead to unstable resonant states in ^{5}Li and

^{8}Be which decay extremely fast. However, the lifetime of the ^{8}Be ground state resonance ($\sim 10^{-16}$ s) is long enough to establish a small ^{8}Be equilibrium abundance under helium burning conditions ($T \sim 10^8$ K, $\rho \sim 10^5$ g/cm^3 for a sun-like star), which amounts to about $\sim 5 \times 10^{-10}$ of the equilibrium ^{4}He abundance. As pointed out by Salpeter [24], this small ^{8}Be equilibrium abundance allows the capture of another ^{4}He nucleus to form the stable ^{12}C nucleus. The second step of the triple-alpha fusion reaction is highly enlarged by the presence of an s-wave resonance in ^{12}C at a resonance energy of 287 keV. To derive the triple-alpha reaction rate under helium burning conditions, it is sufficient to know the properties of this resonance. These quantities have been determined experimentally and it is generally assumed that the triple-alpha rate is known with an accuracy of about 15% for helium burning in Red Giants. A current estimate for the uncertainty of the triple-α rate is given in Ref. [25] which discusses also the influence of the rate on some aspects of subsequent stellar evolution. An improved triple-α rate for temperatures higher than in Red Giant helium burning is given in [26].

The second step in helium burning, the ^{12}C$(\alpha,\gamma)^{16}$O reaction, is the crucial reaction in stellar models of massive stars. Its rate determines the relative importance of the subsequent carbon and oxygen burning stages, including the abundances of the elements produced in these stages. The ^{12}C$(\alpha,\gamma)^{16}$O reaction determines also the relative abundance of ^{12}C and ^{16}O, the two bricks for the formation of life, in the Universe. Stellar models are very sensitive to this rate and its determination at the most effective energy in helium burning ($E = 300$ keV) with an accuracy of better than 20% is asked for. Despite enormous experimental efforts in the last 3 decades, this goal has not been achieved yet, as the low-energy ^{12}C$(\alpha,\gamma)^{16}$O reaction cross section is tremendously tricky. Data have been taken down to energies of about $E = 1$ MeV, requiring an extrapolation of the S-factor to $E = 300$ keV. The data are dominated by a $J = 1^-$ resonance at $E = 2.418$ MeV. Unfortunately there is another $J = 1^-$ level at $E = -45$ keV, just below the $\alpha+^{12}$C threshold. These two states interfere. In the data the broad resonance at $E = 2.418$ MeV dominates, while at stellar burning energies it is likely to be the other way around. It turns out to be quite difficult to determine the properties of the particle-bound states, although a major step forward has been achieved using indirect means by studying the β-decay of the ^{16}N ground state to states in ^{16}O above the α-threshold and their subsequent decay into the $\alpha+^{12}$C channel [27]. The γ-decay of the $J = 1^-$ states to the ^{16}O ground state is of dipole (E1) nature. If isospin were a good quantum number, this transition would be exactly forbidden, as all involved nuclei (^{4}He, ^{12}C, ^{16}O) have isospin quantum numbers $T = 0$. The observed dipole transition must then come from small isospin-symmetry breaking admixtures. The data, indeed, suggest that the transitions are suppressed by about 4 orders of magnitude compared to 'normal' E1 transitions. Such a large suppression makes it possible that E2 (quadrupole) transitions can compete with the dipole contributions. This is confirmed by measurements of the ^{12}C$(\alpha,\gamma)^{16}$O angular distributions at low energies which are mixtures of dipole and quadrupole contributions. While both, dipole and quadrupole, cross sections can thus be determined from the data (although the measurement of angular distributions is much more challenging than the determination of the total cross section), the extrapolation of the E2 data to stellar energies at $E = 300$ keV is strongly hampered by the fact that the stellar cross section is dominated by the tail of a particle-bound $J = 2^+$ state at $E = -245$ keV, which, however, is much weaker in the data taken at higher energies.

The latest results for the $^{12}C(\alpha,\gamma)^{16}O$ S-factor are presented in [28] and in [29].

The $^{16}O(\alpha,\gamma)^{20}Ne$ is non-resonant at stellar energies and hence very slow, compared to the $\alpha+^{12}C$ reaction. Thus, helium burning finishes with the $^{12}C(\alpha,\gamma)^{16}O$ reaction.

3.3 Carbon, Neon, Oxygen burning

In the fusion of two ^{12}C nuclei, the α and proton-channels have positive Q-values ($Q = 4.62$ MeV and $Q = 2.24$ MeV). Thus, the fusion produces nuclides with smaller charge numbers, which can then interact with other carbon nuclei or produced elements. The main reactions in carbon burning are: $^{12}C(^{12}C,\alpha)^{20}Ne$, $^{12}C(^{12}C,p)^{23}Na$, $^{23}Na(p,\alpha)^{20}Ne$, $^{23}Na(p,\gamma)^{24}Mg$, $^{12}C(\alpha,\gamma)^{16}O$, which determine the basic energy generation. However, many other reactions can occur, even producing elements beyond ^{24}Mg like ^{26}Mg and ^{27}Al.

The $^{12}C+^{12}C$ fusion cross section data at low energies show oscillations, which are characteristic for resonances and have been interpreted as evidence for the existence of $^{12}C+^{12}C$ molecules. Astrophysically it is interesting whether the resonant behavior of the fusion data (measured down to $E \sim 2.3$ MeV) continues to lower energies which might have consequences in the simulations of the screening corrections for this reaction in compact objects.

Neon burning occurs at temperatures just above $T = 10^9$ K. At these conditions the presence of high-energy photons is sufficient to photodissociate ^{20}Ne via the $^{20}Ne(\gamma,\alpha)^{16}O$ reaction which has a Q-value of -4.73 MeV. This reaction liberates α particles which react then very fast with other ^{20}Ne nuclei leading to the production of ^{28}Si via the chain $^{20}Ne(\alpha,\gamma)^{24}Mg(\alpha,\gamma)^{28}Si$. Again, many other reactions induced by protons, ^4He and also neutrons, which are produced within the occurring reaction chains, occur.

In the fusion of two ^{16}O nuclei, the α and proton-channels have positive Q-values ($Q = 9.59$ MeV and $Q = 7.68$ MeV). Like in carbon burning, the liberated protons and ^4He nuclei react with other ^{16}O nuclei. Among the many nuclides produced during oxygen burning are nuclei like ^{33}Si and ^{35}Cl. These have quite low Q-values against electron captures making it energetically favorable to capture electrons from the degenerate electron sea (the Fermi energies of electrons during core oxygen burning is of order 1 MeV) via the $^{33}Si(e^-\nu)^{33}P$ and $^{35}Cl(e^-\nu)^{35}S$ reactions. The emitted neutrinos carry energy away, thus cooling the star. In fact, neutrino emission is the most effective cooling mechanism during all advanced burning stages, subsequent to helium burning. As we will see in the next section, electron captures play also a decisive role in core-collapse supernovae.

3.4 Silicon Burning

The nuclear reaction network during silicon burning is initiated by the photodissociation of ^{28}Si, producing protons, neutrons and α-particles. These particles react again with ^{28}Si or the nuclides produced. During silicon burning, the temperature is already quite high ($T \sim 3 - 4 \times 10^9$ K). This makes the nuclear reactions mediated by the strong and electromagnetic force quite fast and a chemical equilibrium between reactions and their inverse processes establishes. Under such conditions, the abundance distributions of the nuclides present in the network becomes independent of the reaction rates, establishing the Nuclear Statistical Equilibrium (NSE) (see [30, 31]). Then the abundance of a nuclide with proton and neutron numbers Z and N can be expressed in terms of the abundance of free protons and neutrons by:

$$Y_{Z,N} = G_{Z,N}(\rho N_A)^{(A-1)} \frac{A^{3/2}}{2^A} (\frac{2\pi\hbar^2}{m_u kT})^{3/2(A-1)} \exp\{\frac{B_{Z,N}}{kT}\} Y_n^N Y_p^Z \tag{38}$$

where $G_{Z,N}$ is the nuclear partition function (at temperature T), $B_{Z,N}$ the binding excess of the nucleus, m_u the unit nuclear mass and Y_p, Y_n are the abundances of free protons and neutrons. The NSE distribution is subject to the mass and charge conservation, which can be formulated as:

$$\sum_i A_i Y_i = 1 \ ; \sum_i Z_i Y_i = Y_e \tag{39}$$

where Y_e is the electron-to-nucleon ratio. Until the onset of oxygen burning, one has Y_e=0.5 (nuclei like ^{12}C have identical protron and neutron numbers, while the number of electrons equals the proton number). Once electron capture processes start, the Y_e value is reduced (protons are changed into neutrons, while the total number of nucleons is preserved). The value of Y_e changes as reactions mediated by the weak force are not in equilibrium (there is no abundance of neutrinos in the star which can initiate the inverse reactions). However, this changes in the late stage of a core-collapse supernova.

For 'normal' temperatures and densities, NSE favours the nuclei with highest binding energies (among those with $Z/A \sim Y_e$ dictated by charge conservation). However, we observe from Eq. (38) the following two limiting cases which are both relevant for core collapse supernovae. Due to the factor ρ^{A-1} the NSE distribution favors increasingly heavier nuclei with increasing density (at fixed temperature); this happens during the collapse. On the contrary, the factor $T^{-3/2(A-1)}$ implies that, at fixed densities, the NSE distribution drives to nuclei with smaller masses with increasing temperature; in the high-T limit the nuclei get all disassembled into free protons and neutrons; this occurs in the shock-heated material.

We finally note that, during silicon burning, not a full NSE is established. However, the nuclear chart breaks into several regions in which NSE equilibrium is established. The different regions are not yet in equilibrium, as the nuclear reactions connecting them are yet not fast enough. One uses the term 'Quasi-NSE" for these conditions.

4 Core Collapse Supernovae

At the end of hydrostatic burning, a massive star consists of concentric shells that are the remnants of its previous burning phases (hydrogen, helium, carbon, neon, oxygen, silicon). Iron is the final stage of nuclear fusion in hydrostatic burning, as the synthesis of any heavier element from lighter elements does not release energy; rather, energy must be used up. If the iron core, formed in the center of the massive star, exceeds the Chandrasekhar mass limit of about 1.44 solar masses, electron degeneracy pressure cannot longer stabilize the core and it collapses starting what is called a type II supernova. In its aftermath the star explodes and parts of the iron core and the outer shells are ejected into the Interstellar Medium. Although this general picture has been confirmed by the various observations from supernova SN1987a, simulations of the core collapse and the explosion are still far from being completely understood and robustly modelled. To improve the input which goes into the simulation of type II supernovae and to improve the models and their numerical simulations is a very active research field at various institutions worldwide.

The collapse is very sensitive to the entropy and to the number of leptons per baryon, Y_e [32]. In turn these two quantities are mainly determined by weak interaction processes, electron capture and β decay. First, in the early stage of the collapse Y_e is reduced as it is energetically favorable to capture electrons, which at the densities involved have Fermi energies of a few MeV, by (Fe-peak) nuclei. This reduces the electron pressure, thus accelerating the collapse, and shifts the distribution of nuclei present in the core to more neutron-rich material. Second, many of the nuclei present can also β decay. While this process is quite unimportant compared to electron capture for initial Y_e values around 0.5, it becomes increasingly competitive for neutron-rich nuclei due to an increase in phase space related to larger Q_β values.

Electron capture, β decay and photodisintegration cost the core energy and reduce its electron density. As a consequence, the collapse is accelerated. An important change in the physics of the collapse occurs, as the density reaches $\rho_{\text{trap}} \approx 4 \cdot 10^{11}$ g/cm^3. Then neutrinos are essentially trapped in the core, as their diffusion time (due to coherent elastic scattering on nuclei) becomes larger than the collapse time [33]. After neutrino trapping, the collapse proceeds homologously [34], until nuclear densities ($\rho_N \approx 10^{14}$ g/cm^3) are reached. As nuclear matter has a finite compressibility, the homologous core decelerates and bounces in response to the increased nuclear matter pressure; this eventually drives an outgoing shock wave into the outer core; i.e. the envelope of the iron core outside the homologous core, which in the meantime has continued to fall inwards at supersonic speed. The core bounce with the formation of a shock wave is believed to be the mechanism that triggers a supernova explosion, but several ingredients of this physically appealing picture and the actual mechanism of a supernova explosion are still uncertain and controversial. If the shock wave is strong enough not only to stop the collapse, but also to explode the outer burning shells of the star, one speaks about the 'prompt mechanism' [35]. However, it appears as if the energy available to the shock is not sufficient, and the shock will store its energy in the outer core, for example, by dissociation of nuclei into nucleons. Furthermore, this change in composition results to additional energy losses, as the electron capture rate on free protons is significantly larger than on neutron-rich nuclei due to the smaller Q-values involved. This leads to a further neutronization of the matter. Part of the neutrinos produced by the capture on the free protons behind the shock leave the star carrying away energy.

After the core bounce, a compact remnant is left behind. Depending on the stellar mass, this is either a neutron star (masses roughly smaller than 30 solar masses) or a black hole. The neutron star remnant is very lepton-rich (electrons and neutrinos), the latter being trapped as their mean free paths in the dense matter is significantly shorter than the radius of the neutron star. It takes a fraction of a second [36] for the trapped neutrinos to diffuse out, giving most of their energy to the neutron star during that process and heating it up. The cooling of the proto-neutron star then proceeds by pair production of neutrinos of all three generations which diffuse out. After several tens of seconds the star becomes transparent to neutrinos and the neutrino luminosity drops significantly [37].

In the 'delayed mechanism', the shock wave can be revived by the outward diffusing neutrinos, which carry most of the energy set free in the gravitational collapse of the core [36] and deposit some of this energy in the layers between the nascent neutron star and the stalled prompt shock. This lasts for a few 100 ms, and requires about 1% of the neutrino energy to be converted into nuclear kinetic energy. The energy deposition increases the pressure behing the shock and the respective layers begin to expand, leaving between shock front and neutron star surface a region of low density, but rather high temperature. This region is called the 'hot neutrino bubble'. The persistent energy input by neutrinos keeps the pressure high in this region and drives the shock outwards again, eventually leading to a supernova explosion.

It has been found that the delayed supernova mechanism is quite sensitive to physics details deciding about success or failure in the simulation of the explosion. Very recently, two quite distinct improvements have been proposed (convective energy transport [38, 39] and in-medium modifications of the neutrino opacities [40, 41]) which increase the efficiency of the energy transport to the stalled shock.

Current one-dimensional supernova simulations, including sophisticated neutrino transport, fail to explode [42, 43] (however, see [44]). The interesting question is whether the simulations explicitly require multi-dimensional effects like rotation, magnetic fields or convection (e.g. [38, 39]) or whether the microphysics input in the one-dimensional models is insufficient. It is the goal of nuclear astrophysics to improve on this microphysics input. The next section shows that core collapse supernovae are nice examples to demonstrate how important micro-physics input and progress in nuclear modelling can be.

4.1 The Role of Electron Capture and β Decay During Collapse

Late-stage stellar evolution is described in two steps. In the presupernova models the evolution is studied through the various hydrostatic core and shell burning phases until the central core density reaches values up to 10^{10} g/cm^3. The models consider a large nuclear reaction network. However, the densities involved are small enough to treat neutrinos solely as an energy loss source. For even higher densities this is no longer true as neutrino-matter interactions become increasingly important. In modern core-collapse codes neutrino transport is described self-consistently by spherically symmetric multigroup Boltzmann simulations. While this is computationally very challenging, collapse models have the advantage that the matter composition can be derived from Nuclear Statistical Equilibrium (NSE) as the core temperature and density are high enough to keep reactions mediated by the strong and electromagnetic interactions in equilibrium. This means that for sufficiently low entropies, the matter composition is dominated by the nuclei with the highest Q-values for a given Y_e. The presupernova models are the input for the collapse simulations which follow the evolution through trapping, bounce and hopefully explosion.

The collapse is a competition of the two weakest forces in nature: gravity versus weak interaction, where electron captures on nuclei and protons and, during a period of silicon burning, also β-decay play the crucial roles. Which nuclei are important? Weak-interaction processes become important when nuclei with masses $A \sim 55 - 60$ (pf-shell nuclei) are most abundant in the core (although capture on *sd* shell nuclei has to be considered as well). As weak interactions changes Y_e and electron capture dominates, the Y_e value is successively reduced from its initial value ~ 0.5. As a consequence, the abundant nuclei become more neutron rich and heavier, as nuclei with decreasing Z/A ratios are more bound in heavier nuclei. Two further general remarks are useful. There are many nuclei with appreciable abundances in the cores of massive stars during their final evolution. Neither the nucleus with the largest capture rate nor the most abundant one are necessarily the most relevant for the dynamical evolution: What makes a nucleus relevant is the product of rate times abundance.

For densities $\rho \leq 10^{11}$ g/cm^3, stellar weak-interaction processes are dominated by Gamow-Teller (GT) and, if applicable, by Fermi transitions. At higher densities forbidden transitions have to be included as well. To understand the requirements for the nuclear models to describe these processes (mainly electron capture), it is quite useful to recognize that electron capture is governed by two energy scales: the electron chemical potential μ_e, which grows like $\rho^{1/3}$, and the nuclear Q-value. Importantly, μ_e grows much faster than the Q values of the abundant nuclei. We can conclude that at low densities, where one has $\mu_e \sim Q$ (i.e. at presupernova conditions), the capture rate will be very sensitive to the phase space and requires an accurate as possible description of the detailed GT_+ distribution of the nuclei involved. Furthermore, the finite temperature in the star requires the implicit consideration of capture on excited nuclear states, for which the GT distribution can be different than for the ground state. As we will demonstrate below, modern shell model calculations are capable to describe GT_+ rather well and are therefore the appropriate tool to calculate the weak-interaction rates for those nuclei ($A \sim 50 - 65$) which are relevant at such densities. At higher densities, when μ_e is sufficiently larger than the respective nuclear Q values, the capture rate becomes less sensitive to the detailed GT_+ distribution and is mainly only dependent on the total GT strength. Thus, less sophisticated nuclear models might be sufficient. However, one is facing a nuclear structure problem which has been overcome only very recently. We come back to it below, after we have discussed the calculations of weak-interaction rates within the shell model and their implications to presupernova models.

In recent years it has been possible to derive the electron capture (and other weak-interaction rates) needed for presupernova and collapse models on the basis of microscopic nuclear models. The results are quite distinct from the more empirical rates used before and have lead to significant changes in supernova simulations. This progress and the related changes are discussed in the next two subsections.

4.2 Weak-interaction Rates and Presupernova Evolution

The general formalism to calculate weak interaction rates for stellar environment has been given by Fuller, Fowler and Newman (FFN) [45, 46, 47, 48]. These authors also estimated the stellar electron capture and beta-decay rates systematically for nuclei in the mass range $A = 20 - 60$ based on the independent particle model and on data, whenever available. In recent years this pioneering and seminal work has been replaced by rates based on large-scale shell model calculations. At first, Oda *et al.* derived such rates for *sd*-shell nuclei ($A = 17 - 39$) and found rather good agreement with the FFN rates [49]. Similar calculations for *pf*-shell nuclei had to wait until significant progress in shell model diagonalization, mainly due to Etienne Caurier, allowed calculations in either the full *pf* shell or at such a truncation level that the GT distributions were virtually converged. It has been demonstrated in [50] that the shell model reproduces all measured GT_+ distributions very well and gives a very reasonable account of the experimentally known GT_- distribu-

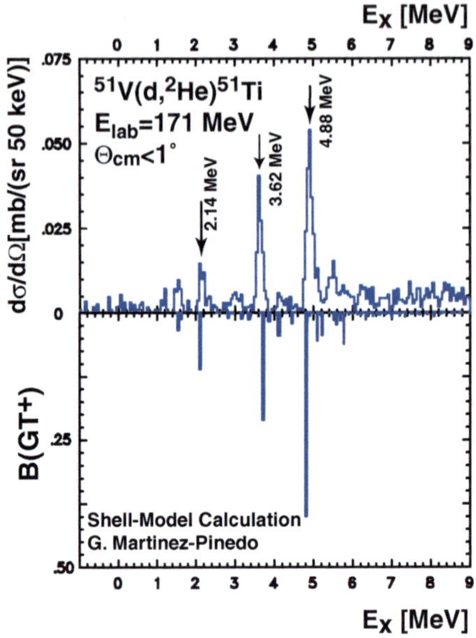

Fig. 2. Comparison of the measured ^{51}V(d,^{2}He)^{51}Ti cross section at forward angles (which is proportional to the GT_+ strength) with the shell model GT distribution in ^{51}V (from [51]).

tions. Further, the lifetimes of the nuclei and the spectroscopy at low energies is simultaneously also described well. Charge-exchange measurements using the (d,^{2}He) reaction at intermediate energies allow now for an experimental determination of the GT_+ strength distribution with an energy resolution of about 150 keV. Fig. 2 compares the experimental GT_+ strength for ^{51}V, measured at the KVI in Groningen [51], with shell model predictions. It can be concluded that modern shell model approaches have the necessary predictive power to reliably estimate stellar weak interaction rates. Such rates have been presented in [52, 53]. for more than 100 nuclei in the mass range A = 45-65. The rates have been calculated for the same temperature and density grid as the standard FFN compilations [46, 47]. An electronic table of the rates is available [53]. Importantly one finds that the shell model electron capture rates are systematically smaller than the FFN rates. The difference is particularly large for capture on odd-odd nuclei which have been previously assumed to dominate electron capture in the early stage of the collapse [54]. The differences are related to an insufficient treatment of pairing in the FFN parametrization, as discussed in [52].

To study the influence of the shell model rates on presupernova models Heger *et al.* [55, 56] have repeated the calculations of Weaver and Woosley [57] keeping the stellar physics, except for the weak rates, as close to the original

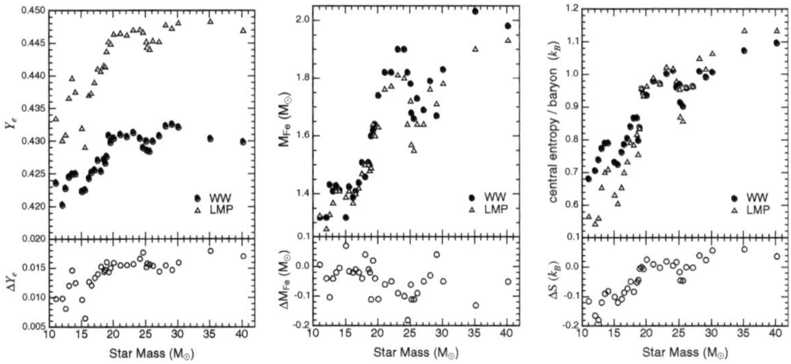

Fig. 3. Comparison of the center values of Y_e (left), the iron core sizes (middle) and the central entropy (right) for $11-40M_\odot$ stars between the WW models, which used the FFN rates, and the ones using the shell model weak interaction rates (LMP).

studies as possible. Fig. 3 exemplifies the consequences of the shell model weak interaction rates for presupernova models in terms of the three decisive quantities: the central Y_e value and entropy and the iron core mass. The central values of Y_e at the onset of core collapse increased by 0.01-0.015 for the new rates. This is a significant effect. We note that the new models also result in lower core entropies for stars with $M \leq 20M_\odot$, while for $M \geq 20M_\odot$, the new models actually have a slightly larger entropy. The iron core masses are generally smaller in the new models where the effect is larger for more massive stars ($M \geq 20M_\odot$), while for the most common supernovae ($M \leq 20M_\odot$) the reduction is by about 0.05 $M_\odot$.

Electron capture dominates the weak-interaction processes during presupernova evolution. However, during silicon burning, β decay (which increases Y_e) can compete and adds to the further cooling of the star. With increasing densities, β-decays are hindered as the increasing Fermi energy of the electrons blocks the available phase space for the decay. Thus, during collapse β-decays can be neglected.

We note that the shell model weak interaction rates predict the presupernova evolution to proceed along a temperature-density-Y_e trajectory where the weak processes are dominated by nuclei rather close to stability. Thus it will be possible, after radioactive ion-beam facilities become operational, to further constrain the shell model calculations by measuring relevant beta decays and GT distributions for unstable nuclei. Ref. [55, 56] identify those nuclei which dominate (defined by the product of abundance times rate) the electron capture and beta decay during various stages of the final evolution of a $15M_\odot$, $25M_\odot$ and $40M_\odot$ star.

4.3 The Role of Electron Capture During Collapse

In collapse simulations a very simple description for electron capture on nuclei has been used until recently, as the rates have been estimated in the spirit of the independent particle model (IPM), assuming pure Gamow-Teller (GT) transitions and considering only single particle states for proton and neutron numbers between $N = 20$–40 [58]. In particular this model assigns vanishing electron capture rates to nuclei with neutron numbers larger than $N = 40$, motivated by the observation [59] that, within the IPM, GT transitions are Pauli-blocked for nuclei with $N \geq 40$ and $Z \leq 40$. However, as electron capture reduces Y_e, the nuclear composition is shifted to more neutron rich and to heavier nuclei, including those with $N > 40$, which dominate the matter composition for densities larger than a few 10^{10} g cm^{-3}. As a consequence of the model applied in the previous collapse simulations, electron capture on nuclei ceases at these densities and the capture is entirely due to free protons. This employed model for electron capture on nuclei is too simple and leads to incorrect conclusions, as the Pauli-blocking of the GT transitions is overcome by correlations [62] and temperature effects [59, 60] (see also [61]).

At first, the residual nuclear interaction, beyond the IPM, mixes the pf shell with the levels of the sdg shell, in particular with the lowest orbital, $g_{9/2}$. This makes the closed $g_{9/2}$ orbit a magic number in stable nuclei ($N = 50$) and introduces, for example, a very strong deformation in the $N = Z = 40$ nucleus ^{80}Zr. Moreover, the description of the B(E2,$0^+ \rightarrow 2_1^+$) transition in ^{68}Ni requires configurations where more than one neutron is promoted from the pf shell into the $g_{9/2}$ orbit [63], unblocking the GT transition even in this proton-magic $N = 40$ nucleus. Such a non-vanishing GT strength has already been observed for ^{72}Ge ($N = 40$) [64] and ^{76}Se ($N = 42$) [65]. Secondly, during core collapse electron capture on the nuclei of interest here occurs at temperatures $T \geq 0.8$ MeV, which, in the Fermi gas model, corresponds to a nuclear excitation energy $U \approx AT^2/8 \approx 5$ MeV; this energy is noticeably larger than the splitting of the pf and sdg orbitals ($E_{g_{9/2}} - E_{p_{1/2}, f_{5/2}} \approx 3$ MeV). Hence, the configuration mixing of sdg and pf orbitals will be rather strong in those excited nuclear states of relevance for stellar electron capture. Furthermore, the nuclear state density at $E \sim 5$ MeV is already larger than 100/MeV, making a state-by-state calculation of the rates impossible, but also emphasizing the need for a nuclear model which describes the correlation energy scale at the relevant temperatures appropriately. This model is the Shell Model Monte Carlo (SMMC) approach [66, 67] which describes the nucleus by a canonical ensemble at finite temperature and employs a Hubbard-Stratonovich linearization [68] of the imaginary-time many-body propagator to express observables as path integrals of one-body propagators in fluctuating auxiliary fields [66, 67]. Since Monte Carlo techniques avoid an explicit enumeration of the many-body states, they can be used in model spaces far larger than those accessible to conventional methods. The Monte Carlo results are in principle exact and are in practice subject only to controllable sampling and discretization errors.

To calculate electron capture rates for nuclei $A = 65$–112 SMMC calculations have been performed in the full pf-sdg shell, considering upto more than 10^{20} configurations. From the SMMC calculations the temperature-dependent occupation numbers of the various single-particle orbitals have been determined. These occupation numbers then became the input in RPA calculations of the capture rate, considering allowed and forbidden transitions up to multipoles $J = 4$ and including the momentum dependence of the operators. The model is described in [62]; first applications in collapse simulations are presented in [69, 74]. There have been several nuclear structure studies performed with the SMMC in the same model space and with the same interaction which give confidence that the model is capable of described correlations across the $N = 40$ gap quite well. These studies dealt with the magicity of the nucleus ^{68}Ni [70], the origin of deformation in the $N \sim Z \sim 40$ nuclei [71] and the competition of pairing and deformation degrees of freedom as function of temperature [72].

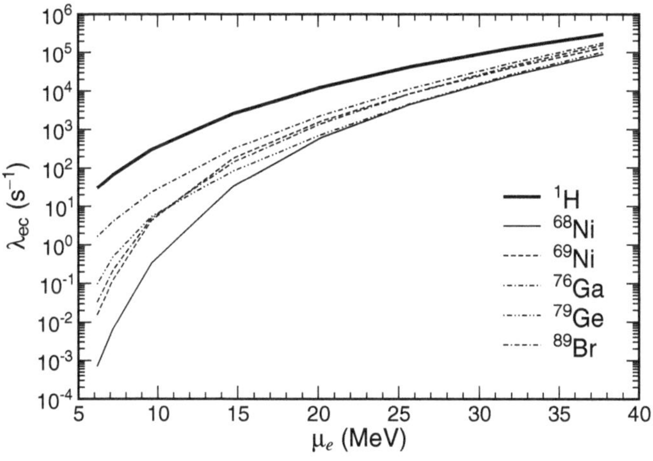

Fig. 4. Comparison of the electron capture rates on free protons and selected nuclei as function of the electron chemical potential along a stellar collapse trajectory taken from [43]. Neutrino blocking of the phase space is not included in the calculation of the rates.

For all studied nuclei one finds neutron holes in the (pf) shell and, for $Z > 30$, non-negligible proton occupation numbers for the sdg orbitals. This unblocks the GT transitions and leads to sizable electron capture rates. Fig. 4 compares the electron capture rates for free protons and selected nuclei along a core collapse trajectory, as taken from [43]. Dependent on their proton-to-nucleon ratio Y_e and their Q-values, these nuclei are abundant at different stages of the collapse. For all nuclei, the rates are dominated by GT transitions at low densities, while forbidden transitions contribute sizably at $\rho \geq 10^{11}$ g/cm^3.

Simulations of core collapse require reaction rates for electron capture on protons, $R_p = Y_p\lambda_p$, and nuclei $R_h = \sum_i Y_i\lambda_i$ (where the sum runs over all the nuclei present and Y_i denotes the number abundance of a given species), over wide ranges in density and temperature. While R_p is readily derived from [58], the calculation of R_h requires knowledge of the nuclear composition, in addition to the electron capture rates described earlier. In [69, 74] NSE has been adopted to determine the needed abundances of individual isotopes and to calculate R_h and the associated emitted neutrino spectra on the basis of about 200 nuclei in the mass range $A = 45 - 112$ as a function of temperature, density and electron fraction. The rates for the inverse neutrino-absorption process are determined from the electron capture rates by detailed balance. Due to its much smaller $|Q|$-value, the electron capture rate on the free protons is larger than the rates of abundant nuclei during the core collapse (Fig. 4). However, this is misleading as the low entropy keeps the protons significantly less abundant than heavy nuclei during the collapse. Fig. 5 shows that the reaction rate on nuclei, R_h, dominates the one on protons, R_p, by roughly an order of magnitude throughout the collapse when the composition is considered. Only after the bounce shock has formed does R_p become higher than R_h, due to the high entropies and high temperatures in the shock-heated

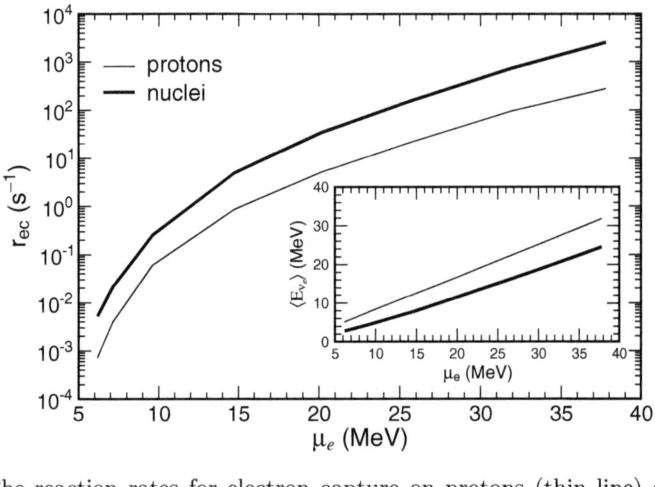

Fig. 5. The reaction rates for electron capture on protons (thin line) and nuclei (thick line) are compared as a function of electron chemical potential along a stellar collapse trajectory. The insert shows the related average energy of the neutrinos emitted by capture on nuclei and protons. The results for nuclei are averaged over the full nuclear composition (see text). Neutrino blocking of the phase space is not included in the calculation of the rates.

matter that result in a high proton abundance. The obvious conclusion is that electron capture on nuclei must be included in collapse simulations.

It is also important to stress that electron capture on nuclei and on free protons differ quite noticeably in the neutrino spectra they generate. This is demonstrated in Fig. 5 which shows that neutrinos from captures on nuclei have a mean energy 40–60% less than those produced by capture on protons. Although capture on nuclei under stellar conditions involves excited states in the parent and daughter nuclei, it is mainly the larger $|Q|$-value which significantly shifts the energies of the emitted neutrinos to smaller values. These differences in the neutrino spectra strongly influence neutrino-matter interactions, which scale with the square of the neutrino energy and are essential for collapse simulations [43, 42] (see below).

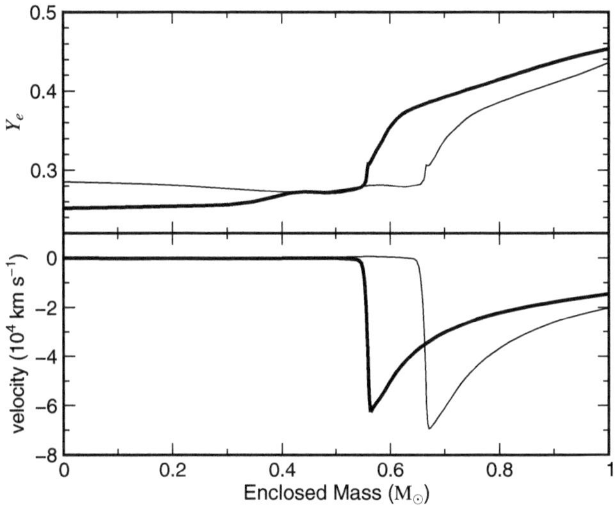

Fig. 6. The electron fraction and velocity as functions of the enclosed mass at bounce for a 15 $M_\odot$ model [55]. The thin line is a simulation using the Bruenn parameterization while the thick line is for a simulation using the combined LMP [73] and SMMC+RPA rate sets. Both models were calculated with Newtonian gravity.

The effects of this more realistic implementation of electron capture on heavy nuclei have been evaluated in independent self-consistent neutrino radiation hydrodynamics simulations by the Oak Ridge and Garching collaborations [74, 75]. The basis of these models is described in detail in Refs. [43] and [42]. Both collapse simulations yield qualitatively the same results. The changes compared to the previous simulations, which adopted the IPM rate estimate from Ref. [58] and hence basically ignored electron capture on nuclei, are significant. Fig. 6 shows a key result: In denser regions, the additional electron capture on heavy nuclei results in more electron capture in the new models. In lower density regions, where nuclei with $A < 65$ dominate, the shell model rates [52] result in less electron capture. The results of these competing effects can be seen in the first panel of Figure 6, which shows the distribution of Y_e throughout the core at bounce (when the maximum central density is reached). The combination of increased electron capture in the interior with reduced electron capture in the outer regions causes the shock to form with 16% less mass interior to it and a 10% smaller velocity difference across the shock. This leads to a smaller mass of the homologuous core (by about 0.1 $M_\odot$). In spite of this mass reduction, the radius from which the shock is launched is actually displaced slightly outwards to 15.7 km from 14.8 km in the old models. If the only effect of the improvement in the treatment of electron capture on nuclei were to launch a weaker shock with more of the iron core overlying it, this improvement would seem to make a successful explosion more difficult. However, the altered gradients in density and lepton fraction also play an important role in the behavior of the shock. Though also the new models fail to produce explosions in the spherically symmetric limit, the altered gradients allow the shock in the case with improved capture rates to reach 205 km, which is about 10 km further out than in the old models.

These calculations clearly show that the many neutron-rich nuclei which dominate the nuclear composition throughout the collapse of a massive star also dominate the rate of electron capture. Astrophysics simulations have demonstrated that these rates have a strong impact on the core collapse trajectory and the properties of the core at bounce. The evaluation of the rates has to rely on theory as a direct experimental determination of the rates for the relevant stellar conditions (i.e. rather high temperatures) is currently impossible. Nevertheless it is important to experimentally explore the configuration mixing between *pf* and *sdg* shell in extremely neutron-rich nuclei as such understanding will guide and severely constrain nuclear models. Such guidance is expected from future radioactive ion-beam facilities.

4.4 Neutrino-induced Processes During a Supernova Collapse

While the neutrinos can leave the star unhindered during the presupernova evolution, neutrino-induced reactions become more and more important during the subsequent collapse stage due to the increasing matter density and neutrino energies; the latter are of order a few MeV in the presupernova models, but increase roughly approximately to the electron chemical potential [58, 62]. Elastic neutrino scattering off nuclei and inelastic scattering on electrons are the two most important neutrino-induced reactions during the collapse. The first reaction randomizes the neutrino paths out of the core and, at densities of a few 10^{11} g/cm^3, the neutrino diffusion time-scale gets larger than the collapse time; the neutrinos are trapped in the core for the rest of the contraction. Inelastic scattering off electrons thermalizes the trapped neutrinos then rather fastly with the matter and the core collapses as a homologous unit until it reaches densities slightly in excess of nuclear matter, generating a bounce and launching a shock wave which traverses through the infalling material on top of the homologous core. In the currently favored explosion model, the shock wave is not energetic enough to explode the star, it gets stalled before reaching the outer edge of the iron core, but is then eventually revived due to energy transfer by neutrinos from the cooling remnant in the center to the matter behind the stalled shock.

Neutrino-induced reactions on nuclei, other than elastic scattering, can also play a role during the collapse and explosion phase [76]. Note that during the collapse only ν_e neutrinos are present. Thus, charged-current reactions $A(\nu_e, e^-)A'$ are strongly blocked by the large electron chemical potential [77, 78]. Inelastic neutrino scattering on nuclei can compete with $\nu_e + e^-$ scattering at higher neutrino energies $E_\nu \geq 20$ MeV [77]. Here the cross sections are mainly dominated by first-forbidden transitions. Finite-temperature effects play an important role for inelastic $\nu + A$ scattering below $E_\nu \leq 10$ MeV. This comes about as nuclear states get thermally excited which are connected to the ground state and low-lying excited states by modestly strong GT transitions and increased phase space. As a consequence the cross sections are significantly increased for low neutrino energies at finite temperature and might be comparable to inelastic $\nu_e + e^-$ scattering [79]. Thus, inelastic neutrino-nucleus scattering, which is so far neglected in collapse simulations, should be implemented in such studies. This is in particular motivated by the fact that it has been demonstrated that electron capture on nuclei dominated during the collapse and this mode generates significantly less energetic neutrinos than considered previously. Examples for inelastic neutrino-nucleus cross sections are shown in Fig. 7. A reliable estimate for these cross sections requires the knowledge of the GT_0 strength (see below). Shell model predictions imply that the GT_0 centroid resides at excitation energies around 10 MeV and is independent of the pairing structure of the ground state [80, 79]. Finite temperature effects become unimportant for stellar inelastic neutrino-nucleus cross sections once the neutrino energy is large enough to reach the GT_0 centroid, i.e. for $E_\nu \geq 10$ MeV.

The trapped ν_e neutrinos will be released from the core in a brief burst shortly after bounce. These neutrinos can interact with the infalling matter just before arrival of the shock and eventually preheat the matter requiring less energy from the shock for dissociation [76]. The relevant preheating processes are charged- and neutral-current reactions on nuclei in the iron and also silicon mass range. So far, no detailed collapse simulation including preheating has been performed. The relevant cross sections can be calculated on the basis of shell model calculations for the allowed transitions and RPA studies for the forbidden transitions [80]. The main energy transfer to the matter behind the shock, however, is due to neutrino absorption on free nucleons. The efficiency of this transport depends strongly on the neutrino opacities in hot and very dense neutron-rich matter [82]. It is likely also supported by convective motion, requiring multidimensional simulations [83].

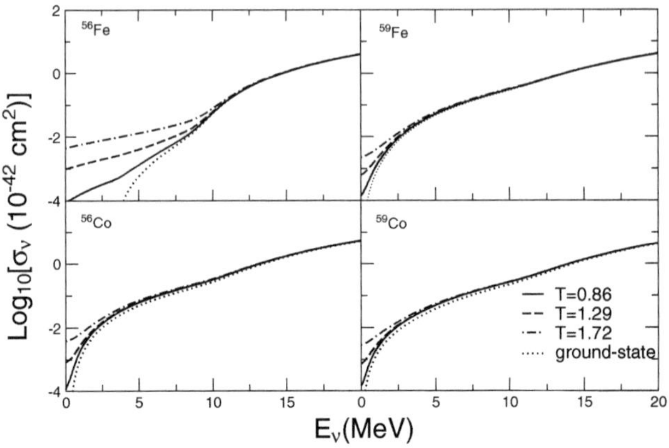

Fig. 7. Cross sections for inelastic neutrino scattering on nuclei at finite temperature. The temperatures are given in MeV (from [81]).

4.5 Explosive Nucleosynthesis

When in an successful explosion the shock passes through the outer shells, its high temperature induces an explosive nuclear burning on short time-scales. This explosive nucleosynthesis can alter the elemental abundance distributions in the inner (silicon, oxygen) shells. Recently explosive nucleosynthesis has been investigated consistently within supernova simulations, where a successful explosion has been enforced by slightly enlargening the neutrino absorption cross section on nucleons or the neutrino mean-free path, which both increase the efficiency of the energy transport to the stalled shock. The results presented in [84, 85] showed that in an early phase after the bounce the ejected matter is actually proton-rich. The proton-to-nucleon ratio Y_e is determined by the competition of the neutrino and anti-neutrino absorption on free nucleons where in this early phase, the ν_e neutrinos have sufficiently large energies to drive the matter proton-rich. In later stages, when neutrinos are produced by pair emission due to cooling of the proto-neutron star, the neutrino opacities in the neutron-rich matter ensures that $\bar{\nu_e}$ have larger average energies than ν_e and $\bar{\nu_e}$ absorption on protons dominates driving the matter neutron-rich (allowing for the r-process to occur).

Nucleosynthesis in the proton-rich environment allows for the generation of certain elements like Sc, Cu, Zn, which have been strongly underproduced in previous studies. Another interesting result is shown in Fig. 8 which shows the importance of neutrino-induced reactions during the nucleosynthesis process. Thus, neutrino interactions do not only determine the Y_e value of the matter, they also strongly influence the matter flow. In the proton-rich matter, neutrons are basically incorporated into wellbound $N = Z$ nuclei like ^{56}Ni or ^{64}Ge, with some free protons available. These nuclei have halflives (beta decay or against proton capture) which are longer than the duration time of the process. Thus, the network would come to a hold, if neutrino interactions were to be ignored. While reactions induced by ν_e are suppressed, as all neutrons are in nuclei, $\bar{\nu_e}$ absorption on the free protons are a source of free neutrons which are then captured by heavy nuclei. Thus the network can continue well to nuclei heavier than ^{64}Ge. It is currently investigated how much this new nucleosynthesis process (called νp process by Martinez-Pinedo) contributes to the production of proton-rich nuclei in the mass range $A \sim 64 - 100$.

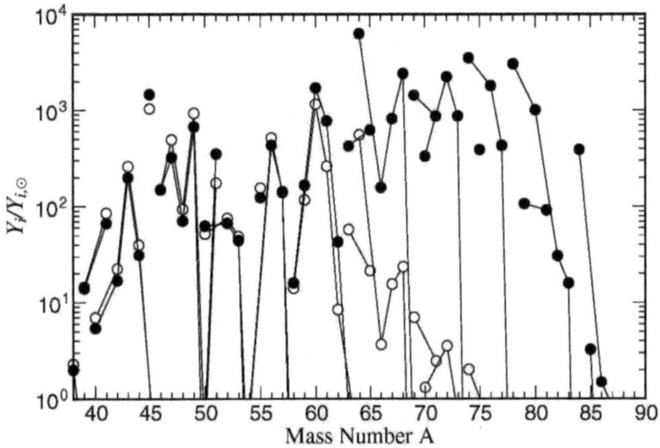

Fig. 8. Elemental abundance yields (normalized to solar) for elements produced in the proton-rich environment shortly after the supernova shock formation. The matter flow stops at nuclei like ^{56}Ni and ^{64}Ge (open circles), but can proceed to heavier elements if neutrino reactions are included during the network (full circles) (from [86]).

4.6 Constraining Neutrino-nucleus Cross Sections from Data

Currently no data for inelastic neutral-current neutrino-nucleus cross sections are available for supernova-relevant nuclei ($A \sim 60$). However, as has been demonstrated in [87], precision M1 data, obtained by inelastic electron scattering, supply the required information about the Gamow-Teller GT_0 distribution which determines the inelastic neutrino-nucleus cross sections for supernova neutrino energies. The argumentation is built on the observation that for M1 transitions the isovector part dominates and the respective isovector M1 operator is given by

$$\mathcal{O}_{iv} = \sqrt{\frac{3}{4\pi}} \sum_k \left[\mathbf{l}(k) bft_z(k) + 4.706\sigma(k)\mathbf{t_z}(k)\right] \mu_N \tag{40}$$

where the sum is over all nucleons and μ_N is the nuclear magneton, and the spin part of the isovector M1 operator is proportional to the desired zero-component of the GT operator. Thus, experimental M1 data yield the needed GT_0 information to determine supernova neutrino-nucleus cross sections, to the extent that the isoscalar and orbital pieces present in the M1 operator can be neglected. First, it is wellknown that the major strength of the orbital and spin M1 responses are energetically well separated. Furthermore, the orbital part is strongly related to deformation and is suppressed in spherical nuclei, like ^{50}Ti, ^{52}Cr, ^{54}Fe. These nuclei have the additional advantage that M1 response data exist from high-resolution inelastic electron scattering experiments [88]. Satisfyingly, large-scale shell model calculations reproduce the M1 data quite well, even in details [87]. The calculation also confirms that the orbital and isoscalar M1 strengths are much smaller than the isovector spin strength. Thus, the M1 data represent, in a good approximation, the needed GT_0 information (upto a constant factor). Fig. 9 compares the inelastic neutrino-nucleus cross sections for the 3 studied nuclei, calculated from the experimental M1 data and from the shell model GT_0 strength. The agreement is quite satisfactory. It is further improved, if one corrects for possible M1 strength outside of the explored experimental energy window. Also differential neutrino-nucleus cross sections as functions of initial and final neutrino energies, calculated from the M1 data and the shell model, agree quite well [87]. On the basis of this comparison, one can conclude that the shell model is validated for the calculations of inelastic neutral-current supernova neutrino-nucleus cross sections. This model can then also be used to calculate these cross sections at the finite temperature in the supernova environment [87]. Juodagalvis et al. have calculated double-differential inelastic neutrino-nucleus cross sections for many nuclei in the $A \sim 56$ mass range, based on shell model treatment of the GT transitions and on RPA studies of the forbidden contributions [81]. A detailed table of the cross sections as function of initial and final neutrino energies and for various temperatures is available from the authors of [81].

4.7 Supernova Neutrino Observation and Constraints

Supernova neutrinos from SN1987a have been observed by the Kamiokande and IMB detectors and have confirmed the general supernova picture. The observed events were most likely due to $\bar{\nu}_e$ neutrinos. To detect the predicted differences in the distributions for the various neutrino families and thus more restrictive tests of current supernova models requires the ability of neutrino spectroscopy by the neutrino detectors. Current and future detectors (e.g. Superkamiokande, SNO, KamLAND, ICARUS, OMNIS) have this capability and will be able to distinguish between the different neutrino types and determine their individual spectra. The various detectors need neutrino-induced cross sections for ^{16}O (Superkamiokande), ^{40}Ar (ICARUS) and ^{208}Pb (OMNIS) (see [89]).

While supernova ν_μ, ν_τ and their antiparticles (combined called ν_x) and ν_e neutrinos are not directly observed yet, Heger et al. [90] point out that

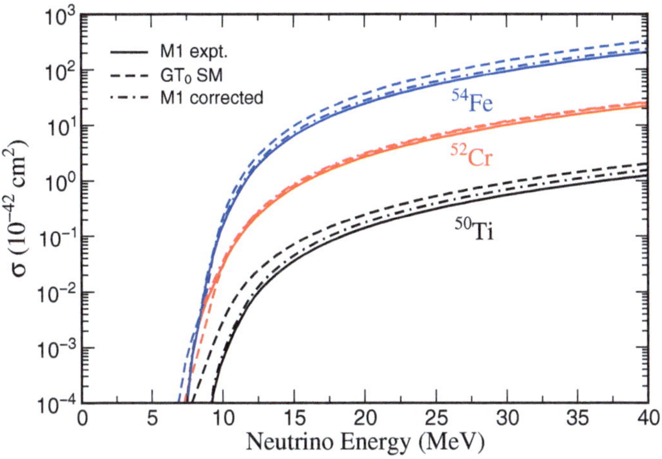

Fig. 9. Neutrino-nucleus cross sections calculated from the M1 data (solid lines) and the shell-model GT_0 distributions (dotted) for ^{50}Ti (multiplied by 0.1), ^{52}Cr, and ^{54}Fe (times 10). The long-dashed lines show the cross sections from the M1 data, corrected for possible strength outside the experimental energy window.

constraints on their spectra might be attainable from nucleosynthesis considerations within the ν process [91]. In the ν process certain nuclides can be made in a large fraction of their observed solar abundance by neutrino-induced spallation off nuclei in the outer shells of a massive star during a supernova explosion. The main observation is that the huge fluence of neutrinos in a supernova can overcome the tinyness of the neutrino-nucleus cross sections and in situations where, as a rule of thumb [92], the solar abundance of the daughter is smaller than the parent abundance by three orders of magnitude or more, neutrino nucleosynthesis can significantly contribute to the solar production of the daughter. Woosley et al. [91] showed that the ν-process, i.e. nucleosynthesis by neutrino-induced reactions, can be responsible for most or a large fraction of the solar ^{11}B, ^{19}F, ^{138}La and ^{180}Ta abundance. For ^{11}B and ^{19}F the ν-process excites the parent nuclides ^{12}C and ^{20}Ne above the particle thresholds where they then decay by proton or neutron emission. As both nuclides have rather large particle thresholds, the neutrino nucleosynthesis of ^{11}B and ^{19}F is dominated by neutral-current reactions induced by supernova ν_x neutrinos which have the higher energy spectrum compared to ν_e and $\bar{\nu}_e$ neutrinos. However, as found in detailed stellar evolution studies [90] the rare odd-odd nuclide ^{138}La is mainly made by the charged-current reaction $^{138}\mathrm{Ba}(\nu_e, e^-)^{138}\mathrm{La}$. Hence, the ν-process is potentially sensitive to the spectra and luminosity of ν_e and ν_x neutrinos, which are the neutrino types not observed from SN1987a. The GT strength on ^{138}Ba has recently been measured by the (^{3}He,t) reaction in Osaka [88]; the $^{138}\mathrm{Ba}(\nu_e, e^-)^{138}\mathrm{La}$

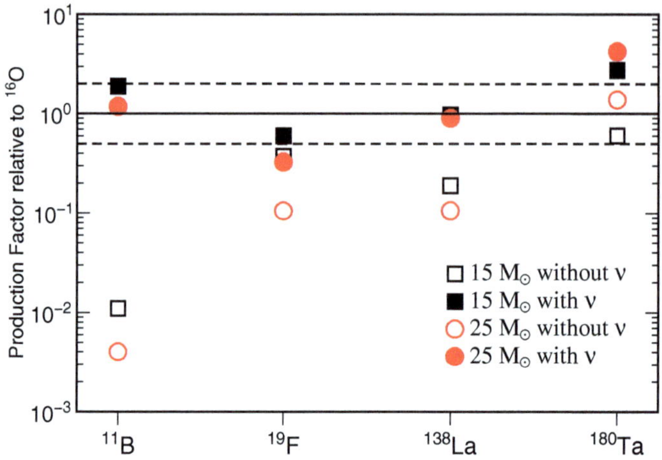

Fig. 10. Abundance yields (normalized to solar) for selected nuclei in 15 $M_\odot$ and 25 $M_\odot$ stars with and without the consideration of neutrino nucleosynthesis reactions. (from [90]).

cross section derived from these data agrees quite well with the calculated one adopted in [90].

We mention that neutrino nucleosynthesis studies are quite evolved requiring state-of-the-art stellar models with an extensive nuclear network. In the first step stellar evolution and nucleosynthesis is followed from the initial hydrogen burning up to the presupernova models. The post-supernova treatment then includes the passage of a neutrino fluence through the outer layers of the star, followed by the shock wave which heats the material and also induces noticeable nucleosynthesis, mainly by photodissociation (the γ process). Modelling of the shock heating is quite essential as the associated γ process destroys many of the daughter nuclides previously produced by neutrino nucleosynthesis.

Acknowledgement

The work presented here has benefitted from fruitful discussions and collaborations with Gabriel Martinez-Pinedo, David Dean, Carla Fröhlich, Alexander Heger, Raphael Hix, Thomas Janka, Andrius Juodagalvis, Tony Mezzacappa, Bronson Messer, Matthias Liebendörfer, Peter von Neumann-Cosel, Achim Richter, Jorge Sampaio, F.-K. Thielemann and Stan Woosley.

References

1. C.E. Rolfs and W.S. Rodney *Cauldrons in the Cosmos*, Chicago Press, Chicago (1988)
2. R.L. Macklin and J.H. Gibbons, Rev. Mod. Phys. **37** (1965) 166
3. H.A. Bethe, Rev. Mod. Phys. **9** (1937) 69
4. W.A. Fowler, G.R. Caughlan and B.A. Zimmerman, Ann. Rev. Astr. Astrophys. **5** (1967) 525
5. J.N. Bahcall, Phys. Rev. Lett. **17** (1966) 398
6. W.A. Fowler, G.R. Caughlan and B.A. Zimmerman, Ann. Rev. Astr. Astrophys. **13** (1975) 69
7. G.R. Caughlan, W.A. Fowler, M.J. Harris and B.A. Zimmerman, At. Nucl. Data Tables **32** (1985) 197
8. G.R. Caughlan and W.A. Fowler, At. Nucl. Data Tables **40** (1988) 238
9. D.D. Clayton, *Principles of Stellar Evolution and Nucleosynthesis* (McGraw-Hill), New York (1968)
10. C. Angulo *et al.*, Nucl. Phys. **A656** (1999) 3
11. E.E. Salpeter, Austr. J. Phys. **7** (1954) 373
12. H.J. Assenbaum, K. Langanke and C.E. Rolfs, Z. Phys. **A327** (1987) 461
13. T. Shoppa, S.E. Koonin, K. Langanke and R. Seki, Phys. Rev. **C48** (1993) 837
14. M. Aliotta *et al.*, Nucl. Phys. **A690** (2001) 790
15. S.E. Woosley, in: *Proceedings of the Accelerated Radioactive Beam Workshop*, eds. L. Buchmann and J.M. D'Auria, TRIUMF, Canada (1985)
16. E. Adelberger *et al.*, Rev. Mod. Phys. **70** (1998) 1265
17. J.N. Bahcall, *Neutrino Astrophysics* (Cambridge University Press), Cambridge, 1989
18. R. Bonetti *et al.*, Phys. Rev. Lett. **82** (1999) 5205
19. A.R. Junghans *et al.*, Phys. Rev. **C68** (2003) 065803
20. G. Imbriani *et al.*, Eur. Phys. J. **A25** (2005) 455
21. R.C. Runkle *et al.*, Phys. Rev. Lett. **94** (2005) 082503
22. S. Degl'Innocenti *et al.*, Phys. Lett. **B590** (2004) 13
23. S.N. Ahmed *et al.*, Phys. Rev. Lett. **87** (2001) 071301; **89** (2002) 011301; preprint nucl-ex/0309004
24. E.E. Salpeter, Phys. Rev. **88** (1952) 547; **107** (1957) 516
25. S.M. Austin, Nucl. Phys. **A758** (2005) 375
26. H. Fynbo *et al.*, Nature **433** (2005) 135
27. L. Buchmann *et al.*, Phys. Rev. Lett. **70** (1993) 726
28. R. Kunz, M. Jaeger, A. Mayer, J. W. Hammer, G. Staudt, S. Harissopulos, and T. Paradellis, Phys.Rev.Lett. **86** (2001), 3244
29. P. Tischhauser *et al.*, Phys. Rev. Lett. **88** (2002) 072501
30. S.L. Shapiro and S.A. Teukolsky, *Black Holes, White Dwarfs and Neutron Stars* (Wiley-Interscience), New York (1983)
31. B.S. Meyer and J.H. Walsh *Nuclear Physics in the Universe*, eds. M.W. Guidry and M.R. Strayer (Institute of Physics), Bristol 1993
32. H.A. Bethe, G.E. Brown, J. Applegate and J.M. Lattimer, Nucl. Phys. **A324** (1979) 487
33. H.A. Bethe, Rev. Mod. Phys. **62** (1990) 801
34. P.Goldreich and S.V. Weber, Ap.J. **238** (1980) 991
35. J.R. Wilson, in *Numerical Astrophysics*, ed. by J.M. Centrella, J.M. LeBlanc and R.L. Bowers, (Jones and Bartlett, Boston, 1985) p. 422

36. A. Burrows, Ann.Rev.Nucl.Sci. **40** (1990) 181.
37. A. Burrows, Ap.J. **334** (1988) 891.
38. A. Burrows and B.A. Fryxell, Science **258** (1992) 430.
39. E. Müller and H.-T. Janka, Astron.Astrophys. **317** (1994) 140
40. S. Reddy, M. Prakash and J.M. Lattimer, Phys.Rev. D **58** (1998) 3009.
41. A. Burrows and R.F. Sawyer, Phys.Rev. C **58** (1998)
42. H.Th. Janka and M. Rampp, Astrophys. J. **539** (2000) L33
43. A. Mezzacappa *et al.*, Phys. Rev. Lett. **86** (2001) 1935
44. J.R. Wilson, in *Fermi and Astrophysics*, ed. R. Ruffini, to be published in Nuovo Cimento
45. G.M. Fuller, W.A. Fowler and M.J. Newman, ApJS **42** (1980) 447
46. G.M. Fuller, W.A. Fowler and M.J. Newman, ApJS **48** (1982) 279
47. G.M. Fuller, W.A. Fowler and M.J. Newman, ApJ **252** (1982) 715
48. G.M. Fuller, W.A. Fowler and M.J. Newman, ApJ **293** (1985) 1
49. T. Oda, M. Hino, K. Muto, M. Takahara and K. Sato, At. Data Nucl. Data Tabl. **56** (1994) 231
50. E. Caurier, K. Langanke, G. Martínez-Pinedo and F. Nowacki, Nucl. Phys. A **653** (1999) 439.
51. C. Bäumer *et al.*, Phys. Rev. **C68** (2003) 031303
52. K. Langanke and G. Martínez-Pinedo, Nucl. Phys. **A673** (2000) 481
53. K. Langanke and G. Martínez-Pinedo, At. Data Nucl. Data Tables **79** (2001) 1
54. M.B. Aufderheide, I. Fushiki, S.E. Woosley and D.H. Hartmann, Ap.J.S. **91** (1994) 389
55. A. Heger, K. Langanke, G. Martinez-Pinedo and S.E. Woosley, Phys. Rev. Lett. **86** (2001) 1678
56. A. Heger, S.E. Woosley, G. Martinez-Pinedo and K. Langanke, Astr. J. **560** (2001) 307
57. S.E. Woosley and T.A. Weaver, Astrophys. J. Suppl. **101** (1995) 181
58. S. W. Bruenn, Astrophys. J. Suppl. **58**, 771 (1985); A. Mezzacappa and S. W. Bruenn, Astrophys. J. **405**, 637 (1993); **410**, 740 (1993)
59. G. M. Fuller, Astrophys. J. **252**, 741 (1982)
60. J. Cooperstein and J. Wambach, Nucl. Phys. **A420** (1984) 591
61. J. Pruet and G.M. Fuller, Astrophys. J. Suppl. **149** (2003) 189
62. K. Langanke, E. Kolbe and D.J. Dean, Phys. Rev. C **63**, 032801 (2001)
63. O. Sorlin *et al.*, Phys. Rev. Lett. **88**, 092501 (2002)
64. M.C. Vetterli *et al.*, Phys. Rev. C **45**, 997 (1992)
65. R.L. Helmer *et al.*, Phys. Rev. C **55**, 2802 (1997)
66. C.W. Johnson, S.E. Koonin, G.H. Lang and W.E. Ormand, Phys. Rev. Lett. **69** (1992) 3157
67. S.E. Koonin, D.J. Dean and K. Langanke, Phys. Rep. **278** (1997) 1
68. J. Hubbard, Phys. Lett. **3** (1959) 77; R.D. Stratonovich, Sov. Phys. - Dokl. **2** (1958) 416
69. K. Langanke *et al.*, Phys. Rev. Lett. **90** (2003) 241102
70. K. Langanke *et al.*, Phys Rev. **C67** (2003) 044314
71. K. Langanke, D.J. Dean and W. Nazarewicz, Nucl. Phys. **A728** (2003) 109
72. K. Langanke, D.J. Dean and W. Nazarewicz, Nucl. Phys. **A757** (2005) 360
73. K. Langanke and G. Martinez-Pinedo, Rev. Mod. Phys. **75** (2003) 819
74. R.W. Hix *et al.*, Phys. Rev. Lett. **91** (2003) 201102

75. H-Th. Janka *et al.*, to be published
76. W.C. Haxton, Phys. Rev. Lett. **60** (1988) 1999
77. S.W. Bruenn and W.C. Haxton, Astr. J. **376** (1991) 678
78. J.M. Sampaio, K. Langanke and G. Martinez-Pinedo, Phys. Lett. **B511** (2001) 11
79. J.M. Sampaio, K. Langanke, G. Martinez-Pinedo and D.J. Dean, Phys. Lett. **B529** (2002) 19
80. J. Toivanen *et al.*, Nucl. Phys. **A694** (2001) 395
81. A. Juodagalvis *et al.*, Nucl. Phys. **A747** (2005) 87
82. S. Reddy, M. Prakash, J.M. Lattimer and S.A. Pons, Phys. Rev **C59** (1999) 2888
83. A. Mezzacappa, Nucl. Phys. **A688** (2001) 158c
84. J. Pruet *et al.*, Astr. J. **623** (2005) 1
85. C. Fröhlich *et al.*, astro-ph/0410208, Astr. J. in press
86. C. Fröhlich *et al.*, submitted to Phys. Rev. Lett.
87. K. Langanke, G. Martinez-Pinedo, P. v. Neumann-Cosel and A. Richter, Phys. Rev. Lett. **93** (2004) 202501
88. P. v. Neumann-Cosel *et al.*, Phys. Lett. **B443** (1998) 1
89. E. Kolbe *et al.*, J. Phys. **29** (2003) 2569
90. A. Heger *et al.*, Phys. Lett. **B606** (2005) 258
91. S.E. Woosley *et al.*, Astr. J. **356** (1990) 272
92. S.E. Woosley, private communication

Pulsars as Probes of Relativistic Gravity, Nuclear Matter, and Astrophysical Plasmas

James M. Cordes

Cornell University, Ithaca, NY 14853 USA cordes@astro.cornell.edu

1 Introduction

Pulsars are rotating neutron stars (NSs) that provide unique information about their interiors, about the relativistic plasma physics of their magnetospheres, and about the intervening media through which pulses emitted by these objects must propagate. NS spins represent the clock mechanism that underlies usage of pulsars as precise timekeepers for their motions. This last topic is the final aim of this chapter, namely using pulsars to probe gravity in binary stellar systems — particularly those involving a pulsar and another compact star (white dwarf [WD], NS, or black hole [BH])— and also to probe gravitational wave backgrounds that affect the location of both the pulsar and the Earth. Pulsars also provide exquisite opportunities for probing neutron stars and their magnetospheres through a variety of techniques.

To set the stage, I will describe the historical and scientific background and the phenomenology needed to understand the capabilities and limitations of using pulsars as clocks. The four main sections of this chapter are:

1. Pulsar basics that are relevant to their use as laboratories.
2. Pulsar populations in the Milky Way Galaxy, their high velocities and mechanisms for inducing them, and modeling the interstellar medium (ISM) using pulsars;
3. Pulsar surveys and how they will yield rare objects that can serve as gravitational laboratories;
4. Using pulsar timing to probe gravity and the internal workings of neutron stars.

The chapter finishes with a list of "big questions" that define the forefront of NS and related science.

The reader is referred to other recent articles that discuss in greater detail some of the topics presented here, including the tremendous capabilities of the next generation radio telescope, the Square Kilometer Array [1, 2, 3, 4, 5, 6], and books on compact objects [7, 8].

A. Carramiñana et al. (eds.), Solar, Stellar and Galactic Connections between Particle Physics and Astrophysics, 43–76.

2 Neutron Stars' Greatest Hits

Since the pulsar discovery in 1967 [9], neutron star astronomy has provided many discoveries and important insights into the way stars evolve, and provided opportunities for probing nuclear matter and making precision measurements of neutron star masses and constraints on relativistic gravity. A short list of important milestones in this history includes:

1. 1967: Discovery of pulsars by Jocelyn Bell, Tony Hewish et al. using a radio array telescope designed to find compact radio sources that displayed interplanetary scintillation [9].
2. 1967: Discovery of gamma-ray bursts (GRBs) with the Vela satellite network, whose purpose was to detect gamma-rays from nuclear explosions. GRBs were unannounced until 1973 [10].
3. 1968: Discovery of the pulsar in the Crab Nebula and measurement of the secular increase in spin period, verifying a prediction by T. Gold about the spinning NS concept [11, 12, 13].
4. 1969: Detection of a rapid spinup in the Vela pulsar [14], signifying that the NS interior contained neutrons in a superfluid state.
5. 1973: Nobel Prize in Physics to A. Hewish for the discovery of pulsars.
6. 1974: Discovery of the first binary pulsar, B1913+16, with a NS companion in an 8-hr orbit [15].
7. 1979: The March 5, 1979 gamma-ray burst event from a source in the Large Magellanic Cloud was detected by many satellites. The source turned out to be the first *soft-gamma repeater* (SGR), now recognized to be a NS with very large magnetic field, a *magnetar*.
8. 1982: Discovery of the first millisecond pulsar, B1937+21, with a period of 1.56 ms [16].
9. 1992: Discovery of the first extrasolar planets. Remarkably, these orbit a millisecond pulsar and were identified using pulse timing methods [17].
10. 1993: Nobel Prize to R. Hulse and J. Taylor for the discovery of the binary neutron star system containing the pulsar B1913+16, which led to a demonstration that gravitational waves are emitted by the system in accord with Einstein's General Theory of Relativity.
11. 1990s-2000s: Identification of pulsars as a runaway population through constraints on their space velocity distribution and discovery of an individual object moving fast enough to escape the Galaxy. About 50% of the pulsar population will escape [18, 19].
12. 2003: Discovery of the first double *pulsar* [20], J0737−3939, a 2.4 hr binary with a relatively slow, 2.8 s pulsar and an old, recycled pulsar with 23 ms period. This system is the best gravitational laboratory so far, showing GR apsidal advance of 17 deg yr^{-1} as compared to Mercury's $\sim$ 43 arc sec $\mathrm{century}^{-1}$.
13. 2004: A giant flare from the magnetar SGR 1806–20 on 27 December 2004 induced an ionospheric perturbation comparable to that produced

by the Sun, even though the object is halfway across the Milky Way Galaxy [21, 22, 23].

14. 2005: Analysis of the short-duration GRB 050709 and its afterglow, suggesting a lower total event energy and a location offset from the source's birth galaxy, consistent with a merger event of two NS or a NS and a black hole [24].

3 Pulsar Basics

Pulsars are important in physics and astrophysics because they serve as

(1) Laboratories for gravity near the strong field limit, especially binary pulsars with NS or BH companions in compact orbits with orbital periods of a few hours or less;

(2) Gravitational wave detectors of long-wavelength gravitational waves (wavelengths $\sim$ light-years);

(3) Fossil information for the physics of core-collapse supernovae via their characteristic high velocities in the Galaxy;

(4) Venues for nonlinear, relativistic plasma physics; and

(5) Probes of the interstellar medium, including magnetic fields and turbulence and as a foreground through which we view the radio universe.

In order to understand and use NS as gravitational laboratories, etc., we need to understand a great deal of pulsar phenomenology.

3.1 Who Discovered Neutron Stars?

It is well known that Baade and Zwicky hypothesized, in 1933, the existence of neutron stars as remnant cores of stars that explode as supernovae and they did so only one year after the neutron itself was discovered by Chadwick and only two years after Chandrasekhar published an analysis of the limiting mass for white dwarf (WD) stars. Contributing to this fast pace of events were the models for neutron stars by Oppenheimer and his colleagues in the late 1930s that incorporated General Relativity. Then followed nearly thirty years of "dark ages" until pulsars were discovered by accident by Jocelyn Bell and Tony Hewish using a low-frequency array telescope near Cambridge, UK. Since the 1967 discovery of pulsars there has been a virtual explosion of activity over the entire electromagnetic spectrum and also neutrino and gravitational wave detectors.

However, this synopsis begs the question that heads this section. Directly or indirectly who, indeed, discovered neutron stars? I think great credit should be given to the Chinese and Native Americans. The Chinese detected — and recorded — the explosion of the star on July 4, 1054 that became the Crab

Nebula and left behind a rapidly rotating neutron star, the Crab pulsar, that was discovered in 1968, 954 years after the Chinese observations. Chinese records indicate that the explosion was visible for 23 days during daytime and were able to monitor it for 653 days until it became invisible to the naked eye. Less quantitatively but equally dramatic is the pictograph from the Anasazis in Chaco Canyon, New Mexico that shows a bright star near a cresent moon that is consistent with the lunar phase on the dates given by the Chinese.

The Crab Nebula itself was discovered by Charles Messier in 1758, becoming the first entry in his catalog of "uninteresting" nebulous objects to be avoided if you were a serious searcher of comets. The Crab appellation was given by Lord Rosse in 1844. In modern times both the Crab nebula and its pulsar have been the targets of thousands of astronomical observations. While sometimes considered Rosetta stones for supernova remnants and pulsars owing to the large body of empirical data and their easy detection across the electromagnetic spectrum, both may be anomalous and thus *not* representative of their object classes.

3.2 Endstates of Stellar Evolution

The cores of ordinary stars — by which we mean main sequence stars in whose centers hydrogen is fused to helium — eventually collapse because gravity wins the tug of war it has with gas pressure. What then happens depends on the mass of the collapsing core. For low-enough masses, the collapse is halted by degeneracy pressure. This pressure arises from the Pauli exclusion principle for fermions, which states that no two particles may have the same quantum state, and which may be viewed in terms of the uncertainty principle, $\Delta x \Delta p \geq \hbar$. Gravitational collapse requires non-zero particle momenta and, hence, pressure that exceeds that associated with random thermal motions. When particles become relativistic, the star cannot find an equilibrium state at finite radius. Low mass cores become white dwarf stars while intermediate masses become neutron stars supported, respectively, by degenerate electron and degenerate neutron pressure. The Chandrasekhar mass is the maximum mass that can be supported and depends on the mean particle mass and is in the ballpark of 1 $M_{\odot}$ for white dwarfs and $\sim$ 1.4 to 2 $M_{\odot}$ for NSs. The uncertainty on the maximum NS mass is both theoretical and empirical, the former because of our lack of precise knowledge about the equation of state of nuclear matter, the latter because of residual model dependences on inferences from binary pulsar timing observations. Still larger masses lead to black holes.

While the three mass regimes for WDs, NSs and BHs are defined primarily with respect to fundamental constants and secondarily to composition, astrophysical factors complexify how the original masses of progenitor main sequence stars lead to particular compact objects. These are associated with the fact that most stars are born in multiple star systems (binary, triple stars, etc.). During the evolution of a binary, for example, mass exchange can take place as one of the stars evolves off the main sequence and later its companion. In sufficiently compact systems, mass exchange can change the ultimate endstate from a WD to a NS, for example, of the acceptor object if it receives a significant amount of mass. Some binaries are disrupted if the more massive, faster-evolving object explodes as a supernova and carries away sufficient energy to unbind the system. Others become unbound after a second explosion. Those massive, rare binaries that remain bound and are compact will inspiral owing to energy loss from gravitational wave emission, ultimately leading to a gravitational catastrophe associated with the merger of the two stars.

3.3 Why Pulsars are Highly Magnetized Neutron Stars

Radio pulsars were discovered by accident during a survey of the sky at 81 MHz (just below the FM band) whose goal was to identify compact quasi-stellar objects that displayed interplanetary scintillation[1]. The key technical element of the survey that allowed the discovery of pulsars was high time resolution of less than one second, an unprecedented short time scale for astronomical samples. On a human level, it was the persistence of Jocelyn Bell in pursuing the origin of the "bit of scruff" on literally miles of chart recordings that led to the discovery.

In a pulsar, radio pulsations are caused by the rotation of the NS, which swings a beam of radiation into and out of the line of sight to the observer. This light house model emerged in the late 1960s as the overwhelmingly favored picture for the pulsar phenomenon for the following reasons:

1. Other causes of the periodicity, such as radial vibrations of a star or orbital motion, were not fast enough to account for the range of periods seen. Moreover, the increase of the pulse period with time was much more readily explainable as the spindown of a rotating object. Decaying orbital motion would lead to a decrease in pulse period, contrary to observation.
2. The energetics of spindown were consistent with the rate at which energy is pumped into the surroundings of the Crab pulsar [25].

Stability of Spinning Objects

Some simple calculations can verify these statements. First, consider the stability of a spinning object, by analyzing the forces on a test mass m on the surface of a star of mass M, radius R_* and angular spin frequency Ω. To stay bound, the gravitational force must exceed the centrifugal force:

[1] IPS is a propagation effect similar to the twinkling of optical starlight. Twinkling is caused by turbulent, refractive index variations in the Earth's atmosphere associated with temperature variations on spatial scales ~ 1 cm. Refractive index variations in the ionized solar wind (a.k.a. the interplanetary medium) are caused by turbulent electron density variations.

$$\frac{GMm}{R_*^2} > \frac{m(\Omega R_*)^2}{R_*}.$$

By defining the mean mass density as $\bar{\rho} = M/\left(4\pi R_*^3/3\right)$ and the period $P = 2\pi/\Omega$, we can write the inequality as

$$\bar{\rho} > \frac{3\pi}{GP^2}.$$

Table 1 evaluates this expression for several pulsars and compares the implied density lower bounds with densities of ordinary stars and WDs. While the first pulsar discovery could have been accounted for by a spinning WD, the discovery of the Crab pulsar ruled out spinning WDs and the discovery of the first millisecond pulsar, B1937+21, implied a mean density comparable to that of an atomic nucleus.

Table 1. Constraints on Average Densities of Spinning Objects

Object	Year	P (ms)	Implied Average Density $\bar{\rho}$ (gm cm^{-3})
Ordinary star	—	—	~ 1
White Dwarf	—	—	$\sim 10^6$
First Pulsar Discovered (CP1919)	1967	1337	$> 10^{5.9}$
Crab Pulsar	1968	33.1	$> 10^{11}$
Millisecond pulsar (B1937+21)	1982	1.56	$> 10^{13.8}$
Fastest spin allowed	—	0.5	$> 10^{14}$

Spindown, Magnetic Torques and the Braking Index

A second calculation yields the spindown energy loss rate and relates it to the surface magnetic field. We assume that the neutron star is highly conducting and that a dipolar magnetic field is frozen into it. For a magnetic moment $\boldsymbol{m}$ inclined at an angle α from the spin axis, the energy loss rate from magnetic dipole radiation (MDR) is

$$\dot{E}_{\mathrm{MDR}} = I\Omega\dot{\Omega} = \frac{2|\boldsymbol{m}|^2\Omega^4 \sin^2\alpha}{3c^3},$$

where Ω is the spin frequency and $\dot{\Omega}$ is its time derivative. Both quantities are observables that are derived readily from pulsar timing measurements. The moment of inertia is $I = 10^{45} I_{45}$ gm cm^2 typical of NS models to within a factor of two. The surface magnetic field B is

$$B = 10^{12}\,\text{Gauss}\left(P\dot{P}_{-15}\right)^{1/2}.$$

where the temporal period derivative $\dot{P} = 10^{-15}\,\text{s}\,\text{s}^{-1}\,\dot{\text{P}}_{-15}$. It was well known before the discovery of pulsars that the Crab Nebula's expansion was accelerating and energy input was needed at the level $\sim 10^{38}$ erg s^{-1}. With $P = 0.33$s and $\dot{P}_{-15} = 422.7$, the implied $\dot{E}$ accounts for that needed to drive the acceleration.

The characteristic time scale for spindown is given by

$$\tau_S = \frac{\Omega}{2|\dot{\Omega}|} = \frac{P}{2\dot{P}},$$

which is related to the chronological age t by

$$t = \tau_S\left[1 - (P_0/P)^{n-1}\right].$$

If a pulsar was born spinning with a period $P_0 \ll P$ then the characteristic and chronological time scales are nearly identical. How does this work for the Crab pulsar? The Crab Nebula and pulsar were born in 1054 AD, so their chronological age of 952 yr (in 2006) is smaller than the spindown age, $\tau_S = 1239$ yr, calculated from current spindown parameters. If the Crab pulsar was born with P_0 a non-negligible fraction of its present period, the spindown time can be reconciled with the true age. However, there are likely to be other factors, such as non-constancy of the braking index or time-variation of the magnetic field.

Do pulsars really spin down in accordance with MDR? The short answer is no. But one must qualify this by stating that the implied torques are not really all that different from what we expect from MDR. We generalize the spindown law by writing

$$\dot{\Omega} = k_\Omega \Omega^n,$$

where k_Ω is a coefficient that may or may not depend on time and n is the braking index. For MDR, $n = 3$. Detailed measurements of a few pulsars, particularly young, highly magnetized pulsars with large Ω and large $\dot{\Omega}$, it is also possible to measure the second derivative, $\ddot{\Omega}$. If we *assume* that k_Ω is time independent, then the braking index is

$$\hat{n} = \frac{\Omega\ddot{\Omega}}{\dot{\Omega}^2},$$

where the caret signifies that the value obtained is merely an estimate, which may not equal the true braking index. What has been found is that $\hat{n} \approx 2.5$ for a few objects and $\hat{n} \approx 1.1$ for the Vela pulsar.

Such values of n can be interpreted in various ways. First, they are not all that different from $n = 3$ for MDR. That suggests that the use of the MDR formula to estimate surface magnetic fields may be ok. Second, the differences from $n = 3$ may result from several factors, including

1. The true braking index may be different from $n = 3$ if there are other contributions to $\dot{E}$ that combine with the magnetic torque. Examples are mass loss from a wind or gravitational radiation.
2. The coefficient k_Ω may vary with time. If so, then
$$\hat{n} = n + (n-1)\frac{\tau_S}{\tau_k},$$
where $\tau_k \equiv -\dot{k}/\ddot{k}$. For $n > 1$, a decaying (growing) coefficient yields $\hat{n} > n$ ($\hat{n} < n$). At present, there is conflicting evidence about whether magnetic fields decay on time scales that are relevant to the lifetimes of pulsars.
3. The size of the magnetosphere influences the torque mechanism, as the finite size of the conducting region relative to the light cylinder radius (defined below) can reduce the apparent braking index to values < 3 [26] . The braking index is $n \sim 2$ for a very short-period pulsar and asymptotes to $n = 3$ as the pulsar spins down.
4. Internal torques involving interactions between the crust and a more fastly rotating superfluid can alter the effective moment of inertia of the crust as well as transferring angular momentum from the superfluid to the crust.
5. Pulsars that show X-ray jets such as the Crab and Vela pulsars may also have conducting equatorial disks that alter the spindown torques, tending to reduce the braking index [27].

It is possible that there are other contributions to the torque besides losses from MDR. Internal torques from interactions with a more fastly-rotating neutron superfluid are thought to underly *glitches* [28], which appear as sudden spinups of the pulsar. An additional external torque may arise from any influx of neutral material into the magnetosphere. Cheng [29] writes the total spindown energy loss as

$$\dot{E}_{\rm total} = \dot{E}_{\rm MDR} + \frac{\Omega^4 B^2 R^6}{c^3}\left(\frac{I_*}{I_{\rm GJ}}\right),$$

where the first term is the contribution from low-frequency dipole radiation and the second term results from the torque exerted by magnetospheric currents; I_* is the actual current and $I_{\rm GJ}$ is the Goldreich-Julian current [30], the integral over the magnetic polar cap of the current $-c\Omega \cdot \mathrm{B}/2\pi c$ that is implied if the magnetosphere is nearly "force free." External gains in Cheng's model contribute to the current flow and thus alter the energy loss rate from that expected purely from MDR.

3.4 Manifestations of Neutron Stars

Originally, in the 1930s, the prospects appeared slim for detecting neutron stars because the envisioned method was to detect thermal X-rays from the blackbody emission from a hot neutron star. The small radius of a a neutron star (~ 10 km) worked against the success of such a plan, at least at those times. As discussed, neutron stars were first detected by accident through radio emission from their magnetospheres. Only later in the 1960s were neutron stars detected in X-rays and that was from magnetospheric emission or from hot, accreting gas, not surface blackbody radiation, A short list of the ways in which neutron stars have been detected is:

1. **Rotation-driven pulsars:** Radiation has been seen across the entire electromagnetic spectrum from the Crab pulsar that derives from spin energy losses. How the spin energy is converted to photons is not well understood but involves radiation that appears to be coupled to the magnetic field, as in curvature and synchrotron radiation. Inverse Compton radiation may also contribute to high-energy emission. Radio emission is necessarily *coherent*, i.e. the intensity levels seen require collective radiation rather than randomly superposed radiation from independent electrons.
2. **Accretion driven pulsars:** NS in binary systems usually accrete material from a companion star at one or more stages. The X-ray luminosity is $L_X = \varepsilon \dot{M} c^2$, where $\dot{M}$ is the accretion rate and $\varepsilon \approx 0.1$ is an efficiency factor. Gas falling into the magnetosphere, if free falling, would approach a few tenths of the speed of light. Collisional heating drives the gas to X-ray emitting temperatures. Low-mass X-ray binaries (LMXB) are NS with stellar companions that typically will form a WD. High-mass X-ray binaries are NS and BH with companions that also will eventually form another NS or BH.
3. **Magnetic-driven emission from Magnetars:** some objects appear to radiate from dissipation of magnetic energy rather than spin or accretion energy. Magnetars radiate pulsed X-rays in excess of their spin energy loss rates, $\dot{E}$ and generally appear to be isolated (as opposed to binary) objects. Such emission may derive from heating of the NS crust from dissipation of magnetic energy. Energetic bursts from anomalous X-ray pulsars (AXPs) and soft-gamma repeaters (SGRs) probably derive from crust quakes driven by magnetic stresses.
4. **Gamma-ray emission from gravitational catastrophes:** We have known for many years that some NS will spiral-in and coalesce with their companions (WD, NS or BH) on time scales $\sim 10^8$ yr. A notable example is the Hulse-Taylor binary pulsar. What happens during coalescence? During the 1990s people thought that inspiraling NS-NS, NS-BH or BH-BH binaries might account for most gamma-ray bursts (GRBs). However localization of the longer GRB events ($\gg$ sec) and linking their afterglows with star-forming regions in galaxies and association of a few GRBs with extragalactic supernovae suggests that such GRBs are associated with *hypernovae*. A hypernova is an explosion that appears to involve energy releases larger than that of a canonical Type II supernova (10^{51} erg in baryonic matter and photons, 10^{53} erg in neutrinos). There may yet be a role for coalescences in the GRB story, however. Short-duration bursts may in fact be associated with coalescences, an idea that receives support

from the recent discovery of a burst source well outside the confines of an optical galaxy. Coalescing binaries will have high space velocities due to momentum kicks imparted by the explosions that produced the compact objects and will either oscillate to large distances from their host galaxy or escape the galaxy altogether (particularly at high redshift when galaxies tended to have lower masses). Detection of gravitational waves from merging binaries is to be expected in the not-too-distant future, perhaps when LIGO II comes on line or when LISA is launched.

3.5 Pulsar Classes

The distribution of pulsars vs. P and $\dot{P}$ in the so-called "$P - \dot{P}$" diagram allows us to identify groupings that reflect the formation and evolution of NS.

1. *Canonical pulsars:* These pulsars, like those first discovered, have present-day spin periods ranging from tens of milliseconds to 8 s and surface magnetic field strengths $B \sim 10^{12\pm1}$ G. They are often thought to be born with periods $\sim$ 10 ms, though evidence suggests that some objects are born with periods longer than 0.1 s. In the standard picture of NS formation, *all* pulsars start as canonical pulsars. In the $P - \dot{P}$ diagram of Fig. 1 most of these pulsars are located at $P \sim 1$s and $\dot{P} \sim 10^{-15}$. *Young pulsars* are especially important members of this class as they are associated with supernova remnants and often show large numbers of glitches.
2. *Modestly recycled pulsars:* are objects in binaries that survived a first SN explosion and subsequently accreted matter that spun-up the pulsar and reduced the effective dipolar component of the magnetic field. Accretion is terminated in these objects by a second supernova explosion that usually, but not always, unbinds the binary. Those that survive are seen today as relativistic NS-NS binaries. Evolutionarily, it is possible that some surviving binaries include black-hole companions. In the $P - \dot{P}$ diagram of Fig. 1 these pulsars are typically located around $P \sim 30$ ms and $\dot{P} \sim 10^{-18}$.
3. *Millisecond pulsars (MSPs):* objects in binaries that survive the first SN explosion and in which the companion object eventually evolves into a white dwarf. The long preceding accretion spins the pulsar up to millisecond periods while attenuating the (apparent) dipolar field component to $10^8 - 10^9$ G. The consequent small spin-down rates seem to underly the high timing precision of these objects and imply spin-down time scales that often exceed a Hubble time. In the $P - \dot{P}$ diagram (Fig. 1) these pulsars are typically located around $P \sim 5$ ms and $\dot{P} \sim 10^{-20}$. Evolutionary scenarios that produce recycled pulsars and MSPs are discussed in [31].
4. *Strong-magnetic-field pulsars:* Recently discovered radio pulsars have inferred fields $\gtrsim 10^{14}$ G [32, 33], rivalling those inferred for "magnetar" objects identified through their X-and-γ radiation that seems to derive from non-rotational sources of energy. The relationship between magnetars and these high-field radio emitting pulsars, whose radiation derives

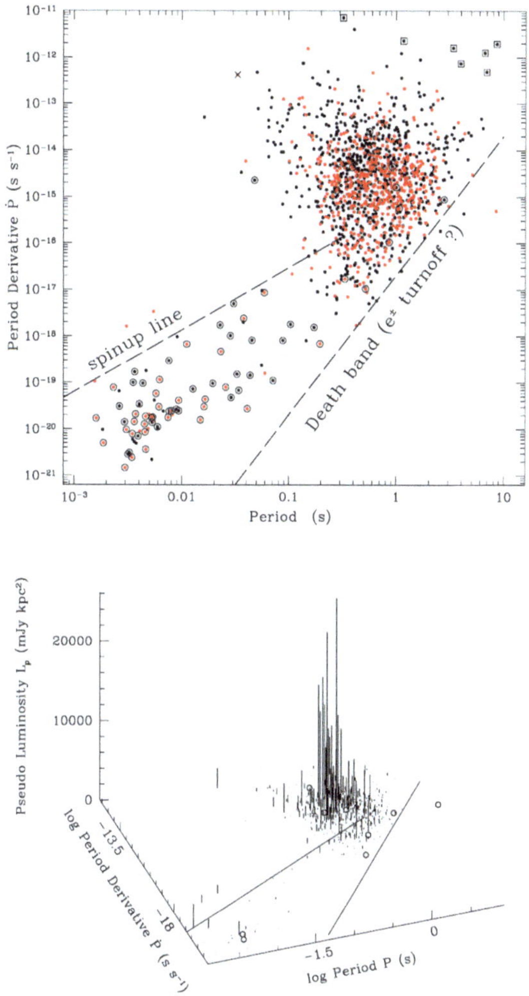

Fig. 1. *Top:* The $P - \dot{P}$ diagram for radio pulsars and magnetars using 1394 pulsars from the ANTF pulsar catalog (http://www.atnf.csiro.au/research/pulsar/psrcat/) that have $\dot{P} > 0$ and $S_{400} > 1$ mJy kpc^2. Pulsars in the largest grouping have surface magnetic field strengths $\sim 10^{12\pm1}$ Gauss. Objects in the bottom left quadrant are "recycled" pulsars, many of which are in binary systems (designated by circles around the points) and have field strengths $\sim 10^9$ Gauss. As pulsars spin down they move downward toward the right such that $\dot{P} \propto P^{2-n}$ where n is the braking index (see text). The spinup line is an estimate of the asymptotic spin state that a pulsar can reach as it spins up from accretion. The death line signifies the empirical fact that pulsars either become much weaker or shut off entirely in their radio emission as they age. *Bottom:* A 3D plot that adds a luminosity axis to the $P - \dot{P}$ plane. The third axis is actually the "pseudo-luminosity" commonly used in radio pulsar studies, $L_p = S_{400}D^2$, where S_{400} is the time-averaged flux density at 400 MHz and D is the distance in kpc. The distribution of L_p indicates that the most luminous pulsars are at intermediate spin periods and that pulsars become dimmer before they reach the death line region.

solely from spin energy, is not yet known. In the $P - \dot{P}$ diagram of Fig. 1 these radio pulsars are typically located around $P \sim 5$ s and $\dot{P} \sim 10^{-13}$.

Death or Dearth Line?

There is clearly a dearth of pulsars in the bottom right corner of the $P - \dot{P}$ diagram. Some pulsar surveys discriminate against long-period objects in the signal processing. Also, longer period pulsars have narrower beams and thus a smaller beaming fraction, defined as the fraction of objects whose beams will intersect the line of sight. Nonetheless, selection effects alone cannot explain the deficit of longer period pulsars. One explanation for the dearth is that the $e^{\pm}$ cascade thought to be required for radio emission (see below) cannot be sustained for long periods and/or small magnetic fields (and hence small $\dot{P}$s). The bounding line is not unlike that drawn in Figure 1 but one needs to invoke surface magnetic field topologies that have much smaller radii of curvature near the magnetic axis than does a dipolar field. Such may be accounted for by higher-order multipolar components near the surface. However, the distribution of pulsars does not conform to that expected if there is a *bona fide* threshold effect that is the same for all pulsars. Pulsars pile up at longer periods because $\dot{P} \propto P^{2-n} \sim P^{-1/2}$ whereas the peak density of pulsars in the $P - \dot{P}$ plane is well to the left of the drawn death line. Rather than shutting off suddenly, pulsars may simply fade out as the spin energy available for driving radiation wanes. This interpretation is consistent with the fact that the radio luminosity becomes a sizable fraction of $\dot{E}$ for long-period objects [19]. Thus the nearly empty region in the lower-right of the $P - \dot{P}$ diagram may simply result from fading away of the radio luminosity rather than a dramatic shutoff.

Spinup Line for Recycled Pulsars

NSs that remain bound in a binary after the supernova explosion that formed them often accrete material as their companions expand and overflow their Roche lobes. Accreting material produces X-ray emission and also carries angular momentum that spins up the NS because the angular momenta in the system will have aligned in the pre-supernova stages. Accreting material effectively applies a torque to the NS at the Alfven radius defined by balance of magnetic and gas pressure. Once the NS spins faster than the Keplerian orbital speed at this radius, the accretion will no longer spin up the star. This constitutes the spinup line, the asymptotic period that a NS can reach as it moves across the $P - \dot{P}$ plane to shorter periods. Another consequence of accretion is attenuation of the effective dipolar field strength, which may arise from burial and dissipation of the field or by reconfiguration of the magnetic field topology. These effects appear to account for the existence and spin properties of millisecond pulsars. That recycling can turn on radio emission as a pulsar recrosses the death-line region implies that the "death" process is reversible and does not involve dissipation of consumables, such as particular elements on the surface of the NS.

Many but not all MSPs are currently in binary systems with WD companions. The most economical explanation for the existence of isolated MSPs is that the re-activated pulsar energy loss evaporates its companion, a process that is ongoing in the MSP B1957+20. Theoretical work has trouble obtaining short-enough evaporation time scales so there appears to be some mechanism that enhances the evaporation rate.

3.6 Pulsar Magnetospheres

Neutron stars have gravitational fields $\sim 10^{11}$ larger than Earth's. However, canonical pulsars (10^{12} Gauss with 1-s spin periods) induce electric forces that are 10^9 larger than gravity.

Figure 2 shows cartoons of pulsar magnetospheres and particle acceleration regions inspired by [30, 34] along with a semi-realistic display of the radio beam pattern. Key elements of the magnetosphere include:

1. *Light Cylinder (LC):* The surface where the corotation velocity would be c is given by $R_{\mathrm{LC}} = c/\Omega = cP/2\pi = 10^{9.68}$ cm.
2. *Dipolar Magnetic Field:* Much work is consistent with the dominant field structure being dipolar. Spindown is dominated by the dipolar component because relevant torques are applied at the LC at radius $r_{\mathrm{LC}} \gg R_*$. Radio polarization measurements probe the structure of the magnetosphere and indicate in some cases that the topology at radii where *conal* radio emission originates is close to dipolar in form. However, some pulsars show departures from the dipolar form either in polarization measurements or in the appearance of "extra" pulse components. Magnetospheric physics indicates that the topology should include a toroidal component near the LC because of field line inertia.
3. *Closed Magnetosphere:* Field lines that close within the LC are nearly equipotentials and thus do not support particle acceleration.
4. *Open Field Line Region:* The open field lines extend through the LC and reconnect through unknown, complex topologies. Particle acceleration takes place because the field lines are not equipotentials.
5. *Magnetic Polar Cap:* The foot of the last open field line bordering the closed magnetosphere is at an angle θ_{PC} relative to the dipole axis given by $\sin\theta_{\mathrm{PC}} = (R_*/r_{\mathrm{LC}})^{1/2}$. For large enough P, the diameter of the magnetic polar cap is

$$2\theta_{\mathrm{PC}} \approx 2(R_*/r_{\mathrm{LC}})^{1/2} \approx 1.66^\circ\, P^{-1/2} \left(\frac{R_*}{10\,\mathrm{km}}\right)^{1/2} .$$

6. *Radio Emission Altitude:* Empirical work strongly supports the view that radio emission occurs in the open field line region at a range of altitudes, some small, some large with respect to the NS radius. Radio

pulse components include core and cone components [35]. Core emission (Fig. 2) has an angular width comparable to $2\theta_{PC}$ and so probably originates close to the NS surface. For a dipolar field, the opening angle of the open field line region at radius r_{em} is a factor $(r_{em}/R_*)^{1/2}$ larger than the polar cap size, thus yielding pulse widths (in angular units) $W = 1.66^\circ P^{-1/2} (r_{em}/10\,\mathrm{km})^{1/2}$. Rankin [36] determined a scaling law for core beam widths $\sim 2.45^\circ P^{-1/2}$ by investigating observed pulse shapes and taking into account orientation angles of spin and dipole axes relative to the line of sight. Taken at face value, this suggests an emission altitude of 22 km or about two stellar radii for $R_* = 10$ km. Alternatively, NS radii could be larger than 10 km if the equation of state is relatively hard; however, other constraints and considerations suggest that 10 km is a good estimate for R_*. It is also recognized that gravitational bending of ray paths plays a role both in the beaming of outgoing radiation and in the pair production cascade that also depends on photon trajectories. Conal emission is significantly broader than core emission and is likely produced at many NS radii (10 to 100) above the surface. Rankin [35] argues convincingly that there are probably two, essentially conal beams that vary in relative strength for different pulsars.

7. *Outer gap emission:* an outer acceleration region is associated with the null surface where $\boldsymbol{\Omega} \cdot \boldsymbol{B} = 0$ and the Goldreich-Julian charge density needed to short-out parallel electric fields would vanish. Such a region is unstable and thus gives rise to parallel fields that can accelerate particles. The implied beam components and, thus, pulse components are skewed in angle and pulse phase, respectively, from polar-cap components. Outer-gap components appear to underly high-energy and other non-radio emission from the Crab and Vela pulsars and a few other objects. Interestingly, a few objects show "giant" radio pulses that align with the high-energy components. These objects include the Crab pulsar ($P = 0.033$ s) and the MSP B1937+21 ($P = 1.56$ ms), which have drastically different light cylinder sizes and surface magnetic field strengths but, remarkably, have nearly identical magnetic field strengths at their light cylinders if we assume a dipolar scaling $\propto r^{-3}$.

4 Pulsar Distances, Velocities and Kicks

Pulsars were first recognized to be a high-velocity population by Gunn and Ostriker in 1970 [53]. Since then our knowledge of NS kinematics has grown enormously owing to empirical work on the pulsar distance scale and on proper motion measurements and other measures of pulsar velocities.

As shown in Table 4, pulsar distances have been measured directly through parallax determinations using pulse timing or interferometry. Associations of pulsars with supernova remnants and other objects yield additional distances.

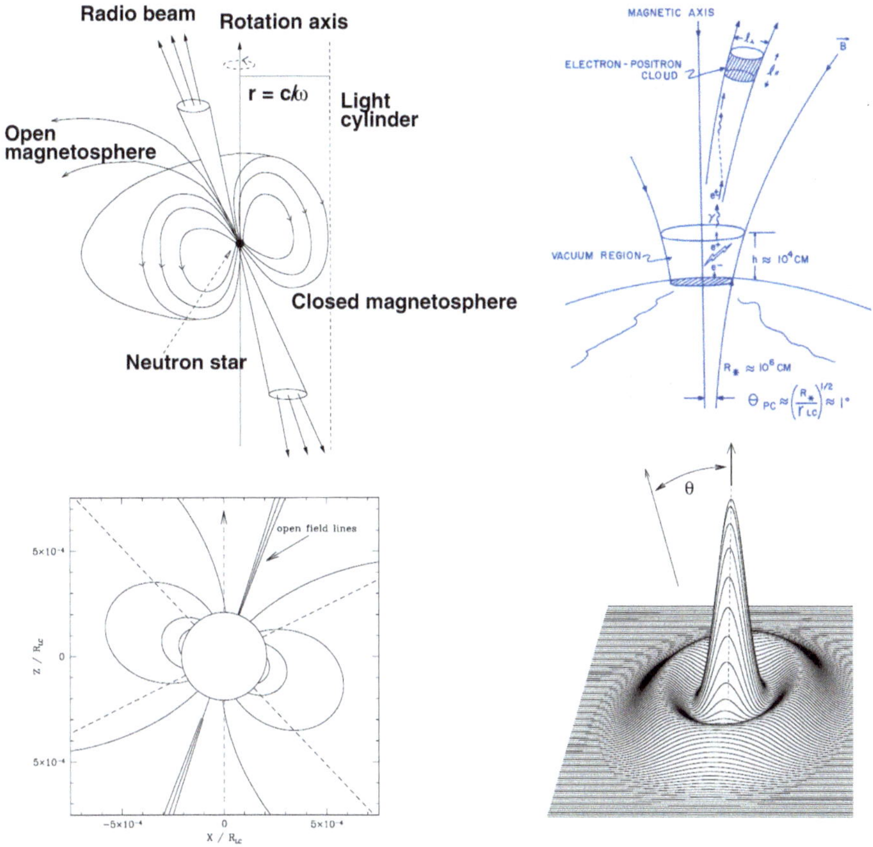

Fig. 2. The workings of radio pulsars. *Top left:* schematic picture of a highly magnetized neutron star with dipole moment skewed from the rotation axis. The light cylinder with radius $c/\Omega = cP/2\pi$ defines the boundary between the open and closed portion of the magnetosphere by the last field line that closes inside it. Radio emission is thought to arise from the relativistic particle flow along the open field lines near the dipole axis. Some radio emission may originate from additional acceleration regions that are associated with a surface defined by $\boldsymbol{\Omega} \cdot B = 0$, the so called "outer gap" regions. *Top right:* Detail of activity at the surface of the neutron star according to the picture of [34]. The size of the magnetic polar cap is as shown. Boundary conditions at the NS surface and at the last closed-field-line surface require $\boldsymbol{E} \cdot \boldsymbol{B} n_e 0$, thus leading to acceleration of particles. Under conditions that require large Ω and/or large B, an electron-positron ($e^{\pm}$) cascade can be sustained that drives a two-stream instability, which provides coherent radio emission. Turn-off of the cascade may account for the "death line" in the $P - \dot{P}$ diagram. *Bottom left:* Field line structure near the neutron star. The open field lines (those that do not close inside the light cylinder) are shown. *Bottom right:* The radio beam pattern for a typical pulsar, showing an inner core and outer, hollow-cone beam component. The strengths of these components appear to depend on P and possibly on $\dot{P}$. Additional components are needed for some pulsars, including a second hollow-cone component and components associated with outer-gap emission regions.

Atomic hydrogen (HI) absorption at 21 cm wavelength combined with a kinematic model for Galactic rotation yields distance constraints. By far, most pulsar distances are obtained using the dispersion measure combined with a Galactic model for the electron density.

Table 2. Pulsar Distance Estimates

Type	Number	Comments & Limitations
Parallaxes:		
Interferometry	16	1 mas @ 1kpc, ionosphere
Timing	8	1.6 μs @ 1kpc, timing noise
Optical	1	HST point-spread function
Associations:		
Supernova Remnants	8	ISM perturbations
Globular Clusters	16 clusters	Spectroscopic distances
Large/Small Magellanic Clouds	8	
HI Absorption:	74	bright pulsars, Galactic rotation model
DM + Electron Density Model (NE2001):	all radio pulsars	ISM perturbations, Galactic structure

The distance scale is obviously important for establishing luminosities of pulsars across the electromagnetic spectrum. More importantly, good distance estimates are needed to correct pulsar timing measurements for acceleration effects that are a function of distance, such as the Shklovsky effect[2] and Galactic acceleration. These contribute terms to pulse arrival times that are covariant with the long-term evolution of compact orbits and thus affect GR tests.

Recent very long baseline interferometry [47, 48, 50, 49, 51] has yielded important new parallax measurements and over the next few years, a large number of parallaxes should emerge. Parallaxes provide anchor points for electron density models for the Galaxy which then can be used to the much larger sample of pulsars where parallaxes are lacking.

[2] An object moving with speed $v_\perp$ across the line of sight will show an apparent $\dot{P} = V_\perp^2 P/Dc$, where $V_\perp$ is the transverse pulsar speed and D is the distance. even if the pulse period is intrinsically steady. This simple geometric effect arises because the distance to the pulsar steadily increases with time. Note that the motion along the line of sight is already absorbed into the nominal period.

4.1 Dispersion Measure Distances

The *dispersion measure* (DM) is the observable for estimating pulsar distances that is obtained routinely for all radio pulsars. Figure 3 shows a single pulse from the Crab pulsar that displays differential time of arrival (TOA) vs. frequency. Pulsed flux evidently is emitted simultaneously (or nearly so) at the pulsar, so the systematic variation with frequency is caused by propagation through the interstellar plasma.

The plasma frequency in the ISM $\nu_p \approx 1.56\,\mathrm{kHz}\,(n_e/0.03\,\mathrm{cm}^{-3})^{1/2}$ and the gyrofrequency $\nu_\mathrm{B} \approx 2.8\,\mathrm{Hz} B_{\mu G}$. Magnetic fields introduce birefringence that is most easily detected as Faraday rotation. The index of refraction in a cold magnetized plasma like the ISM for $\nu \gg \nu_p$ and $\nu \gg \nu_{\mathrm{B}\|}$ is (e.g. Thomson, Moran & Swenson 2001)

$$n_{\ell,r} \approx 1 - {\nu_p}^2/2\nu^2 \mp {\nu_p}^2 \nu_{\mathrm{B}\|}/2\nu^3, \tag{1}$$

where $\nu_p = (n_e e^2/\pi m_e)^{1/2}$ is the plasma frequency and $\nu_{\mathrm{B}\|} = eB\cos\theta/m_e c$ is the electron gyrofrequency calculated for the magnetic field component along the line of sight; the $\mp$ cases apply for LHCP and RHCP waves.

The frequency dependent time delay is calculated by integrating (group velocity)$^{-1} = dk/d\omega = c^{-1}\,(n_{\ell,r} + \nu dn_{\ell,r}/d\nu)$ along the line of sight, giving

$$t = t_\mathrm{DM} \pm t_\mathrm{RM}. \tag{2}$$

The dispersive time delay and the small correction due to birefringence are

$$t_\mathrm{DM} = \frac{e^2}{2\pi m_e c}\frac{1}{\nu^2}\int_0^D ds\, n_e = 4.15\,\mathrm{ms}\ \mathrm{DM}\ \nu_\mathrm{GHz}^{-2} \tag{3}$$

where the dispersion and rotation measures and their standard units (for D in pc, n_e in cm^{-3}, and $\mathrm{B}_\|$ in μG) are

$$\mathrm{DM} = \int_0^D ds\, n_e(s) \qquad (\mathrm{pc\,cm^{-3}}), \tag{4}$$

$$\mathrm{RM} = \frac{e^3}{2\pi m_e^2 c^4}\int_0^D ds\, n_e \mathrm{B}_\| = 0.81 \int_0^D ds\, n_e \mathrm{B}_\| \qquad (\mathrm{rad\,m^{-2}}). \tag{5}$$

The variation of TOA vs ν is thus $\mathrm{TOA}(\nu) \propto \mathrm{DM}\nu^{-2}$, so multifrequency TOA measurements readily yield DM. An electron density model such as NE2001 [52] can be used to invert the integral that defines DM. Such distances are on average good to perhaps 25%. However, any given pulsar can be nearer or farther by factors of two owing to fine-scale structure in the ionized ISM.

4.2 Pulsar Space Velocities

Proper motions of pulsars (their angular motion on the sky) have been measured using the same techniques as for parallaxes (see Table 4). Work over the last 10 years has refined the original conclusion by Gunn and Ostriker [53] that pulsars are a "runaway" population. Specifically, we know that

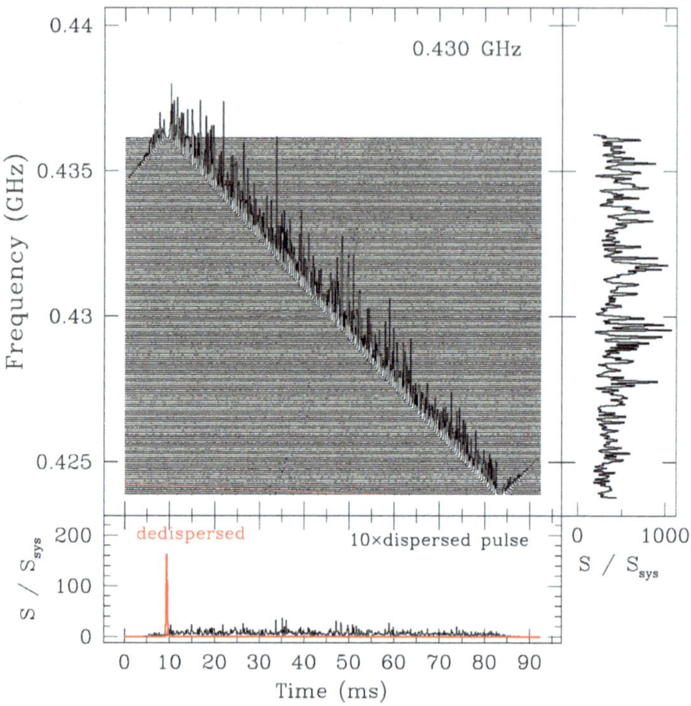

Fig. 3. Plot of intensity against time and frequency, showing a single dispersed pulse as it arrives at different frequencies centered at 0.43 GHz. The right-hand panel shows the pulse amplitude vs. frequency while the bottom panel shows the pulse shape with and without compensating for dispersion delays. This pulse is the largest in one hour of data, has S/N $\sim 1.1 \times 10^4$, and a pulse peak that is 130 times the flux density of the Crab Nebula, or $\sim$ 155 kJy. Note that the segments at either end of the bandpass where the pulse arrival time is opposite the trend at most frequencies is caused by aliasing of the signal.

1. Canonical ($\sim 10^{12}$ Gauss) pulsars have a mean 3D peculiar speed (adding to Galactic rotation) $\sim$ 400-500 km $^{-1}$ [18, 54, 19, 55]. For comparison, solar-type stars have rms peculiar speeds $\sim$ 10 km s^{-1} and so-called OB runaways have speeds up to $\sim$ 100 km s^{-1}.
2. The velocity distribution extends to high velocities ($> 10^3$ km s^{-1}) [51].
3. The velocity distribution is inferred to be *bimodal* [54, 19] with a low velocity component $\sim$ 100 km s^{-1} and high velocity component $\sim$ 500 km s^{-1}. Selection biases undoubtedly affect both the low-velocity and high-velocity pulsars. Recent work [55] reports a good fit to the proper motions of 233 pulsars using only a unimodal, Maxwellian distribution with a 1D rms $\sim$ 265 km s^{-1}. However, this work did not include corrections for

selection effects. With the increase in source sample and the continued output of precision parallaxes, a reanalysis that takes into account the parallax sample as well as the most recent electron density model needs to be done to improve the inferences about the shape of the velocity distribution.

4. Millisecond pulsars are a low-velocity population (by pulsar standards): their typical 3D speed is $\sim$ 50-100 km s^{-1} [56, 55]. Such low speeds are consistent with a *velocity selection effect* in the formation of MSPs: such objects are formed only if their binarity is preserved after the supernova explosion that formed the NS.

Pulsar Recoils and Kicks

Several processes have been suggested for the large peculiar motions of pulsars. Here we describe them and report the implied orientation of the velocity and the spin axis:

1. **Recoil from supernova disruption of a binary** ($\boldsymbol{\Omega} \perp \boldsymbol{V}$): Sudden mass loss from a supernova causes the individual stars in a compact binary to recoil at high velocities [57]. Blaauw's original explanation was for OB runaway stars but also applies to NS formed in binaries.
2. **Recoil from evanescent NS binaries** ($\boldsymbol{\Omega} \perp \boldsymbol{V}$): A rapidly rotating proto-NS may fragment to form a double proto-NS, one of which will lose mass and explode as it reaches the minimum stable mass. The resulting explosion in the very compact binary imparts a large recoil velocity to the surviving NS [61, 62].
3. **Natal kicks from asymmetrical supernova explosions** (random $\boldsymbol{\Omega} \cdot \boldsymbol{V}$ or $\boldsymbol{\Omega} \parallel \boldsymbol{V}$ with rotational averaging): Momentum thrust(s) occurring during the $\sim$ 1 sec of core collapse can lead to even larger space velocities than the recoil mechanism. Candidate effects include neutrino and mass rocket effects associated either with advection in the collapsing core or, if magnetic fields are very large ($> 10^{15}$ Gauss), neutrino asymmetry due to the magnetic field.
4. **Slow acceleration after the supernova explosion** ($\boldsymbol{\Omega} \parallel \boldsymbol{V}$): A magnetic dipole offset from the center of a NS would not only spin down the star but also accelerate the NS translationally [58]. Large enough velocities require short spin periods at birth ($\sim$ 1 ms), and a sizable offset, a few tenths of a NS radius [59].

Several lines of evidence indicate that recoil alone cannot account for the largest pulsar velocities, requiring either natal kicks or the dipole rocket effect. Other evidence for favoring natal kicks involves specific objects for which either geodetic precession or orbital precession occurs, processes that necessitate there having been a sudden misalignment of the spin and orbital angular momenta. It is expected that accreting binaries will have aligned angular momenta because the time scales for alignment are relatively short. Geodetic precession comprises wobble of the spin vector about the total (spin + orbital) angular momentum. A pulsar's magnetic axis and thus its radiation beam(s) will also wobble, producing secular changes in pulse shape. Such changes are seen in several of the NS-NS binaries, including the Hulse-Taylor binary (B1913+16), B1534+12, and the double pulsar, J0737–3939. Statistically, the small fraction of binary radio pulsars also appears to require kicks in addition to recoil. Population synthesis studies indicate that too many pulsars remain in binaries if there is no kick.

Corroborating evidence for natal kicks comes from the orientation of X-ray jets relative to the proper motions of the Crab and Vela pulsars. X-ray jets are aligned with the spin axis, thus implying alignment of the velocity vectors with the spin axes in these objects. Such alignment can occur if the time scale $t_{\rm kick}$ for the kick process is larger than the spin period $P(t)$ of the proto-NS, thus inducing spin averaging of the kick force into a direction along the spin axis [60]. Not all objects show such alignments so either these are chance alignments or mis-aligning effects occur, like the combination of kicks from two supernovae in cases where the binary remains bound after the first.

Given that natal kicks combine with recoil to produce the net velocity distribution, it is a bit of a conundrum to explain the evidently bimodal shape of the velocity distribution. Suppose there are at least two contributions to the peculiar velocity (kicks + recoil) and there may be multiple kick processes (neutrino + matter rocket effects). If these processes act independently each with its own contributing velocity distribution, then the sum of the contributions should have a distribution that is the *convolution* of the various distributions. If these are all individually unimodal then so too will be the sum. The bimodal velocity distribution therefore suggests that the various effects are *not* independent. An example would be if kick amplitudes were somehow influenced by evolution in binary systems. Another suggestion is that all pulsars are not NS but rather some are strange stars that receive kicks from the phase transition to quark matter [63].

Bow Shocks

As the mean pulsar speed V is highly supersonic and super-Alfvénic, shocks are expected at the interface between the pulsar wind and the ISM. The momentum of the relativistic NS wind is $\dot{E}/c$ under the assumption that the spindown energy loss is carried away by the relativistic wind. Balanced with ISM ram pressure, the stand-off radius is

$$R_0 = \left(\frac{\dot{E}}{4\pi c \rho V^2}\right)^{1/2} \approx 266\,{\rm AU} n_H^{-1/2} \left(\frac{\dot{E}}{10^{33}\,{\rm erg\,s^{-1}}}\right)^{1/2} \left(\frac{100\,{\rm km\,s^{-1}}}{V}\right), \quad (6)$$

where ρ is the ISM mass density and n_H is the equivalent effective hydrogen number density.

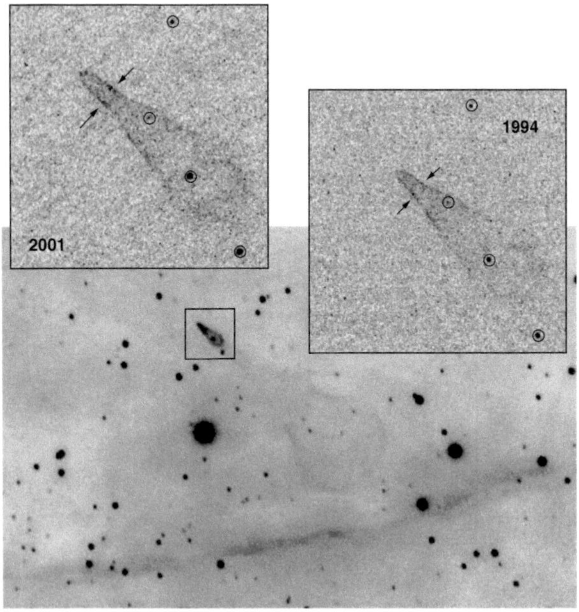

Fig. 4. Hα images of the head of the Guitar nebula. The bottom panel shows a wide-field image of the Guitar nebula obtained at the 5-m Hale Telescope at Palomar [64]. High resolution images of the region marked with a box ($\sim$ 16 arcsec in size) were obtained with the HST PC in 1994 (right) and 2001 (left). North is upward and east is to the left.

This expression conforms to the observed bow shocks, which are rare owing to the requirements needed for the pulsar environment in order to see a detectable tracer, such as Hα emission [64].

Figure 4 shows the Guitar Nebula, a spectacular bow shock seen in Hα. The length of the nebula corresponds to $\sim$ 300 yr of travel. The spindown age of the pulsar is 10^6 yr, so it has moved about 1 kpc from its birth site. The "nose" of the bow shock is unresolvable from the ground because the small $\dot{E}$ and large space velocity $\sim$ 1600 km s^{-1} (using the proper motion and nominal DM distance) imply a small R_0. The nose is resolved in Hubble Space Telescope images, two epochs of which are shown in the figure, which demonstrate the secular advance of the bow shock through the ISM. This and other pulsar bow shocks thus show that (a) the pulsar wind is relativistic and (b) does not manifest any anisotropy that would affect the bow shock contours; such anisotropy should exist but evidently is not sufficient to be observable once rotational averaging takes place. In addition, bow shocks corroborate the velocities inferred for pulsars for which we have only a DM-based distance.

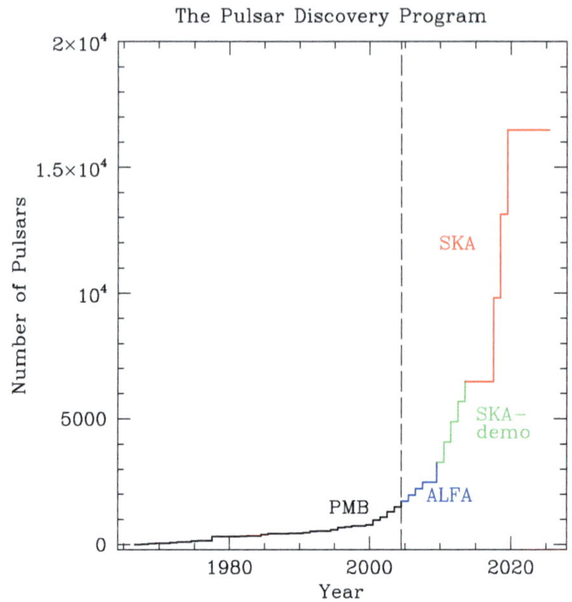

Fig. 5. Pulsar discoveries vs time, past, present and future. PMB = Parkes multibeam survey from the late 1990s to early 2000s; ALFA = Arecibo L-band Feed Array being used in an ongoing pulsar survey that will take the next 5 to 10 yr; SKA-demo = demonstrator array for the Square Kilometer Array that may begin construction around 2011 and have 10% of the eventual collecting area of the SKA. SKA = "complete" Galactic census for pulsars to be conducted with the SKA. The number of active radio pulsars in the Galaxy that are detectable in *periodicity searches* is the birth rate $\sim 10^{-2}$ yr^{-1} multiplied by the typical radio lifetime for canonical pulsars $\sim 10^7$ yr, or $\sim 10^5$ objects. Of these $\sim 20\%$ are beamed toward us, yielding $\sim 2 \times 10^4$ potentially detectable pulsars. Recent work [65] has identified a population of pulsars detectable only or primarily through *single-pulse* searches; the number of such "pulsar transients" is comparable to the number detectable in periodicity surveys.

5 Pulsar Surveys

Pulsar surveys have been ongoing since the original 1967 discovery and have made ever-growing use of innovations in computer technology to increase the sensitivity. Figure 5 shows the pace of discovery, which is expected to increase dramatically with the ongoing Arecibo ALFA survey and surveys expected with the Square Kilometer Array and demonstrator arrays for the SKA, leading to a full-Galactic census of pulsars.

Why conduct a full pulsar census of the Galaxy? The first reason for proposing a complete Galactic census is obvious: the larger the number of pulsar detections, the more likely it is to find rare objects that provide the greatest opportunities for use as physical laboratories. These include binary pulsars as described above and also those with black hole companions; MSPs that can be used as detectors of cosmological gravitational waves; MSPs spinning faster than 1.5 ms, possibly as fast as 0.5 ms, that probe the equation-of-state under extreme conditions; hypervelocity pulsars with translational speeds in excess of 10^3 km s^{-1}, which constrain both core-collapse physics and the gravitational potential of the Milky Way; and objects with unusual spin properties, such as those showing discontinuities ("glitches") and apparent precessional motions (including "free" precession in isolated pulsars and binary pulsars showing geodetic precession).

The second reason for a full Galactic census is that the large number of pulsars can be used to delineate the advanced stages of stellar evolution that lead to supernovae and compact objects. In particular, with a large sample we can determine the branching ratios for the formation of canonical pulsars and magnetars. We can also estimate the effective birth rates for MSPs and for those binary pulsars that are likely to coalesce on time scales short enough to be of interest as sources of periodic, chirped gravitational waves (e.g. [66]).

The third reason is that a maximal pulsar sample can be used to probe and map the ISM at an unprecedented level of detail. Measurable propagation effects include dispersion, scattering, Faraday rotation, and HI absorption that provide, respectively, line-of-sight integrals of the free-electron density n_e, of the fluctuating electron density, δn_e, of the product $B_{\|} n_e$, where $B_{\|}$ is the LOS component of the interstellar magnetic field, and of the neutral hydrogen density. The resulting dispersion measures (DM), scattering measures (SM), rotation measures (RM) and atomic hydrogen column densities ($N_{\rm HI}$) obtained for a large number of directions will enable us to construct a much more detailed map of the Galaxy's gaseous and magnetic components, including their fluctuations.

5.1 Where Should We Search?

From the above discussion, the target classes that comprise the forefront are:

1. *Sub-millisecond pulsars:* Finding any pulsar with $P < 0.5$ ms would rule out all current equations of state (EOS) for neutron star matter and would require the role of strange matter.
2. *Slow pulsars:* Very slow pulsars (5 to 10 s) require radio pulsars to have field strengths $B > 10^{13}$ Gauss in order to produce radio emission. We do not understand the relationship of the few radio pulsars having such strong fields with magnetars, X-ray and gamma-ray objects from which radio emission has not (yet) been detected. What is the longest period at which we might find radio emission?
3. *High-velocity pulsars:* Selection effects work against finding the highest velocities so we do not know what the largest kick is. Finding this out may help us identify or at least rule out some kick mechanisms.
4. *Relativistic binaries:* NS-NS and NS-BH binaries should exist with orbital periods of hours and less. The more compact the orbit, the greater the

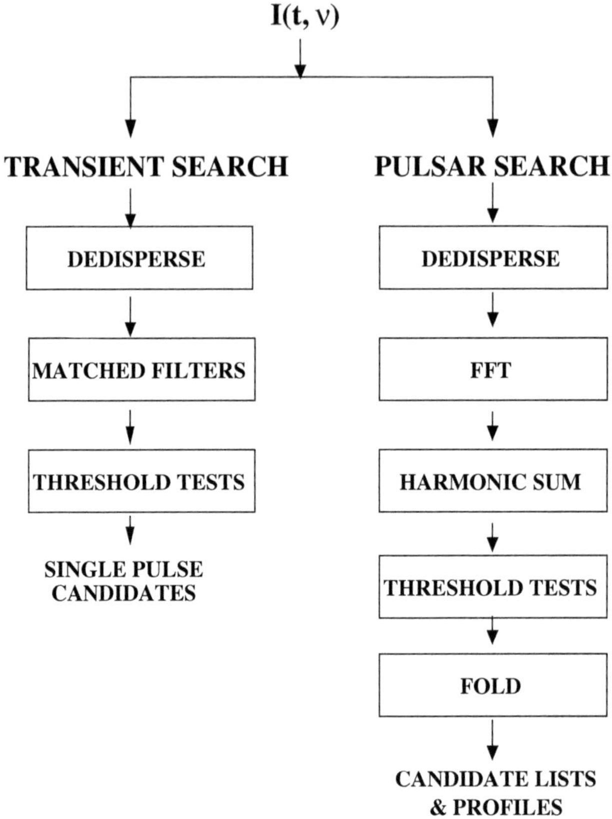

Fig. 6. Flow diagram for pulsar searches. The raw data are a "dynamic spectrum" $I(t, \nu)$, intensity vs. time and frequency that is sampled with resolutions $\delta t \sim 100\,\mu s$ and $\delta\nu \sim 10 - 10^3$ kHz. The first step is to dedisperse the signal (see text) with trial values of DM when the DM is not known (which is typical). Next in the periodicity search is an FFT of the resulting dedispersed time series, followed by trial harmonic sums, which are thresholded to identify candidate signals. The time series is "folded" at candidate periods in order to build up a pulse shape, which is then subjected to additional tests. The acid test in the analysis is to confirm the candidate through a new observation. The other analysis path for a transient search involves cross-correlation of dedispersed time series with a set of matched filters that correspond to different pulse widths. Output correlation functions are subjected to threshold tests to identify candidate single-pulses.

number of relativistic effects that can be detected and the stronger the tests of GR. Also, a comprehensive sample of compact binaries allows better estimates of the merger rate relevant for GRB studies and for detection rates for LIGO and other gravitational wave telescopes.

5. *Globular clusters:* Clusters are prolific factories for recycled pulsars because the stellar density is high enough that stars can exchange partners and thus enhance the prospects for recycling. Some globular clusters may contain medium-mass black holes with orbiting pulsars. Finding them is challenging but will provide great payoff for testing GR.
6. *Pulsars orbiting the massive black hole in the Galactic center:* Pulsars orbiting the $3 \times 10^6\ M_\odot$black hole in the center of the Milky Way are especially interesting for subsequent pulse timing and tests of GR [67, 69, 68]. The GC is similar to, but much larger than, a globular cluster, so we expect many recycled pulsars in binaries to exist in the star cluster around Sgr A*. Presently, no pulsars are known in this region because *radio wave scattering* smears the pulses from any pulsar that happens to be there. The pulse broadening time $\sim$ 300 s at a frequency of 1 GHz but scales as ν^{-4}. Thus a search at $\nu > 10$ GHz can yield pulsars with periods $\lesssim 0.3$ s. Yet shorter period objects require even higher frequencies. However, pulsar spectra typically fall off as ν^{-x} with $x \approx 1.5$, so high-frequency surveys will not detect many, if any, pulsars using existing telescopes. Success in this area may require the SKA or, at least, an Arecibo-sized telescope operating in the southern hemisphere.

5.2 Pulsar Search Processing

A two branch flow chart for search processing is shown in Figure 6. The periodicity search that uses Fourier techniques to identify the periodic signal has characterized pulsar signals up until now. The transient search branch assumes nothing about periodicity and simply identifies single pulses in the time series. Both branches have "dedispersion" in common, which compensates the differential arrival times due to dispersive propagation (Fig. 3). In both branches, identification of the pulsar signal strives for *matched filtering* in order to optimize the signal to noise ratio (S/N) of various test statistics. The same is true for detection of gravitational waves.

There are two types of dedispersion techniques:

1. **Coherent Dedispersion** operates on the radio telescope voltage $\propto$ electric field $E(t)$, which contains phase information about the pulsar radiation. The emitted electric field is modified by the phase factor $\exp(ik(\omega)z)$ where $k(\omega)$ contains DM and other quantities. Dedispersion simply involves multiplication of the FFT of the voltage by the inverse filter, $\exp(-ik(\omega)z)$. Coherent dedispersion is most useful for low-DM pulsars or for high-frequency observations. The resulting time resolution is 1 / (total bandwidth) and thus easily can be better than 1 μs. However, another propagation effect — pulse broadening from interstellar scattering (multipath propagation) — may determine the actual time resolution. Coherent dedispersion is computationally very expensive and is typically used for precision timing and single-pulse studies rather than in pulsar searching.

A notable example is the nano-second duration shot pulse sub-structure seen in giant pulses from the Crab pulsar [70].
2. **Post-detection Dedispersion** is an approximate method that operates on the "detected" (i.e. squared) voltage and shifts the time series obtained in each of a large number of individual frequency channels. Such data are obtained with multi-channel spectrometers whose outputs are recorded at fast intervals ($\sim 100\,\mu s$). Dedispersion involves shifting the intensity $I(\nu, t)$ by a time $t_{\rm DM} \propto {\rm DM}/\nu^2$. The example in Figure 3 shows the dispersed and dedispersed pulse in the bottom panel. This method is computationally much less demanding than coherent dedispersion. However, significant investment must be made in hardware spectrometers that have fast recording capabilities.

Dedispersion uses matched filtering of the dispersion signature in the frequency-time plane. Other aspects of matched filtering for pulsar searches include:

1. Single pulse searches: one must match the shape and the width of the pulses, neither of which are known in advance.
2. Periodicity searches: the period as well as the pulse shape and width needs to be matched. An approximation for doing so is *harmonic summing*, which involves using partial sums of candidate harmonics in the power spectrum of the dedispersed time series.
3. Pulsars in compact orbits: acceleration within a binary with orbital period of a few hours can smear the pulse if uncompensated even for data sets of duration T of just a few minutes. Optimal S/N mandates a search in an acceleration parameter for the cases $T \gg P_{\rm orb}$. For $T \gtrsim P_{\rm orb}$, matched filtering involves a large parameter space, although approximate methods can reduce the computational load if orbital sidebands are searched as part of the harmonic summing process [71].

6 Probing Gravity and Neutron Stars with Pulsar Timing

6.1 Arrival Time Analysis

Pulsar timing is conceptually simple: at an observatory a pulsar's signal is processed (i.e. dedisperse and average over some number of pulse periods) to establish a time of arrival (TOA) for the pulse. Measurements of TOAs over long periods of time by definition correspond to integer multiples of pulse periods separating the TOAs. Through appropriate analysis, one can identify the many effects that contribute to the TOA, including those within the pulsar itself, those that arise from binary motion (if any) of the pulsar, interstellar propagation effects, and those that arise from the observatory's motion in the solar system. Added to these are relativistic effects occurring anywhere along the line of sight. A comprehensive model that accounts for all these effects,

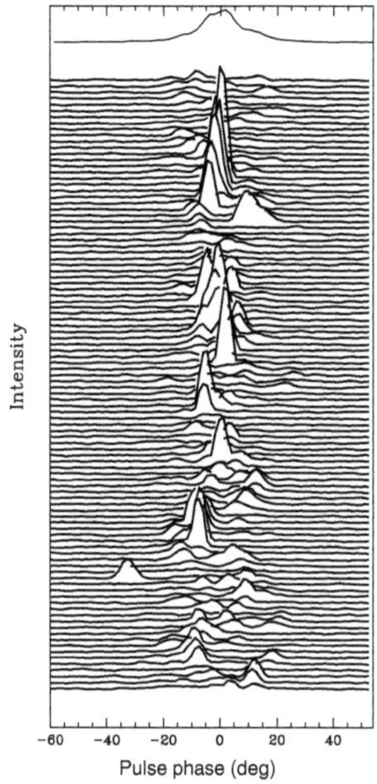

Fig. 7. A sequence of pulses from the pulsar B1740+09 [37] with their average at the top.

when fitted to the TOAs, will yield timing residuals that have zero mean, showing non-zero values only as a result of random measurement errors.

In practice, TOAs are estimated from a measured pulse shape using (again) *matched filtering*. This involves a cross correlation of a template filter, which is the idealized or long-term-average pulse shape, with the data and identifying the time offset between template and pulse. Such a procedure is based on the assumption that the pulsar's pulse shape is invariant with time and that the only source of random error on the shape is additive. These are only approximately true. As seen in Figure 7, single pulses show significant amplitude fluctuations and phase jitter. These induce $N^{-1/2}$ fluctuations when N pulses are combined.

Once a TOA is estimated, it is useful to refer it to an (approximate) inertial frame, the solar system's barycenter, as shown in Figure 8. Barycentric TOAs may then be analyzed with a timing model. The simplest model applies to an isolated pulsar that spins down smoothly on a time scale much longer ($\gg 10^3$ yr) than the set of TOAs (years). For this case a quadratic or cubic phase model suffices:

$$\phi_{\rm spin}(t) = \phi_0 + \nu t + \frac{1}{2}\dot{\nu}t^2 + \frac{1}{6}\ddot{\nu}t^3, \tag{7}$$

where the data are assumed to start at $t = 0$. For most pulsars, even the cubic term from spindown is negligible. Figure 9 shows the residual phases $R(t) = \phi(t) - n_t$, where n_t is the integer number of periods corresponding to time t) (expressed as $P\delta\phi(t)$ in time units) for 14 pulsars. Some objects

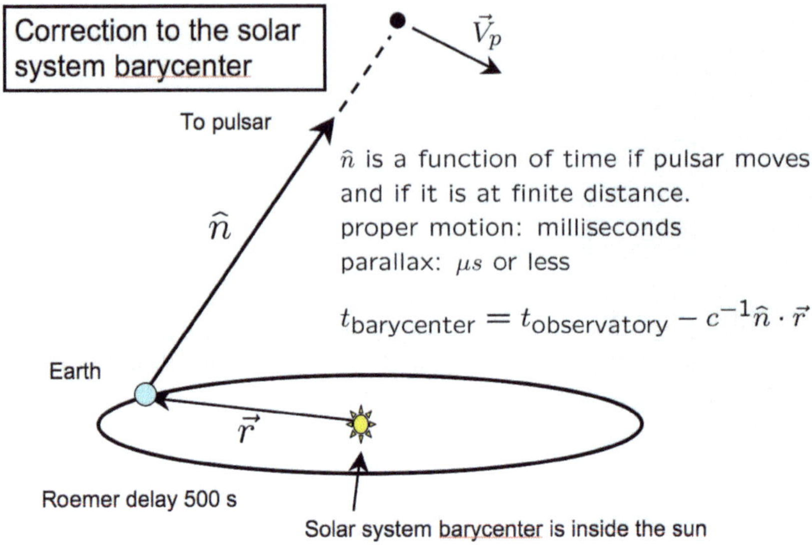

Fig. 8. Correction of topocentric arrival times measured at a terrestrial observatory to the solar system's barycenter, located inside the Sun.

show $R(T)$ consistent with random errors while others clearly show large, correlated residuals. In all of these cases, the excess residuals are caused by "timing noise," which appears to be stochastic with a "red" spectrum, meaning that power is concentrated at low fluctuation frequencies. A better name for this timing noise would be spin noise because it all appears to derive from activity within the NS, such as crustquakes combined with noisy superfluid interactions with the NS crust. Spin noise correlates with large $\dot{P}$ and thus appears to have something to do with the rate at which the crust spins down in response to the magnetic torque and how the spindown is communicated to the internal, more fastly rotating superfluid. Spin noise is of astrophysical interest in and of itself. However, our discussion here will treat it as a limitation to how well we can measure other effects, such as gravitational effects.

6.2 Arrival Time Contributions

Contributions to arrival times (or, equivalently, pulse phase) include effects that are imposed in one or more locations along the line of sight. Ideally, we would like to estimate the actual *emission time* t_e from the TOA or *reception time* t_r of a pulse:

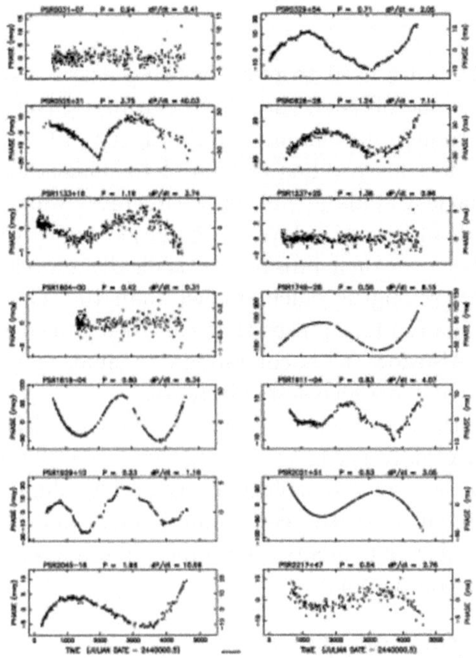

Fig. 9. Pulse phase residuals from quadratic fits to arrival times.

$$
\begin{array}{ll}
t_e = t_r - D/c & \text{Path length} \\
\quad + 4.15\,\mathrm{ms}\,\mathrm{DM}/\nu^2 & \text{Plasma dispersion (ISM)} \\
\quad + \Delta_{R\odot} + \Delta_{E\odot} + \Delta_{S\odot} & \text{Roemer, Einstein, Shapiro delays (solar system)} \\
\quad - \Delta_{R\mathrm{psr}} + \Delta_{E\mathrm{psr}} + \Delta_{S\mathrm{psr}} & \text{Roemer, Einstein, Shapiro delays (binary pulsar)} \\
\quad + \delta TOA_{\mathrm{ISM}} & \text{ISM scattering fluctuations} \\
\quad + \delta TOA_{\mathrm{orbit\ noise}} & \text{Orbital fluctuations} \\
\quad + \delta TOA_{\mathrm{spin\ noise}} & \text{Spin(timing) noise} \\
\quad + \delta TOA_{\mathrm{grav.\ waves}} & \text{Gravitational backgrounds,}
\end{array}
$$

where the various terms are discussed in detail in [6]. The second line is the dispersion delay associated with the ISM. The Roemer term is the delay associated with the separation and relative motion of the observatory and the solar system barycenter; the Einstein term is associated with the gravitational redshift and time dilation of the Sun's (or pulsar companion's) gravitational field; and the Shapiro delay is associated with the gravitational potential of the Sun or the pulsar's companion (if any). Stochastic variations in the TOA include noise from the ISM, orbital noise from (e.g.) an ensemble of asteroids, spin noise within the pulsar, and the last term represents variations associated with gravitational wave backgrounds.

6.3 Binary Pulsars

For pulsars in binary systems, TOAs are used to monitor the pulsar's orbit with extrarordinary precision. For some objects, the TOAs can be measured to a precision of less than 100 ns, though for canonical pulsars, the precision may be as large as 1 ms. Nature fortunately has conspired to make those pulsars in the most interesting binary systems — pulsars with white dwarf or NS companions with orbital periods of hours to days — the objects with the best TOA precision. This arises because such pulsars are typically recycled objects with short periods and field strengths much smaller than 10^{11} G. In a fixed amount of time, there are many more pulses that can be summed to improve the signal-to-noise ratio. Also low fields correspond to small values of $\dot{P}$ that appear to be associated with low levels of spin noise.

Compact orbits need to be described by General Relativity or some other gravitational theory. Therefore TOA analysis can be used to test how well GR works [38, 39] and determine the masses of the orbiting objects. The Shapiro delay provides a combined determination of the companion's mass and the inclination of the orbit, since the effect depends on the impact parameter of the line of the sight to the pulsar's companion:

$$s = \Omega_{\rm orb}^{2/3} M^{2/3} m_c^{-1} \cdot \left(\frac{a_p \sin i}{c}\right) \left(\frac{GM_\odot}{c^3}\right)^{-1/3}, \tag{8}$$

where $M = m_p + m_c$ is the total mass, $\Omega_{\rm orb}$ is the angular frequency of the orbit, and a_p is the semi-major axis of the pulsar's orbit. Apsidal motion — precession of the (approximately) elliptical orbit at a rate $\dot{\omega}$ provides another combination of parameters involving the mass,

$$\dot{\omega} = 3\frac{(GM)^{2/3}\,\Omega_{\rm orb}^{5/3}}{(1-e^2)\,c^2}. \tag{9}$$

By combining measurements of s and $\dot{\omega}$, the individual masses m_p and m_c and the orbital inclination can be determined. Through this approach, masses of a sizable number of neutron stars have been determined (Figure 10).

Radiation of gravitational waves causes the two objects to spiral in, yielding a increase in orbital frequency at a rate, $\dot{\Omega}_{\rm orb}$. The measured increase has been shown to be consistent with GR for several pulsars, most notably for the Hulse-Taylor binary and for the double pulsar.

6.4 Pulsars as Gravitational Wave Detectors

A gravitational wave causes a test mass to move at the frequency of the wave. The amount of motion is very small. In a pulse timing context, both the Earth and the pulsar serve as test masses (Figure 11). Long-period waves (years) induce *cumulative* changes in pulse phase (and, thus, TOAs), so the greatest sensitivity is to waves that are comparable in period to the length

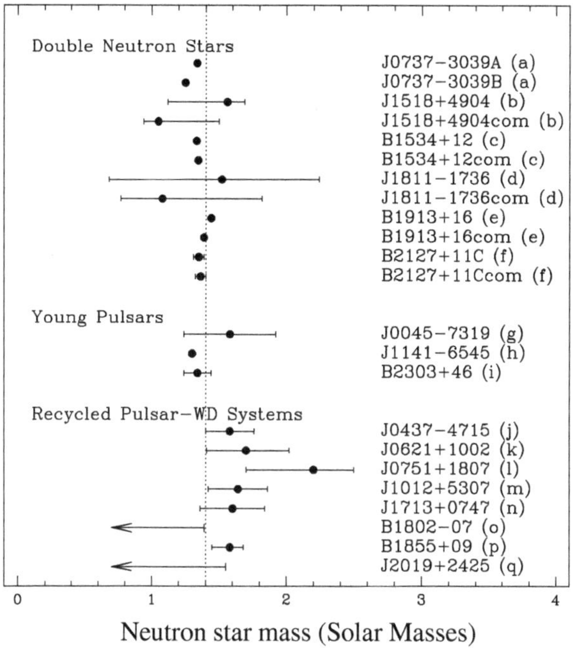

Fig. 10. Mass estimates of NS obtained from pulse-timing techniques.

of a many-year data set of TOAs. The technique is well discussed in [6] and was originally proposed by [40]. Measurements [41, 42, 43] yield upper bounds on the effective mass-energy density of a cosmological background of waves at the level of $\Omega_g h^2 \lesssim 10^{-8}$, where Ω_g is the density in units of the closure density of the universe. A good discussion of one kind of background is in [44]. Various workers have proposed a "pulsar timing array" [45, 46] which would use multiple pulsars to discriminate between other sources of timing perturbations while also detecting a correlated signal that is present because the Earth is in common to all lines of sight. The pulsar timing array is a key science project for the Square Kilometer Array [4].

7 Big Questions in the Physics of Neutron Stars and their Use as Cosmic Laboratories

Here is a list of questions related to the material of this chapter that I think illustrate the forefront of neutron star science.

1. **Formation and Evolution of Neutron Stars:**
   a) What determines if a NS is born as a magnetar or as a canonical pulsar?

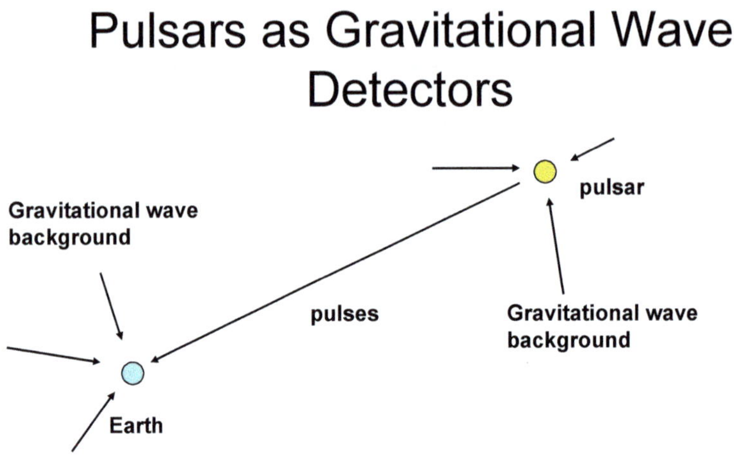

Fig. 11. Effects of gravitational wave backgrounds on the timing of pulsars.

   b) How fast do NS spin at birth?
   c) How fast can recycled pulsars spin?
   d) What is the role of instabilities and gravitational radiation in determining the spin state?
   e) How do momentum thrusts during core collapse affect the resulting spin state and translational motion of the NS?
   f) What processes determine the high space velocities of NS? (i) Neutrino emission, (ii) Matter rocket effects, (iii) Electromagnetic rocket effect from off-center magnetic-dipole radiation (Harrison-Tademaru), (iv) Gravitational wave rocket effect.
   g) Are orbital spiral-in events at all related to high-energy bursts, such as short-duration GRBs?
2. **NS Structure and the Equation of State:**
   a) Are NS really NS or might some of them be strange stars?
   b) What comprises the core of a NS?
   c) What is the distribution of masses among pulsars and other NS?
   d) In what regions of a NS are the neutrons (protons) in a superfluid (superconducting) state?
   e) How large are interior magnetic fields?
3. **Magnetospheres and Emission Physics:**
   a) What quantum electrodynamic processes are relevant for electromagnetic emissions?
4. **NS as Gravitational Laboratories:**

   a) Can departures from General Relativity be identified by monitoring the orbits of compact binary pulsars?
   b) Does the Strong-Equivalence Principle hold to high precision in pulsars with WD or BH companions?
5. **NS as Gravitational Wave Detectors:**
   a) Can we use pulsars as GW detectors or do spin noise and pulse-phase jitter limit the precision of their timing?
   b) Can we detect long-period gravitational waves to detect gravitational wave backgrounds from: (i) The early universe? (ii) Mergers of supermassive black holes? (iii) Topological defects (cosmic strings)?
6. **Pulsars as Probes of Galactic Structure:**
   a) What is the spiral structure of the Milky Way Galaxy?
   b) What is the nature of turbulence in the warm ionized ISM?

References

1. Cordes, J.M.: Pulsar Observations I. – Propagation Effects, Searching, Distance Estimates, Scintillations and VLBI. In: *Single-Dish Radio Astronomy: Techniques and Applications, ASP Conf Ser* 278, eds S. Stanimirovic, D. Altschuler, P. Goldsmith, C. Salter (Astron. Soc. Pacific, San Francisco 2002) pp 227-250
2. Stairs, I. H.: Pulsar Observations II. – Coherent Dedispersion, Polarimetry, and Timing. In: *Single-Dish Radio Astronomy: Techniques and Applications, ASP Conf Ser* 278, eds S. Stanimirovic, D. Altschuler, P. Goldsmith, C. Salter (Astron. Soc. of the Pacific, San Francisco 2002) pp 251–269
3. Cordes, J. M. et al. 2004, New Astronomy Reviews, 48, 1413
4. Kramer, M. et al. 2004, New Astronomy Reviews, 48, 993
5. Cordes, J. M. et al. 2004, New Astronomy Reviews, 48, 1459
6. Backer, D. C. & Hellings, R. W. 1986, AnnRevAstAp, 24, 537
7. Shapiro, S. L. & Teukolsky, S. A. 1983, Neutron Stars, Black Holes and White Dwarfs, New York, Wiley-Interscience
8. Glendenning, N. K. 2000, Compact Stars, Springer
9. Hewish, A. et al. 1968, Nature 217, 709
10. Klebesadel, R. W., Strong, I. B., & Olson, R. A. 1973, ApJLetters 182, L85
11. Staelin, D. H., & Reifenstein, E. C. 1968, Science 162, 1481
12. Cocke, W. J., Disney, M. J., & Taylor, D. J. 1969, Nature 221, 525
13. Comella, J. M. et al. 1969, Nature 221, 453.
14. Radhakrishnan, V., & Manchester, R. N. 1969, Nature 222, 228
15. Hulse, R. A., & Taylor, J. H. 1975, ApJLetters, 195, L51
16. Backer, D. C. et al. 1982, Nature 300, 615
17. Wolszczan, A., & Frail, D. A. 1992, Nature 355, 145
18. Lyne, A. G. & Lorimer, D. R. 1994, ApJ 369, 127
19. Arzoumanian, Z., Chernoff, D. F. & Cordes, J. M. 2002, ApJ 568, 289
20. Lyne, A. G. et al. 2004, Science, 303, 1153
21. Palmer, D. M. 2005, Nature 434, 1107
22. Gaensler, B. M. et al. 2005, Nature 434, 1104
23. Cameron, P. B. et al. 2005, Nature 434, 1112

24. Fox, D. B. et al. 2005, Nature 437, 845
25. Gold, T. 1969, Nature 221, 25
26. Melatos, A. 1997, MNRAS 288, 1049.
27. Blackman, E. G., & Perna, R. 2004, ApJLetters, 601, L71
28. Shemar, S. L. & Lyne, A. G. 1996, MNRAS, 282, 677
29. Cheng, A. F. 1985, ApJ 299, 917
30. Goldreich, P. & Julian, W. H. 1969, ApJ 157, 869
31. Bhattacharya, D. & van den Heuvel, E. P. J. 1991, Phys. Reports, 203, 1
32. Camilo, F. et al. 2000, ApJ 541, 367
33. McLaughlin, M. A. et al. 2003, ApJ 591, L135
34. Ruderman, M. A., & Sutherland, P. G. 1975, ApJ 196, 51
35. Rankin, J. M. 1983, ApJ 274, 333
36. Rankin, J. M. 1990, ApJ 352, 247
37. McLaughlin, M. A. et al. 2002, ApJ 564, 333
38. Blandford, R. & Teukolsky, S. 1976, ApJ 205, 580
39. Damour, T. & Taylor, J. H. 1992, Phys. Rev. D, 45, 1840
40. Detweiler, S. 1979, ApJ 234, 1100
41. Romani, R. W. & Taylor, J. H. 1983, ApJLetters, 265, L35
42. Stinebring, D. R. 1990, Phys Rev Letters, 65, 285
43. Kaspi, V., Taylor, J. H. & Ryba, M. F. 1994, ApJ 428, 713
44. Jaffe, A. H. & Backer, D. C. 2003, ApJ 583, 616
45. Foster, R. S. & Backer, D. C. 1990, ApJ 361, 300
46. Lommen, A. N. et al. 2003, ASP Conf. Ser. 302: Radio Pulsars, 302, 81
47. Brisken, W. F. et al. 2003, Astron J 126, 3090.
48. Chatterjee, S. et al. 2001, ApJ 550, 287
49. Vlemmings, W. H. T., Cordes, J. M., & Chatterjee, S. 2004, ApJ 610, 402.
50. Chatterjee, S. et al. 2004, ApJ 604, 339
51. Chatterjee, S. et al. 2005, ApJLetters, 630, L61
52. Cordes, J. M. & Lazio, T. J. W. 2002, arXiv: astro-ph/0207156; see also http://www.astro.cornell.edu/~cordes/NE2001
53. Gunn, J. E. & Ostriker, J. P. 1970, ApJ 160, 979
54. Cordes, J. M. & Chernoff, D. F. 1998, ApJ 505, 315
55. Hobbs, G. et al. 2005, MNRAS, 360, 974
56. Cordes, J. M. & Chernoff, D. F. 1997, ApJ 482, 971
57. Blaauw, A. 1961, BAN, 15, 265
58. Harrison, E. R. & Tademaru, E. 1975, ApJ 201, 447
59. Lai, D., Chernoff, D. F. & Cordes, J. M. 2001, ApJ 549, 1111
60. Spruit, H. C. & Phinney, E. S. 1998, Nature 393, 139
61. Colpi, M. & Wasserman, I. 2002, ApJ 581, 1271
62. Colpi, M. & Wasserman, I. 2003, arXiv:astro-ph/0302332
63. Bombaci, L. & Popov, S. B. 2004, AstAp, 424, 627
64. Chatterjee, S. & Cordes, J. M. 2002, ApJ 575, 407
65. McLaughlin, M. et al. 2006, Nature 439, 817
66. Burgay, M. et al. 2003, Nature 426, 531
67. Cordes, J. M. & Lazio, T. J. W. 1997, ApJ 475, 557
68. Pfahl, E. & Loeb, A. 2004, ApJ 615, 253
69. Wex, N. & Kopeikin, S. M. 1999, ApJ 514, 388
70. Hankins, T. H. et al. 2003, Nature 422, 141
71. Ransom, S. M., Cordes, J. M. & Eikenberry, S. S. 2003, ApJ 589, 911
72. Weisberg, J. M. et al. ApJS 150, 317.
73. Chatterjee, S., & Cordes, J. M. 2004, ApJLett, 600, L51
74. Nice, D. J. et al. 2005, ApJ 634, 1242

Theory of Gamma-Ray Burst Sources

Enrico Ramirez-Ruiz

Institute for Advanced Study, Einstein Drive, Princeton, NJ 08540:
enrico@ias.edu

In the sections which follow, we shall be concerned predominantly with the theory of γ-ray burst sources. If the concepts there proposed are indeed relevant to an understanding of the nature of these sources, then their existence becomes inextricably linked to the *metabolic pathways* through which gravity, spin, and energy can combine to form collimated, ultrarelativistic outflows. These threads are few and fragile, as we are still wrestling with trying to understand non-relativistic processes, most notably those associated with the electromagnetic field and gas dynamics. If we are to improve our picture-making we must make more and stronger ties of physical theory. But in reconstructing the creature, we must be guided by our eyes and their extensions. In this introductory section we have therefore attempted to summarise the observed properties of these ultra-energetic phenomena.

1 A Field Guide to Gamma-Ray Bursts

1.1 A Burst of Progress

The first sighting of a γ-ray burst, alias *GRB*, came on July 2, 1967, from the military *Vela* satellites monitoring for nuclear explosions in verification of the Nuclear Test Ban Treaty [52]. These γ-ray flashes proved to be rather different from the man-made explosions that the satellites were designed to detect. Over the next 30 years, hundreds of GRB detections were made. Frustratingly, they continued to vanish too soon to get an accurate angular position to permit follow-up observations. The reason for this is that γ-rays are notoriously hard to focus, so γ-ray images are generally not very sharp.

Before 1997, most of what we knew about GRBs was based on observations from the Burst and Transient Source Experiment (BATSE) on board the *Compton Gamma Ray Observatory* (*CGRO*), whose results have been summarised in [30]. BATSE, which measured about 3000 events, revealed that between two or three bursts occur somewhere in the observable universe on a

A. Carramiñana et al. (eds.), Solar, Stellar and Galactic Connections between Particle Physics and Astrophysics, 77–113.

typical day. While they are on, they outshine every other source in the γ-ray sky, including the sun. Although each is unique, the bursts fall into one of two rough categories. Bursts that last less than two seconds are *short*, and those that last longer – the majority – are *long*. The two categories differ spectroscopically, with short bursts having relatively more high-energy γ-rays than long bursts do (Section 1.2).

Arguably the most important result from BATSE concerned the distribution of bursts. They occur isotropically – that is, evenly over the entire sky, suggesting a cosmological distribution with no dipole and quadrupole components. This finding cast doubt on the prevailing wisdom, which held that bursts came from sources within the Milky Way. The uniform distribution led most astronomers to conclude that the instruments were picking up some kind of event happening throughout the universe. Unfortunately, γ-rays alone did not provide enough information to settle the question for sure. The detection of radiation from bursts at other wavelengths would turn out to be essential. Visible light, for example, could reveal the galaxies in which the bursts took place, allowing their distances to be measured. Attempts were made to detect these burst counterparts, but they proved fruitless.

A watershed event occurred in 1997, when the spacecraft *BeppoSAX* succeeded in obtaining high-resolution X-ray images [22] of the predicted fading afterglow of GRB 970228 – so named because it occurred on February 28, 1997. This detection, followed by a number of others at an approximate rate of 10 per year, led to positions accurate to about an arc minute, which allowed the detection and follow-up of the afterglows at optical and longer wavelengths [110]. This paved the way for the measurement of redshift distances, the identification of candidate host galaxies, and the confirmation that they were at cosmological distances [71].

Among the first GRBs pinpointed by *BeppoSAX* was GRB 970508. Radio observations of its afterglow provided an essential clue. The glow varied erratically by roughly a factor of two during the first three weeks, after which it stabilised and then began to diminish. The large variations probably had nothing to do with the burst source itself; rather they involved the propagation of the afterglow light through space. Just as the Earth's atmosphere causes visible starlight to twinkle, interstellar plasma causes radio waves to scintillate. Therefore, if GRB 970508 was scintillating at radio wavelengths and then stopped, its source must have grown from a mere point to a discernible disk. "Discernible" here means a few light-weeks across. To reach this size the source must have been expanding at a considerable rate – close to the speed of light [115]. The *BeppoSAX* and follow-up observations have transformed our view of GRBs. The old concept of a sudden release of energy concentrated in a few brief seconds has been discarded. Indeed, even the term "afterglow" is now recognised as misleading – the energy radiated during both phases is comparable.

The next step for GRB astronomy is to flesh out the data on burst, afterglow and host-galaxy characteristics. One needs to measure many hundreds of bursts of all varieties: long and short, bright and faint, bursts that are mostly γ-rays and bursts that are mostly X-rays. Currently we are obtaining burst positions from the second *High Energy Transient Explorer* satellite, launched in October 2000, and the Interplanetary Network, a series of small γ-ray detectors piggybacking on planetary spacecraft. The *Swift* mission, recently launched, now offers multi-wavelength observations of tens of *long* GRBs and their afterglows. On discovering a GRB, the γ-ray instrument triggers automatic onboard X-ray and optical observations. A rapid response determines whether the GRB has an X-ray and/or visible afterglow. A striking development in the last several months has been the measurement and localization of fading X-ray signals from several *short* GRBs by *Swift*. Other missions, though not designed solely for GRB discovery, will also contribute. The *International Gamma-Ray Astrophysics Laboratory*, detects between 10 to 20 GRBs a year. The *Energetic X-ray Imaging Survey Telescope*, planned for launch a decade from now, will have a sensitive gamma-ray instrument capable of detecting thousands of GRBs.

1.2 Gamma Rays: Clues

GRBs are brief flashes of radiation at hard X-ray and soft γ-ray energies that display a wide variety of time histories, though in $\sim 25\%$ of the cases a characteristic single-pulse profile is observed, consisting of a rapid rise followed by a slower decay [30]. GRBs were first detected at soft γ-ray energies with wide field-of-view instruments. Peak soft γ-ray fluxes reach hundreds of photons cm^{-2} s^{-1} in rare cases. The BATSE instrument is sensitive in the 50-300 keV band, and provides the most extensive data base of GRB observations during the prompt phase. It searches for GRBs by examining strings of data for $> 5.5\sigma$ enhancements above the background on the 64 ms, 256 ms, and 1024 ms time scales, and triggers on GRBs as faint as ≈ 0.5 ph cm^{-2} s^{-1}, corresponding to energy flux sensitivities $< 10^{-7}$ ergs cm^{-2} s^{-1}.

The integral size distribution of BATSE GRBs in terms of peak flux ϕ_p is very flat below ~ 3 ph cm^{-2} s^{-1}, and becomes steeper than the $-3/2$ behaviour expected from a Euclidean distribution of sources at $\phi_p > 10$ ph cm^{-2} s^{-1} [111]. GRBs typically show a very hard spectrum in the hard X-ray to soft γ-ray regime, with a photon index breaking from ≈ -1 at photon energies $E_{ph} < 50$ keV to a -2 to -3 spectrum at $E_{ph} >$ several hundred keV [5]. Consequently, the distribution of the peak photon energies E_{pk} of the time-averaged νF_ν spectra of BATSE GRBs are typically found in the 100 keV - several MeV range [64]. The general trend is that the spectrum softens, and E_{pk} decreases, with time. More precise statements must however wait for larger area detectors.

In Figure 1 are plotted representative spectra, in the unconventional coordinates ν and νF_ν, the energy radiated per logarithmic frequency interval. Some obvious points should be emphasised. We measure directly only the specific luminosity $D^2 \nu I_\nu \equiv (1/4\pi)\nu L_\nu$ (the energy radiated in the direction

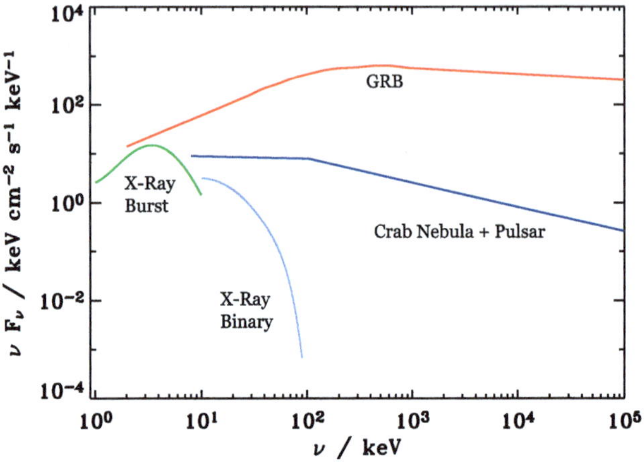

Fig. 1. Representative spectra $\nu F_\nu \propto \nu^2 N(\nu)$ of various high energy phenomena. The GRB spectrum is nonthermal, the number of photons varying typically as $N(\nu) \propto \nu^{-\alpha}$, where $\alpha \sim 1$ at low energies changes to $\alpha \sim 2$ to 3 above a photon energy 0.1-1 MeV.

of the earth per second per steradian per logarithmic frequency interval by a source at luminosity distance D), and its dimensionless distance-independent ratios between two frequencies (BATSE triggers, for example, are based on the count rate between 50 keV and 300keV). The apparent bolometric luminosity $\int_{-\infty}^{\infty} D^2 \nu I_\nu d(\text{In}\nu)$ may be quite different from the *true* bolometric luminosity $\int_{4\pi}\int_0^{\infty} D^2 I_\nu d\nu d\Omega$ if the source is not isotropic. The GRB spectrum shown in Figure 1 is that of 9206022 which was observed simultaneously by BATSE, COMPTEL and *Ulysses*. The time integrated spectrum on those detectors ranges from 25 keV to 10MeV [43].

Most GRBs are accompanied by a high-energy tail which contains a significant amount of energy – $\nu^2 N(\nu)$ is almost constant. The EGRET instrument on the *CGRO* detected 7 GRBs with $>$ 30 MeV emission during the prompt phase [30], including the extraordinary burst GRB 940217, which displayed $>$ 100 MeV emission 90 minutes after the onset of the GRB, including one photon with energy near 20 GeV [49]. This gives unambiguous evidence for the importance of nonthermal processes in GRBs. TeV radiation has been reported to be detected with the Milagrito water Cherenkov detector from GRB 970417a [4]. If correct, this requires that this source be located at $z < 0.3$ in order that $\gamma - \gamma$ attenuation with the diffuse intergalactic infrared radiation be small.

A class of X-ray rich GRBs, with durations of order of minutes and X-ray fluxes in the range 10^{-8}-10^{-7} ergs cm^{-2} s^{-1} in the 2-25 keV band, has been detected with many X-ray satellites, including *Ariel V*, *HEAO-1*, *ROSAT*, *Ginga*, and *Beppo-SAX* [45].

A *typical* GRB (if there is such a thing) lasts about ten seconds. Observed durations vary, however, by six orders of magnitude, from several milliseconds [30] to several thousand seconds [53]. The shortest BATSE burst had a duration of 5ms with a 0.2ms structure [10]. The longest so far, GRB940217, displayed GeV activity one and a half hours after the main event [49]. The bursts GRB 961027a, GRB961027b, GRB 961029a and GRB 961029b occurred in the same region of the sky within two days [30]; if this *gang of four* is considered as a single event then the longest duration so far is two days! These observations may indicate that some sources display a continued activity (at a variable level) over a period of days.

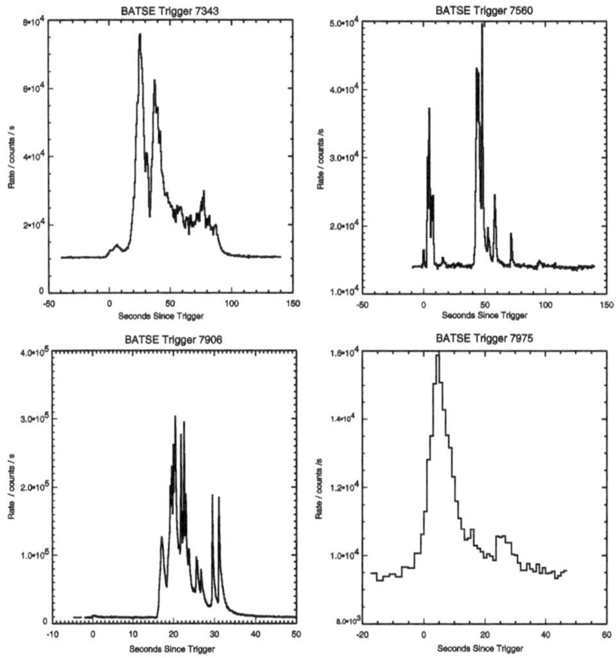

Fig. 2. BATSE lightcurves of various individual bursts. The y axis is the photon count rate in the 0.05 to 0.5 MeV range; the x axis is the time in seconds since the burst trigger. Both before and after the burst trigger, no γ-rays are detected, above background, from the same direction. Clockwise, from top left: GRB 990123, GRB 990510, GRB 000131, and GRB 000131.

The bursts have complicated and irregular time profiles which vary drastically from one burst to another. They range from smooth, fast rise and quasi-exponential decay, through curves with several peaks, to variable curves with many peaks. Various profiles, selected from the BATSE catalogue, are shown in Figure 2. About 3% of bursts are preceded by a precursor with a lower peak intensity than the main event [57], while 8% of them contain at least one long period of quiescence [87]. Several bursts observed by *Ginga* showed significant thermal X-ray emission before the main high energy pulse [73]. Some of these bursts were also followed by low energy X-ray tails. These are probably pre-discovery detections of the X-ray afterglows observed now by *Swift* and *HETE-II* and previously uncovered by *BeppoSAX*.

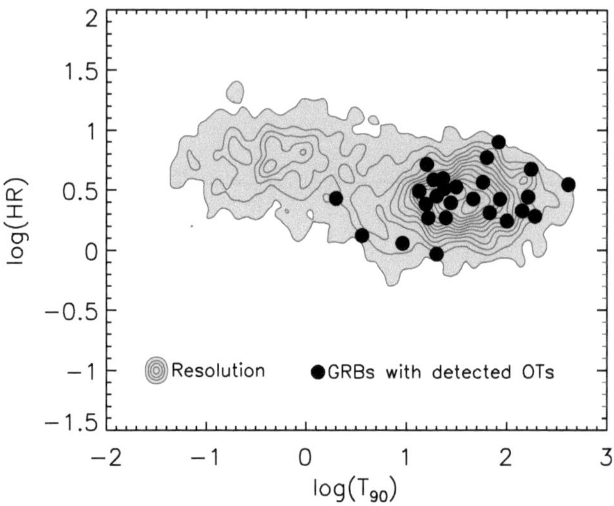

Fig. 3. Hardness ratio versus duration of BATSE bursts. The hardness ratio is a measure of the shape of the spectrum: larger values correspond to harder spectra. Different line levels denote number density contours. The bimodality of the distribution of their duration is confirmed by the associated spectral shape. Indicated are the positions in the diagram of the *BeppoSax* GRBs with detected optical transients. All information derived from the precise localization of *BeppoSax* refers only to long duration GRBs.

The duration of a GRB is defined by the time during which the middle 50% (t_{50}) or 90% (t_{90}) of the counts above background are measured. A bimodal duration distribution is measured, irrespective of whether the t_{50} or t_{90} durations are considered [58]. About two-thirds of BATSE GRBs are long-duration GRBs with $t_{90} > 2$ s, with the remainder comprising the short-duration GRBs. This bimodality is confirmed by the associated spectral shape, since short bursts, on average, appear harder than long GRBs. Figure 3 shows the hardness ratio as a function of the duration of the emission. The hardness ratio is a measure of the slope of the spectrum where larger values means that the flux at high energies dominates.

The BATSE results also showed that the angular distribution of GRBs on the sky is isotropic, and that the size distribution exhibits a strong flattening for faint bursts [65]. This behaviour follows from a cosmological origin of GRB sources, with the decline in the number of faint bursts due to cosmic expansion. Follow-up X-ray observations with the Narrow Field Instrument on *BeppoSAX* has permitted redshift determinations that firmly establish the distance scale to the sources of > 2 s duration GRBs, which are those to which *BeppoSAX* was sensitive (Figure 3).

1.3 A Warm Afterglow

Launched on April 1996, the *BeppoSAX* satellite made breakthrough observations of GRBs, succeeding in positioning them with error boxes of only a few arcminutes through its coded mask Wide Field Camera, sensitive at medium–hard X–ray energies (i.e. 2–25 keV). The *BeppoSAX* GRB observations revealed that essentially all long-duration GRBs have fading X-ray afterglows [23]. The Wide Field Camera on *BeppoSAX* has sensitivity down to $\sim 10^{-10}$ ergs cm^{-2} s^{-1} with $< 10'$ error boxes. Spacecraft slewing requires 6-8 hours, but permits Narrow Field Instrument X-ray observations with sensitivity down to $\sim 10^{-14}$ ergs cm^{-2} s^{-1} and error boxes $< 0.5'$. The first X-ray afterglow was obtained from GRB 970228 [22], which revealed an X-ray source which decayed according to a power-law behaviour $\phi_X \propto t^{\chi}$, with $\chi \sim -1.33$. Typically, $\chi \sim -1.1$ to -1.5 in X-ray afterglows. The spectral shape is remarkably softer than the prompt emission, with $F(\nu)$ roughly proportional to ν^{-1}.

The small X-ray error boxes allow deep optical and radio follow-up studies. GRB 970228 was the first GRB from which an optical counterpart was observed [110], and GRB 970508 was the first GRB for which a redshift was measured [26, 71]. Detection of optical emission lines from the host galaxy, and absorption lines in the fading optical afterglow due to the presence of intervening gas has provided redshifts for about 30 GRBs. No optical counterparts are detected from approximately one-half of GRBs with well-localized X-ray afterglows, and these are termed *dark* bursts. These sources may be undetected in the optical band because of dusty media [44] or because they are optically faint by nature [7].

The monochromatic flux decreases in time as a power law $F_\nu(t) \propto t^{-0.8}$ – t^{-2}. Usually, the magnitudes of the optical afterglow detected $\sim$ one day after the γ–ray event are in the range 19–21. An observation that attracted much attention was the discovery [1] of a prompt and extremely bright (visual magnitude $m_v \sim 9$) optical flash in GRB 990123, 15 s after the GRB started and while it was still going on. The term *flash* is indeed appropriate. A magnitude 9 at a redshift $z = 1.6$ corresponds a power $L \sim 5 \times 10^{49}$ erg s^{-1} in the optical band, meaning that if the event had instead taken place a few thousands light-years away, it would have been as bright as the midday sun, albeit for a short time. Due to the brightness of the afterglow and its prompt optical emission, photometric observations were extensive, making it one of the best-studied afterglows (Figure 4).

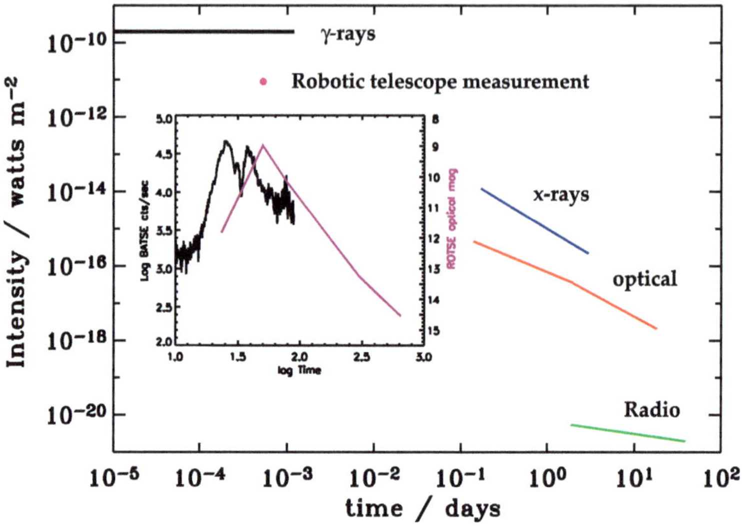

Fig. 4. One of the brightest GRB yet recorded went off on January, 23, 1999. The radiation, which started out concentrated in the γ-ray range during the burst, progressively evolved into an afterglow radiation that peaked in the X-rays, then ultraviolet (UV), optical, infrared, and radio. GRB 990123 is still the only burst detected so far in the optical while the prompt emission was still on, by the robotic telescope ROTSE, 22 seconds after the trigger at $m_v \sim 11.7$, reaching $m_v \sim 9$ a few tens of seconds after [1].

Approximately 40% of GRBs have radio counterparts and the transition from scintillating to smooth behaviour in the radio afterglow of GRB 970508 provides direct evidence for relativistic source expansion [32].

The redshifts of nearly 100 GRBs are now known (early 2006), with the median $\langle z \rangle \sim 1.5$ and the largest measured redshift at $z = 6.295$. The corresponding distances imply apparent isotropic γ-ray energy releases in the range from $\approx 10^{51}$-10^{54} erg (Figure 5). For a solar-mass object, this implies that an unusual large fraction of the energy is converted into γ-ray photon energy. This spread in the inferred luminosities obtained under the assumption of isotropic emission may be reduced if most GRB outflows are jet-like.

X-ray emission lines, possibly due to Fe Kα fluorescence, were detected during a re-brightening phase in the afterglow of GRB 970508 [82], and in the afterglow spectra of GRB 991216 [83] and GRB 000214 [3]. A detection of soft X-ray emission features was also reported from GRB 011211 [92]. The observed frequencies of the lines appear displaced from the laboratory frequency, as expected from the Doppler shift caused by the expansion of the universe, in agreement with the redshift measured in optical lines from the host galaxy. A transient absorption feature was detected from GRB 990705 [2], and X-ray absorption in excess of the Galactic hydrogen column density has also been reported in GRB 980329 [35]. These results offer an alternative method of

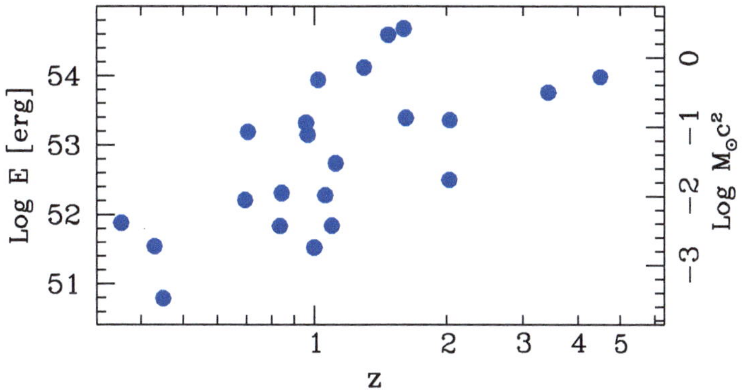

Fig. 5. Apparent isotropic γ-ray energy as a function of redshift. The energy is calculated, assuming isotropic emission in a common comoving bandpass, for 22 *BeppoSax* GRBs with known redshift and spectra, using the compilation of [16]. A particularly useful updated link with all the relevant information about bursts with good localization is maintained by Jochen Greiner at: http://www.aip.de/ jcg/grbgen.html.

redshift determination, and provide important clues about the environment of the progenitor object, showing that a high column density of gas must be present in the vicinity of these sources.

1.4 Hosts, Supernova Family Ties, and Cosmological Setting

Demography

For the *long* GRB afterglows localized so far, a host galaxy has been found in most cases. As of early 2006, plausible or certain host galaxies have been found for all but 1 or 2 of the bursts with optical, radio, or X-ray afterglows localised with arcsecond precision. The median apparent magnitude is $R \approx 25$ mag, with tentative detections or upper limits reaching down to $R \approx 29$ mag. The few missing cases are at least qualitatively consistent with being in the faint tail of the observed distribution of host galaxy magnitudes [20]. We note also that the observations in the visible probe the UV in the rest frame, and are thus especially susceptible to extinction. However, sub-mm detections of dusty GRB hosts are currently limited by the available technology to only a handful of ultraluminous sources [8].

The majority of redshifts so far are from the spectroscopy of host galaxies, but an increasing number are based on the absorption-line systems seen in the spectra of the afterglows (which are otherwise featureless power-law continua). Reassuring overlap exists in several cases; invariably, the highest-z absorption system corresponds to that of the host galaxy, and has the strongest lines. In some cases (a subset of the so-called “dark bursts”) no optical transient is detected, but a combination of the X-ray and radio transient unambiguously pinpoints the host galaxy [27].

Are the GRB host galaxies special in some way? This is hard to answer [27] from their visible ($\sim$ restframe UV) luminosities alone: the observed light traces an indeterminate mix of recently formed stars and an older population, and cannot be unambiguously interpreted in terms of either the total baryonic mass, or the instantaneous star formation rate (SFR).

The magnitude and redshift distributions of GRB host galaxies are typical for the normal, faint field galaxies, as are their morphologies when observed with the *Hubble Space Telescope* (*HST*): they are often compact, and sometimes suggestive of a merging system [17], but that is not unusual for galaxies at comparable redshifts. Within the host galaxies, the distribution of GRB-host offsets follows the light distribution closely [17], which is roughly proportional to the density of star formation (especially for the high-z galaxies). It is thus fully consistent with a progenitor population associated with the sites of massive star formation. Spectroscopic measurements provide direct estimates of recent, massive SFR in GRB hosts [27]. The observed *unobscured* SFRs range from a few tenths to a few $M_{\odot}$ yr^{-1} [8]. All this is entirely typical for the normal field galaxy population at comparable redshifts. However, such measurements are completely insensitive to any fully obscured SFR components.

Equivalent widths of the [O II] 3727 doublet in GRB hosts, which may provide a crude measure of the SFR per unit luminosity (and a poorer measure of the SFR per unit mass), are on average somewhat higher [27] than those observed in magnitude-limited field galaxy samples at comparable redshifts [48]. A larger sample of GRB hosts, and a good comparison sample, matched both in redshift and magnitude range, are necessary before any solid conclusions can be drawn from this apparent difference. One intriguing hint comes from the flux ratios of [Ne III] 3869 to [O II] 3727 lines: they are on average a factor of 4 to 5 higher in GRB hosts than in star forming galaxies at low redshifts [27] . Strong [Ne III] requires photoionization by massive stars in hot H II regions, and may represent indirect evidence linking GRBs with massive star formation.

The interpretation of the luminosities and observed star formation rates is vastly complicated by the unknown amount and geometry of extinction. The observed quantities (in the visible) trace only the unobscured stellar component, or the components seen through optically thin dust. Any stellar and star formation components hidden by optically thick dust cannot be estimated at all from these data, and require radio and sub-mm observations [89]. As of late 2002, radio and/or sub-mm emission powered by obscured star formation has been detected from 4 GRB hosts [34, 8]. The surveys to date are sensitive only to the ultra-luminous ($L > 10^{12} L_{\odot}$) hosts, with SFR of several hundred $M_{\odot}$ yr^{-1} (Figure 6). Modulo the uncertainties posed by the small number statistics, the surveys indicate that about 20% of GRB hosts are objects of this type, where about 90% of the total star formation takes place in obscured regions.

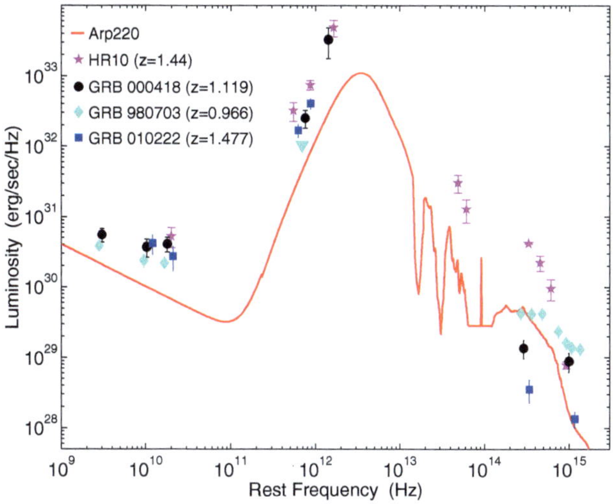

Fig. 6. Spectral energy distributions (SEDs) of host galaxies of GRB 000418, GRB 980703, and GRB 010222 compared to the SED of the local starburst galaxy Arp 220, and the high-z starburst galaxy HR 10. The luminosities are plotted at the rest frequencies to facilitate a direct comparison. These highly star forming galaxies are more luminous than Arp 220, and are similar to HR 10, indicating that their bolometric luminosities exceed 10^{12} $L_{\odot}$, and their star formation rates are of the order of 500 $M_{\odot}$ yr^{-1}. On the other hand, the spectral slopes in the optical regime are flatter than both Arp 220 and HR 10, indicating that the GRB host galaxies are bluer than Arp 220 and HR 10. Panel from [8].

Given the uncertainties of the geometry of optically thin and optically thick dust, optical colours of GRB hosts cannot be used to make any meaningful statements about their net star formation activity. The broad-band optical colours of GRB hosts are not distinguishable from those of normal field galaxies at comparable magnitudes and redshifts [17]. It is notable that the optical/NIR colours of GRB hosts detected in the sub-mm are much bluer than typical sub-mm selected galaxies, suggesting that the GRB selection may be probing a previously unrecognised population of dusty star-forming galaxies [89, 108, 8].

On the whole, the GRB hosts seem to be representative of the normal, star-forming field galaxy population at comparable redshifts, and so far there is no evidence for any significant systematic differences between them.

Supernova Partnership

At cosmological distances, the total energy of GRBs is roughly of the same order of magnitude as that of supernovae; in GRBs, however, the energy is emitted much more rapidly. The reason is that in supernovae the energy (except that in neutrinos) is thermalised by a large amount of mass (several solar masses). It therefore came as something of a surprise when [38] found that the *BeppoSAX* error box of GRB 980425 contained the supernova SN 1998bw. This supernova is located in a spiral arm of the nearby galaxy ESO 184-G82, at a redshift of 2550 km/s, corresponding to a distance of 40 Mpc. On the basis of very conservative assumptions regarding the error box and the time window in which the supernova occurred, [38] determined that the probability that any supernova with peak optical flux a factor of 10 below that of SN 1998bw would be found in the error box by chance coincidence is 10^{-4}.

With respect to apparent properties (peak flux, duration, burst profile) GRB 980425 was not remarkable. Of course, at its distance of 40 Mpc, its total energy 8×10^{47} erg/s is some five orders of magnitude smaller than that of normal GRBs.

Independent of its connection with a GRB, SN 1998bw is extraordinary for its very high radio luminosity near the peak of the SN lightcurve [59]. An analysis of the optical lightcurve [38] and its early spectra [50] showed that it was an extremely energetic event – total explosive energy in the range $2-6\times10^{52}$ erg, i.e. a factor of $\sim$ 30 higher than is typical for a Ib/c supernova. Because the sampling volume for low luminosity bursts such as 980425 is smaller than that of normal GRBs by a factor of 10^6, the rate (per galaxy) of the former events may well exceed those of the latter by a large factor. Due to their small distances they are expected to contribute a $\phi_p^{-3/2}$ component to the log $N(>\phi_p)$ distribution. From the absence of a turn-up at the flux limit ($\phi_p \simeq 0.2$), [55] inferred that such a Euclidean component can contribute at most 10% to the observed BATSE burst sample. With a normal GRB rate of 10^{-8} per galaxy per year [117], the corresponding limit on events like SN 1998bw is thus a few 10^{-4} per galaxy per year. With an observed rate of type Ib/c supernova of a few times 10^{-3} per year per galaxy [112], this rather weak limit serves to show that at most a fraction of the SN Ib/c produce GRBs.

The observational basis for a connection between GRBs and supernovae was greatly enriched with the discovery by [15] of a late time component superposed on the power law optical lightcurve of GRB 980326, which they argue reflects an underlying supernova. The optical lightcurve of GRB 980326 showed an initial rapid decay; the lightcurve flattened after $\sim$ 10 days to a constant value $R = 25.5 \pm 0.5$. Such flattening has been seen in the light curves of other afterglows as well, and has been interpreted as the signature of an underlying host galaxy [111]. Observations by [15] made $\sim$ 3 weeks after the burst revealed a surprising brightening of the afterglow, to a flux level 60 times above that expected from an extrapolation of the power-law decay. At the same time, the spectral energy distribution became very red. Observations made $\sim$ 9 months after the burst showed that any host galaxy is fainter than R=27.3. Using the multicolour light curve of SN 1998bw [38] as a template, [15] found that they can reproduce the observed optical light curve of GRB 980326 by a combination of a power-law and a bright, simultaneous 1998bw-like supernova at a redshift $z \sim 1$.

A number of other GRBs since then have shown similar late deviations from a power-law decline (the best example being GRB 011121 [40]), but they still lacked a clear spectroscopic detection of an underlying supernova. Detection of such signature was recently reported by [106] for GRB 030329. Due to its extreme brightness and slow decay, spectroscopic observations were extensive. The early spectra consisted of a power-law decay continuum ($F_\nu \propto \nu^{-0.9}$) typical of GRB afterglows with narrow emission features identifiable as Hα, [OIII], Hβ and [OII] at $z = 0.1687$, making GRB 030329 the second nearest burst overall (GRB 980425 possibly associated with the nearby SN 1998bw is the nearest at $z = 0.0085$) and the *classical* burst with the lowest known redshift.

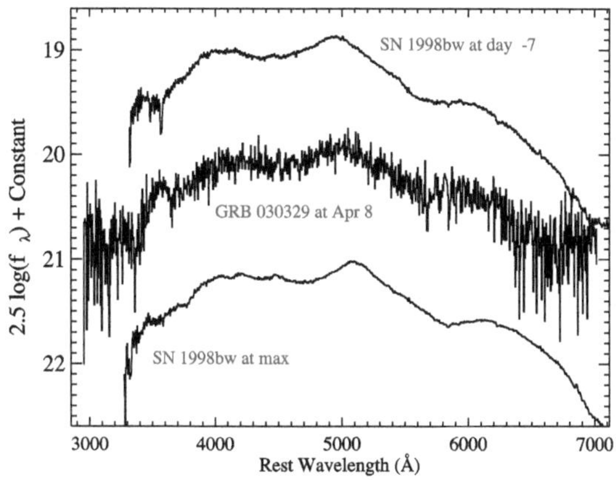

Fig. 7. The supernova dominated spectrum of GRB 030329 obtained in April 8 with the MMT telescope by [106]. The residual spectrum shows broad bumps at approximately 5000 Å and 4200 Å (rest frame), which are similar to those seen in the spectrum of the peculiar type Ic SN 1998bw a week before maximum [79].

Beginning April 6 (i.e. 8 days after the GRB), the spectra showed the development of broad peaks in flux, characteristic of a supernova. The broad bumps are seen at approximately 5000 Å and 4200 Å (rest-frame). Over the next few days the SN features became more prominent as the afterglow faded and the SN brightened towards maximum. The afterglow spectrum of April 8, clearly supernova dominated, is shown in Figure 7. For comparison, spectra of SN 1998bw at maximum and a week before maximum are also displayed [79]. The similarities are striking. While the presence of supernovae has been inferred from the lightcurves and colours of GRB afterglows in the past, this is the first convincing spectroscopic evidence that a supernova was lurking beneath the optical afterglow of a long duration GRB.

GRBs and Cosmology

While interesting on their own, GRBs are now rapidly becoming powerful tools to study the high-redshift universe and galaxy evolution, thanks to their apparent association with massive star formation, and their brilliant luminosities. There are three basic ways of learning about the evolution of luminous matter and gas in the universe. First, a direct detection of sources (i.e., galaxies) in emission, either in the UV/optical/NIR (the unobscured components), or in the FIR/sub-mm/radio (the obscured component). Second, the detection of galaxies selected in absorption along the lines of sight to luminous background sources, traditionally QSOs. Third, diffuse extragalactic backgrounds. Studies of GRB hosts and afterglows can contribute to all three of these methodological approaches, bringing in new, independent constraints for models of galaxy evolution and of the history of star formation in the universe [11, 89, 27].

Already within months of the first detections of GRB afterglows, no optical afterglows were found associated with some well-localised bursts despite deep and rapid searches; the prototype *dark* burst was GRB 970828 [28]. Perhaps the most likely explanation for the non-detections of optical transients when sufficiently deep and prompt searches are made is that they are obscured by dust in their host galaxies. This is an obvious culprit if indeed GRBs are associated with massive star formation. The census of optical afterglow detections for well-localised bursts can thus provide a completely new and independent estimate of the mean obscured star formation rate [89]. There is one possible loophole in this argument: GRBs may be able to destroy the dust in their immediate vicinity up to ~ 10 pc [116, 39], and if the rest of the optical path through their hosts ($\sim$ kpc scale?) was dust-free, the optical afterglow would become visible. Such a geometrical arrangement may be unlikely in most cases, and our argument probably still applies. A more careful treatment of the dust evaporation geometry is needed, but it is probably safe to say that GRBs can provide a valuable new constraint on the history of star formation in the universe.

Absorption spectroscopy of GRB afterglows is now becoming a powerful new probe of the ISM in evolving galaxies, complementary to the traditional studies of quasar absorption line systems. The key point is that the GRBs, almost by definition (that is, if they are closely related to the sites of ongoing or recent massive star formation, as the data seem to indicate), probe the lines of sight to dense, central regions of their host galaxies ($\sim 1-10$ kpc scale). On the other hand, the quasar absorption systems are selected by the gas cross section, and favour large impact parameters ($\sim 10-100$ kpc scale), mostly probing the gaseous halos of field galaxies, where the physical conditions are very different.

The growing body of data on GRB absorption systems shows exceptionally high column densities of gas, when compared to the typical quasar absorption systems; only the highest column density DLA systems (themselves ostensibly star-forming disks or dwarfs) come close [72]. This opens the interesting prospect of using GRB absorbers as a new probe of the chemical enrichment history in galaxies in a more direct fashion than is possible with the quasar absorbers, where there may be very complex dynamics of gas ejection, infall, and mixing at play.

Possibly the most interesting use of GRBs in cosmology is as probes of the early phases of star and galaxy formation, and the resulting reionization of the universe at $z \sim 6-20$. The bursts for which redshifts are known are bright enough to be detectable, in principle, out to much larger distances than those of the most luminous quasars or galaxies detected at present [60]. Within the first minutes to hours after the burst, the optical light from afterglows is known to have a range of visual magnitudes $m_v \sim 10 - 15$, far brighter than quasars, albeit for a short time. Thus, promptly localized GRBs could serve as beacons which, shining through the pregalactic gas, provide information about much earlier epochs in the history of the universe. The presence of iron or other X-ray lines provides an additional tool for measuring GRB distances, which may be valuable for investigating the small but puzzling fraction of bursts which have been detected only in X-rays but not optically, perhaps due to a high dust content in the host galaxy.

Short-Lived Mysteries

Until recently, *short* GRBs were known predominantly as bursts of γ-rays, largely devoid of any observable traces at any other wavelengths. However, a striking development in the last several months has been the measurement and localization of fading X-ray signals from several *short* GRBs, making possible the optical and radio detection of afterglows, which in turn enabled the identification of host galaxies at cosmological distances [18, 31, 41, 86]. The presence in old stellar populations e.g., of an elliptical galaxy for GRB050724, rules out a source uniquely associated with recent star formation [9]. In addition, no bright supernova is observed to accompany *short* GRBs [18, 31, 47], in distinction from most nearby *long* GRBs [46].

The newly launched *Swift* spacecraft is expected to yield localizations for about 100 *long* and 10 *short* bursts per year. *Swift* is equipped with γ-ray, X-ray and optical detectors for on-board follow-up, and capable of relaying to the ground arc-second quality burst coordinates within less than a minute from the burst trigger, allowing even mid-size ground-based telescopes to obtain prompt spectra and redshifts. This will permit much more detailed studies of the burst environment, the host galaxy, and the intergalactic medium between galaxies.

This concludes our compendium of the *facts*. For ease of reference in the chapters that follow, they have been assembled here with a minimum of speculative interpretation.

2 Metabolic Pathways

In this section, we present a partial summary of some general ways by which gravity, angular momentum and the electromagnetic field can couple to power ultrarelativistic outflows, along with reviewing the most popular current models for the central source.

2.1 Bestiary

As is well known, one of the first proposals that was made, soon after the discovery of quasars in 1963, was that they were powered by accretion onto massive black holes[119, 99]. The fundamental reason why this proposal was made was that quasars were known to be prodigiously powerful, with luminosities equivalent to hundreds of galaxies, and that up to $\sim 0.1c^2 \equiv 10^{20}$ erg g^{-1} of energy per unit mass could be released by lowering matter close to a black hole. This efficiency could be over a hundred times that traditionally associated with nuclear power. Since this time, we have also learned about black holes with masses $\sim 5 - 10$ $M_\odot$ in Galactic binary systems and ultra-luminous X-ray sources which, with decreased confidence, we also associate with black holes, primarily on energetic grounds. In these objects, accretion (and the accompanying radiation) is usually thought to be limited by the Eddington rate, a self–regulatory balance imposed by Newtonian gravity and radiation pressure. The standard argument gives a maximum luminosity

$$L_{\rm Edd} = 1.3 \times 10^{38}(M/M_\odot) \text{ erg s}^{-1}. \tag{1}$$

Although this may not be strictly the case in reality – as in the current argument concerning the nature of ULXs – it does exhibit the qualitative nature of the effect of radiation pressure on accreting plasma, in the limit of large optical depth.

The photon luminosity, for the few-second duration of a typical short burst, is of course colossal: it exceeds by many thousands the most extreme output from any active galactic nucleus (thought to involve super massive black holes), and is 12 orders of magnitude above the Eddington limit for a stellar-mass object. The total energy, however, is not out of line with some other phenomena encountered in astrophysics - indeed it is reminiscent of the energy released in the core of a supernova. The Eddington photon limit (1) is circumvented if the main cooling agent is emission of neutrinos rather than electromagnetic waves. This regime requires correspondingly large accretion rates, of the order of one solar mass per second, and is termed hypercritical accretion. Such high accretion rates are never reached for black holes in XRBs or AGN, where characteristic rates are below the Eddington rate. They can, however, be achieved in the process of forming neutron stars and solar-mass black holes in the core collapse of massive stars. In such a situation, the densities and temperatures are so large ($\rho \simeq 10^{12}$ g cm^{-3}, $T \simeq 10^{11}$ K) that photons are completely trapped, and neutrinos are emitted copiously.

Table 1. Estimated rates of GRBs and plausible progenitors in yr^{-1} Gpc^{-3} derived from [36, 80, 104]

Progenitor	Rate($z = 0$)
NS–NS	80
BH–NS	10–300
BH–WD	10
BH–He	1000
SN Ib/c	60000
GRBs	$0.5(4\pi/\Omega)$

Unless they are beamed into less than one percent of the solid angle, the triggers for GRBs are thousands of times rarer than supernovae. The current view is that they arise in a very small fraction $\sim 10^{-6}$ of stars which undergo a catastrophic energy release event toward the end of their evolution. One conventional possibility is the coalescence of binary neutron stars [61, 75, 29, 74, 94, 96]. Double neutron star (NS) binaries, such as the famous PSR1913+16, will eventually coalesce due to angular momentum and energy losses to gravitational radiation. When a neutron star binary coalesces, the rapidly-spinning merged system could be too massive to form a single neutron star; on the other hand, the total angular momentum is probably too large to be swallowed immediately by a black hole [95]. The expected outcome would then be a spinning hole, orbited by a torus of NS debris.

Other types of progenitor have been suggested - e.g. a NS-BH merger, where the neutron star is tidally disrupted before being swallowed by the hole [74, 54, 62]; the merger of a He star (or a WD) with a black hole [120, 36]; or a category labelled as hypernovae [77] or collapsars [118, 63], where the collapsing core is too massive to become a neutron star, but has too much angular momentum to collapse quietly into a black hole (as in a so-called *failed* supernova). Table 1 provides a summary of the various rate estimates for some of these GRB progenitors, while Figure 8 illustrates their different production channels. The reader is referred to the excellent review by [66] for a description of other source models. Aside from the rate of SNe events, the rate of short GRBs and plausible progenitors in Table 1 are uncertain by at least a factor of a few. All of the progenitor scenarios listed roughly scale with the rate of star formation; therefore, the rates at redshift of $z = 1$ are a factor of ~ 10 higher than locally.

It has become increasingly apparent in the last few years that most plausible GRB progenitors suggested so far (e.g. NS-NS or NS-BH mergers, He-BH or WD-BH mergers, and failed SN) are expected to lead to a black hole plus debris torus system, although a possible exemption includes the formation from a stellar collapse of a fast rotating neutron star with an ultrahigh magnetic field. The binding energy of the orbiting debris, and the spin energy of the BH are the two main reservoirs in the former case. The first provides up

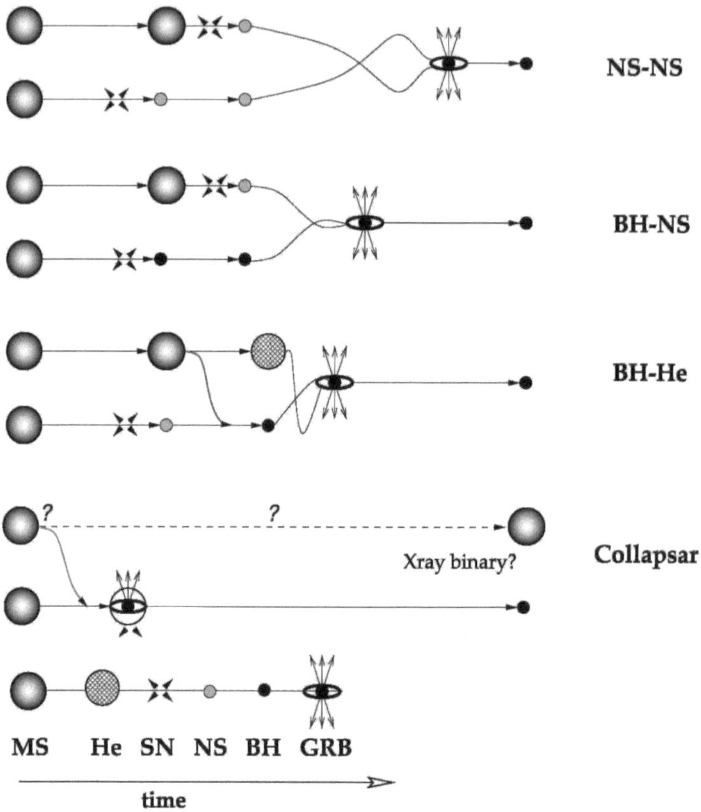

Fig. 8. Schematic scenarios for plausible GRB progenitors. The dominant production channel for each scenario is depicted, where MS denotes the primary main sequence star. The (rough) relative in-spiral times due to gravitational radiation for compact mergers are shown. BH–He mergers occur, in general, much more rapidly than NS–NS or NS–WD mergers.

to 42% of the rest mass energy of the torus, while the latter can grant up to 29% (for a maximal spin rate) of the mass of the black hole itself. How can the energy be transformed into outflowing relativistic plasma? There seem to be two options. The energy released as thermal neutrinos is expected to be reconverted, via collisions outside the dense core, into electron-positron pairs and photons. Alternatively, strong magnetic fields anchored in the dense matter could convert the gravitational binding energy of the system into a Poynting-dominated outflow. A brief summary of the various metabolic pathways is presented in Figure 9.

Neutrino Emission

The $\nu\bar{\nu} \to e^+e^-$ process [29] can tap the thermal energy of the torus produced by viscous dissipation. For this mechanism to be efficient, the neutrinos must escape before being advected into the hole; on the other hand, the efficiency of conversion into pairs (which scales with the square of the neutrino density) is low if the neutrino production is too gradual. An $e^{\pm}, \gamma$ fireball arises from the enormous compressional heating and dissipation which can provide the driving stresses leading to the relativistic expansion.

Typical estimates suggest a fireball of $\leq 10^{51}$ erg [97, 85], except perhaps in the *collapsar* or failed SN Ib/c case where [85] estimate $10^{52.5}$ erg for optimum parameters. If the fireball is collimated into a solid angle $\Omega_{\nu\bar{\nu}}$ then of course the apparent *isotropized* energy would be larger by a factor $(4\pi/\Omega_{\nu\bar{\nu}})$.

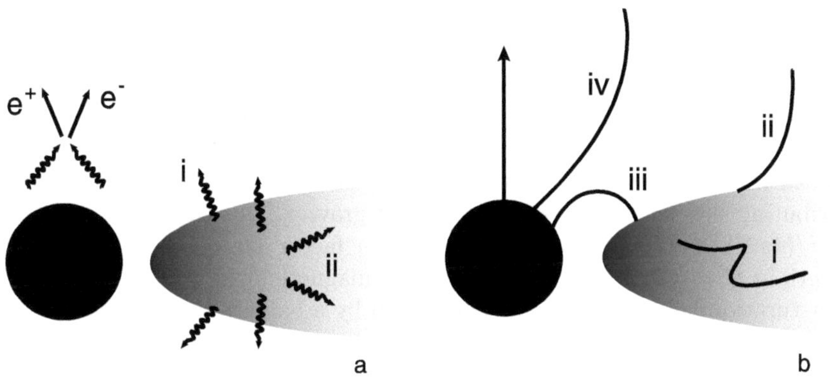

Fig. 9. Metabolic pathways for the extraction of energy. *Panel a:* Energy released as neutrinos is reconverted via collisions outside the dense core into electron-positron pairs or photons (i). The neutrinos that are emitted from the inner regions of the debris deposit part of their energy in the outer parts of the disc (ii), driving a strong baryonic outflow. This wind may be responsible for collimating the jet. *Panel b:* Strong magnetic fields anchored in the dense matter can convert the binding and/or spin energy into a Poynting outflow. A dynamo process of some kind is widely believed to be able to operate in accretion discs, and simple physical considerations suggest that fields generated in this way would have a canonical length-scale of the order of the disc thickness (i). Open field lines can connect the disc outflow and may drive a hydromagnetic wind (ii). The above mechanism can tap the binding energy of the debris torus. A rapidly rotating hole could contain a larger energy reservoir. This energy could be extracted in principle through MHD coupling to the rotation of the hole (iii;iv).

Magnetic Processes

An alternative way to tap the torus energy is through dissipation of magnetic fields generated by the differential rotation in the torus [76, 74, 69, 51].

The orbiting debris with its large magnetic seed fields and its turbulent fluid motion will give rise to a plethora of electromagnetic activity.

If the sun and the simulations, are any guide, then a magnetic field that is growing in the disc will also escape into the corona. We expect coronal arches, as well as larger scale magnetic structures, to be quite common and to be regenerated on an orbital timescale. If the footpoints of an arch are at different orbital radii in the disc then they will separate tangentially in a single period. Field lines will be stretched across the disc surface and will quickly be forced to reconnect. Differential rotation will cause the loop to twist and probably to undergo some topological rearrangement. This provides an alternative mechanism for heating the disc corona and perhaps for driving an outflow through thermal heating (similar to the neutrino heating mechanism described above). Although these tangential flux loops may be regenerated by buoyancy, they will also be stretched and pulled back into the disc.

Field amplification could in principle continue until the magnetic pressure becomes comparable to the gas pressure. This limit can also be written in the form $B^2/(4\pi\Sigma) < c_s\Omega$, where Σ is the surface mass density. The radial force exerted by such a field is predominantly the tension force, the magnitude of which integrated over the disc thickness is $B_\phi B_p/2\pi$. A poloidal field can in principle exist up to strengths such that this radial force starts contributing significantly to support against gravity. This can be formulated as $B^2/(4\pi\Sigma) < \Omega^2 r$. This limit on B^2 is a factor $\Omega r/c_s$ larger than the strength in $B = \sqrt{8\pi\rho c_s^2}$, at which point the magnetic field becomes buoyant. The saturation field amplitude is determined by a balance between nonlinear growth, and dissipative process like reconnection and buoyant escape, and can clearly really only be estimated through careful, three dimensional, numerical simulations (which are just now becoming possible).

If the magnetic fields do not thread the black hole, then a Poynting outflow can at most carry the gravitational binding energy of the torus. For a maximally rotating and a non-rotating black hole this is 0.42 and 0.06 of the torus rest mass, respectively. The torus or disc mass in a NS-NS merger is $M_{\rm t} \sim 0.1 M_\odot$ [98], and for a NS-BH, a He-BH, WD-BH merger or a binary WR collapse it may be estimated at $M_{\rm t} \sim 1 M_\odot$[37]. In the He-BH merger and WR collapse the mass of the disc is uncertain due to lack of calculations on continued accretion from the envelope, so 1 $M_\odot$ is just a rough estimate. The largest energy reservoir is therefore associated with NS-BH, He-BH or binary WR collapse, which have larger discs and fast rotation, the maximum energy being $\sim 8 \times 10^{53}\epsilon(M_{\rm t}/M_\odot)$ erg; for the failed SNe Ib (which is a slow rotator) it is $\sim 1.2 \times 10^{53}\epsilon(M_{\rm t}/M_\odot)$ erg, and for the (fast rotating) NS-NS merger it is $\sim 0.8 \times 10^{53}\epsilon(M_{\rm t}/0.1 M_\odot)$ erg, where ϵ is the efficiency in converting gravitational into MHD jet energy.

Field lines that leave the surface of the inner disc, go up to high latitude, and then connect with either the gas plunging into the hole or the event horizon of the black hole, are likely to be more significant and they, too, can extract energy and angular momentum. They have the further advantage that they are likely to propagate in a region where the Alfvén speed is large and there is causal contact between the inflow and the disc from closer to the hole. In fact it is quite likely that, except in the region close to the disc or the infalling matter, the Alfvén speed will become relativistic, which may allow substantial energy to be extracted from the black hole itself [36]. This is the basis of the proposal that spinning black holes power jets, and possibly GRBs. The power is created as a large scale Poynting flux or equivalently as a battery-driven current flowing around an electrical circuit [12]. This neatly avoids the problem of catastrophic radiative drag close to the jet origin, and allows the terminal jet Lorentz factors to be large.

Rapid rotation is essentially guaranteed in a NS-NS merger. Since the central BH will have a mass of about $2.5 M_{\rm BH}$ [94], the NS-NS system can thus power a jet of up to $\sim 1.3 \times 10^{54} \epsilon (M_{\rm BH}/2.5 M_\odot)$ erg. The scenarios less likely to produce a fast rotating BH are the NS-BH merger (where the rotation parameter could be limited to $a \le M_{\rm NS}/M_{\rm BH}$, unless the BH is already fast-rotating) and the failed SNe Ib (where the last material to fall in would have maximum angular momentum, but the material that was initially close to the hole has less angular momentum). For instance, even allowing for low total efficiency $\epsilon \sim 0.3$, a NS-NS merger whose jet is powered by the torus binding energy would only require a modest beaming of the γ-rays by a factor $(4\pi/\Omega) \sim 20$, or no beaming if the jet is powered by the B-Z mechanism, to produce the equivalent of an isotropic energy of $10^{53.5}$ ergs. The beaming requirements of BH-NS and some of the other progenitor scenarios are even less constraining.

3 Great Balls of Fire

In this section, we examine the consequences of the hypothesis that GRB sources are powered by jets of matter which arise from hydrodynamical expansion of a relativistic hot fluid. The properties of these outflows are strongly constrained by the physical processes occurring near their origin site. The emitted radiation is an observable diagnostic of the microphysical processes of particle acceleration and cooling occurring within the bulk flow. Astrophysicists understand supernova remnants reasonably well, despite continuing uncertainty about the initiating explosion; likewise, we may hope to understand the afterglows of GRBs, despite the uncertainties about the *trigger* that we have already emphasised.

At cosmological distances the observed GRB fluxes imply energies of order of up to a solar rest-mass ($\le 10^{54}$ erg), and from causality these must arise in regions whose size is of the order of kilometres in a time scale of the order of seconds. This implies that an $e^\pm, \gamma$ fireball must form [75, 42, 105], which would expand relativistically. The difficulty with this is that a smoothly expanding fireball would convert most of its energy into kinetic energy of accelerated baryons rather than into luminosity, and would produce a quasi-thermal spectrum, while the typical time scales would not explain events much longer than milliseconds. This problem was solved by postulating that shock waves would inevitably occur in the outflow, after the fireball became transparent, and these would reconvert the kinetic energy of expansion into nonthermal particles and radiation energy [90]. Best–guess numbers are Lorentz factors Γ in the range 10^2 to 10^3, allowing rapidly-variable emission to occur at radii in the range 10^{14} to 10^{16} cm.

The complicated light curves can be understood in terms of internal shocks [91] in the outflow itself, caused by velocity variations induced near or at the source. This is followed by the development of a forward shock or blast wave moving into the external medium ahead of the ejecta, and a reverse shock moving back into the ejecta as the latter is decelerated by the back-reaction from the external medium [67]. In the presence of turbulent magnetic fields built up behind the shocks, the electrons produce a synchrotron power-law radiation spectrum similar to that observed in the afterglow [111]. We shall focus here on the afterglow itself, since the prompt γ-ray emission will be considered in fuller detail in the following section.

3.1 Elementary Blast Wave Physics

The simplest version of the standard blast wave model involves a spherical, uncollimated explosion taking place in a uniform surrounding medium. A relativistic pair fireball is formed when an explosion deposits a large amount of energy into a compact volume. The pressure of the explosion causes the fireball to expand, with the thermal energy of the explosion being transformed into bulk kinetic energy due to strong adiabatic cooling of particles in the comoving frame [68, 81]. Because of the Thomson coupling between the particles and photons, most of the original explosion energy is carried by the baryons that were originally mixed into the explosion. Under certain conditions involving less-energetic, temporally extended, or very baryon-clean explosions, neutrons can decouple from the flow [24]. If this does not occur, then the coasting Lorentz factor $\Gamma_0 \cong E_0/M_b c^2$, where M_b is the baryonic mass and $E_0 = 10^{52} E_{52}$ ergs is the apparent isotropic energy release.

In the simplest version of the model, the blast wave is approximated by a uniform thin shell. A forward shock is formed when the expanding shell accelerates the external medium, and a reverse shock is formed due to deceleration of the cold shell. The forward and reverse shocked fluids are separated by a contact discontinuity and have equal kinetic energy densities. From the relativistic shock jump conditions [100], $4\Gamma(\Gamma-1)n_0 = 4\bar{\Gamma}(\bar{\Gamma}-1)n'_{sh}$, where Γ is the blast wave Lorentz factor, $\bar{\Gamma}$ is the Lorentz factor of the reverse shock in the rest frame of the shell, n'_{sh} is the density of the unshocked fluid in the proper frame of the expanding shell, and n_0 is the density of the circumburst medium (CM), here assumed to be composed of hydrogen. Particle acceleration at the reverse shock is unimportant when the reverse shock is non relativistic, which occurs when

$$\Gamma \ll \sqrt{\frac{n'_{sh}}{n_0}} \ . \tag{2}$$

The unique feature of GRBs is that the coasting Lorentz factor Γ_0 may reach values from hundreds to thousands. By contrast, Type Ia and II SNe have ejecta speeds that are in the range of $\sim$ 5000-30000 km s^{-1}. The relativistic motion of the radiating particles introduces many interesting effects in GRB emissions that must be properly taken into account.

Three frames of reference are considered when discussing the emission from systems moving with relativistic speeds: the stationary frame, which is denoted here by asterisks, the comoving frame, denoted by primes, and the observer frame. The differential distance travelled by the expanding blast wave during differential time dt_* is simply $dr = \beta c dt_*$, where $\beta = \sqrt{1-\Gamma^{-2}}$. Due to time dilation, $dr = \beta \Gamma c dt'$. Because of time dilation, the Doppler effect, and the cosmological redshift z, the relationship between comoving and observer times is $(1+z)\Gamma dt'(1-\beta\cos\theta) = (1+z)dt'/\delta = dt$, where θ is the angle between the emitting element and the observer and $\delta = [\Gamma(1-\beta\cos\theta)]^{-1}$ is the Doppler factor. For on-axis emission from a highly relativistic emitting region, we therefore see that $dt \cong (1+z)dr/\Gamma^2 c$; consequently the blast wave can travel a large distance $\Gamma^2 c\Delta t$ during a small observing time interval. A photon measured with dimensionless energy $\epsilon = h\nu/m_e c^2$ is emitted with energy $\delta\epsilon'/(1+z)$.

A blast wave expanding into the surrounding medium acts as a fluid if there is a magnetic field in the comoving frame sufficiently strong to confine the particles within the width of the blast wave. The blast-wave width depends on the duration of the explosion and the spreading of the blast-wave particles. Because a few seconds duration is regularly observed in GRB light curves, the blast-wave width must be $\sim$ light-seconds, and is probably much thicker in view of the 100 s duration observed in some GRBs. We denote this length scale by Δ_0 cm. The shell will spread radially in the comoving frame by an amount $r'_{\parallel} \simeq v_{spr} t'$, where $t' \cong r/(\beta\Gamma c)$ is the available comoving time and v_{spr} is the spreading speed. This implies a shell width in the observer frame of $\Delta = r/\Gamma_0^2$ due to length contraction, and a spreading radius $r_{spr} = \Gamma_0^2\Delta_0$, assuming that the shell spreads with speed $v_{spr} \cong c$. The width of the unshocked blast-wave fluid in the rest frame of the explosion is therefore $\Delta = \min(\Delta_0, r/\Gamma_0^2)$ [68, 101]. Milligauss fields are sufficient to confine relativistic electrons.

As the blast wave expands, it sweeps up material from the surrounding medium to form an external shock [67]. In a colliding wind scenario, the blast wave intercepts other portions of the relativistic wind [91]. Protons captured by the expanding blast wave from the CM will have total energy $\Gamma m_p c^2$ in the fluid frame, where m_p is the proton mass. The kinetic energy swept into the comoving frame by an uncollimated blast wave at the forward shock per unit time is given by [13]

$$\frac{dE'}{dt'} = 4\pi r^2 n_0 m_p c^3 \beta\Gamma(\Gamma-1). \tag{3}$$

The factor of Γ represents the increase of external medium density due to length contraction, the factor $(\Gamma - 1)$ is proportional to the kinetic energy of the swept up particles, and the factor β is proportional to the rate at which the particle energy is swept. Thus the power is $\propto \Gamma^2$ for relativistic blast waves, and $\propto \beta^3$ for non relativistic blast waves. This process provides internal energy available to be dissipated in the blast wave.

A proton that is captured by the blast wave and isotropized in the comoving frame will have energy $\Gamma m_p c^2$ in the comoving fluid frame, or energy $\Gamma^2 m_p c^2$ in the observer frame. The expanding shell will therefore begin to decelerate when $E_0 = \Gamma_0 M_b c^2 = \Gamma_0^2 m_p c^2 (4\pi r_d^3 n_0/3)$, giving the deceleration radius [90, 67]

$$r_d \equiv \left(\frac{3E_0}{4\pi\Gamma_0^2 c^2 m_p n_0}\right)^{1/3} \cong 2.6 \times 10^{16} \left(\frac{E_{52}}{\Gamma_{300}^2 n_0}\right)^{1/3} \text{ cm}, \tag{4}$$

where $\Gamma_{300} = \Gamma_0/300$. Acceleration at the shock front can inject power-law distributions of particles. In the process of isotropizing the captured particles, magnetic turbulence is introduced [84] that can also accelerate particles to very high energies through a second-order Fermi process.

The deceleration time as measured by an on-axis observer is given by

$$t_d \equiv (1+z)\frac{r_d}{\beta_0 \Gamma_0^2 c} \cong \frac{9.6\,(1+z)}{\beta_0}\left(\frac{E_{52}}{\Gamma_{300}^8 n_0}\right)^{1/3} \text{ s}\,. \tag{5}$$

The factor $\beta_0^{-1} = 1/\sqrt{1-\Gamma_0^{-2}}$ generalises the result of [67] for mildly relativistic and non relativistic ejecta, as in the case of Type Ia and Type II supernova explosions. The Sedov radius is given by

$$\ell_S = \Gamma_0^{2/3} r_d = \left(\frac{3E_0}{4\pi m_p c^2 n_0}\right)^{1/3} \cong 1.2 \times 10^{18} \left(\frac{E_{52}}{n_0}\right)^{1/3} \text{ cm}\,, \tag{6}$$

where $M_\odot$ is the total (rest mass plus kinetic) explosion energy in units of Solar rest mass energy. For relativistic ejecta, ℓ_S refers to the radius where the blast wave slows to mildly relativistic speeds, i.e., $\Gamma \sim 2$. The Sedov radius of a SN that ejects a 10 $M_\odot$ envelope could reach ~ 5 pc or more. The Sedov age $t_S = \ell_S/v_0 \cong 700(M_\odot/n_0)^{1/3}/(v_0/0.01c)$ yr for non relativistic ejecta, and is equivalent to t_d in general.

The evolution of an adiabatic blast wave in a uniform surrounding medium for an explosion with a non relativistic reverse shock is given by [25]

$$P(r) = \frac{P_0}{\sqrt{1+(r/r_d)^3}} \cong \begin{cases} \beta_0\Gamma_0\,, & r \ll r_d \\ \beta_0\left(\frac{r}{\ell_S}\right)^{-3/2}\,, & r_d \ll r \end{cases} \tag{7}$$

where $P = \beta\Gamma$ and $P_0 = \beta_0\Gamma_0$ represent dimensionless bulk momenta of the shocked fluid. Equation (7) reduces to the adiabatic Sedov behaviour for non relativistic ($\beta_0\Gamma_0 \ll 1$) explosions, giving $v \propto r^{-3/2}$, $r \propto t^{2/5}$, and $v \propto t^{-3/5}$, as is well-known. In the relativistic ($\Gamma_0 \gg 1$) limit, $\Gamma \propto r^{-3/2}$ and $t \cong c^{-1} \int dr/\Gamma^2 \propto \int dr\; r^3$, yielding $r \propto t^{1/4}$ and $\Gamma \propto t^{-3/8}$ when $\Gamma \gg 1$. The adiabatic behaviour does not apply when radiative losses are important. A non relativistic supernova shock becomes radiative at late stages of the blast-wave evolution. It is not clear whether a GRB fireball is highly radiative or nearly adiabatic during either its gamma-ray luminous prompt phase or during the afterglow [114, 113]. In the fully radiative limit, energy conservation reads $\beta_0\Gamma_0 M_0 \cong \beta\Gamma[M_0 + 4\pi n_0 r^3/3]$, giving the asymptotes $\Gamma \propto r^{-3}$ when $\Gamma_0 \gg \Gamma \gg 1$ and $\beta \propto r^{-3}$ when $\Gamma - 1 \ll 1$.

Most treatments employing blast-wave theory to explain the observed emission from GRBs assume that the radiating particles are electrons. The problem here is that $\sim m_p/m_e \sim 2000$ of the nonthermal particle energy swept into the blast-wave shock is in the form of protons or ions, unless the surroundings are composed primarily of electron-positron pairs [56]. For a radiatively efficient system, physical processes must therefore transfer a large fraction of the swept-up energy to the electron component. In elementary treatments of the blast-wave model, it is simply assumed that a fraction e_e of the forward-shock power is transferred to the electrons, so that

$$L'_e = e_e \frac{dE'}{dt'} \; . \tag{8}$$

If all the swept-up electrons are accelerated, then joint normalisation to power and number gives

$$\gamma_{\min} \cong e_e \left(\frac{p-2}{p-1}\right)\left(\frac{m_p}{m_e}\right)(\Gamma - 1) \cong e_e k_p \left(\frac{m_p}{m_e}\right)\Gamma \tag{9}$$

for $2 < p < 3$, where the last expression holds when $\Gamma \gg 1$ and $k_p = (p-2)/(p-1)$.

The strength of the magnetic field is another major uncertainty. The standard prescription is to assume that the magnetic field energy density $u_b = B^2/8\pi$ is a fixed fraction e_B of the downstream energy density of the shocked fluid. Hence

$$\frac{B^2}{8\pi} = 4 e_B n_0 m_p c^2 (\Gamma^2 - \Gamma) \; . \tag{10}$$

It is also generally supposed in simple blast-wave model calculations that some mechanism – probably the first-order shock Fermi process – injects electrons with a power-law distribution between electron Lorentz factors $\gamma_{\min} \le \gamma \le \gamma_{\max}$ downstream of the shock front. The electron injection spectrum in the comoving frame is modelled by the expression

$$\frac{dN'_e(\gamma)}{dt'd\gamma} = K_e \gamma^{-p} \; , \text{ for } \gamma_{\min} < \gamma < \gamma_{\max} \; , \tag{11}$$

where K_e is normalised to the rate at which electrons are captured.

The maximum injection energy is obtained by balancing synchrotron losses and an acceleration rate given in terms of the inverse of the Larmor time scale through a parameter $e_{\max}$, giving

$$\gamma_{\max} \cong 4 \times 10^7 e_{\max}/\sqrt{B(\mathrm{G})}\ . \tag{12}$$

A break is formed in the electron spectrum at cooling electron Lorentz factor γ_c, which is found by balancing the synchrotron loss time scale t'_{sy} with the adiabatic expansion time $t'_{adi} \cong r/\Gamma c \cong \Gamma t \cong t'_{sy} \cong (4c\sigma_{\mathrm{T}}B^2\gamma_c/24\pi m_e c^2)^{-1}$, giving [102]

$$\gamma_c \cong \frac{3m_e}{16 e_B n_0 m_p c \sigma_T \Gamma^3 t} \tag{13}$$

For an adiabatic blast wave, $\Gamma \propto t^{-3/8}$, so that $\gamma_{min} \propto t^{-3/8}$ and $\gamma_c \propto t^{1/8}$.

The observed νF_ν synchrotron spectrum from a GRB afterglow depends on the geometry of the outflow. Denoting the comoving spectral luminosity by $L'_{sy}(\epsilon') = \epsilon'(dN'/d\epsilon' dt')$, then $\epsilon' L'_{sy}(\epsilon') \cong \frac{1}{2} u_B c \sigma_{\mathrm{T}} \gamma^3 N_e(\gamma)$, with $\gamma = \sqrt{\epsilon'/\epsilon_B}$. For a spherical blast-wave geometry, the spectral power is amplified by two powers of the Doppler factor δ for the transformed energy and time. The νF_ν synchrotron spectrum is therefore

$$f_\epsilon^{sy} \cong \frac{2\Gamma^2}{4\pi D^2}\,(u_B c \sigma_{\mathrm{T}})\,\gamma^3 N'_e(\gamma)\ ,\ \gamma \cong \sqrt{\frac{(1+z)\epsilon}{2\Gamma \epsilon_B}}\ . \tag{14}$$

where D is the luminosity distance.

For collimated blast waves with jet opening angle θ_j, equation (14) does not apply when $\theta_j < 1/\Gamma$ because portions of the blast wave's radiating surface no longer contribute to the observed emission. In this case, the blast-wave geometry is a localised emission region, and the received νF_ν spectrum is given by

$$f_\epsilon^{sy} \cong \frac{\delta^4}{4\pi D^2}\,(\frac{1}{2} u_B c \sigma_{\mathrm{T}})\,\gamma^3 N'_e(\gamma)\ ,\ \gamma \cong \sqrt{\frac{(1+z)\epsilon}{\delta \epsilon_B}}\ . \tag{15}$$

Here the observed flux is proportional to four powers of the Doppler factor: two associated with solid angle, one with energy and one with time.

From this formalism, analytic and numerical models of relativistic blast waves can be constructed. It is useful to define two regimes depending on whether $\gamma_{\min} < \gamma_c$, which is called the weak cooling regime, or $\gamma_{\min} > \gamma_c$, called the strong cooling regime [102]. If the parameters p, e_e and e_B remain constant during the evolution of the blast wave, then a system originally in the weak cooling regime will always remain in the weak cooling regime. In contrast, a system in the strong cooling regime will evolve to the weak cooling regime (Figure 10)

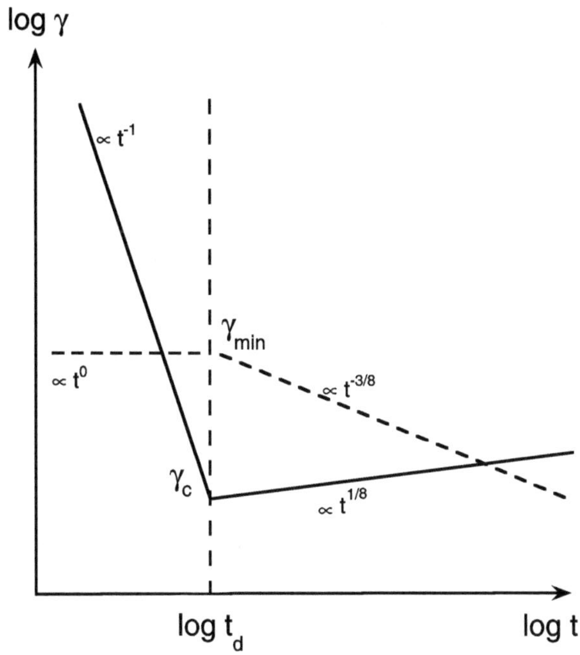

Fig. 10. Characteristic behaviour of the minimum Lorentz factor $\gamma_{\min}$ and the cooling Lorentz factor γ_c with observer time in the non-relativistic reverse shock case. We assume ϵ_ϵ and ϵ_B to remain constant with time and that the blast wave expands into a uniform circumburst medium.

For a power-law injection spectrum given by equation (11), the cooling comoving nonthermal electron spectrum can be approximated by

$$N'_e(\gamma) \cong \frac{N_e^0\,\gamma_0^{s-1}}{s-1}\begin{cases}\gamma^{-s}\,, & \gamma_0<\gamma<\gamma_1\\ \gamma_1^{p+1-s}\gamma^{-(p+1)}\,, & \gamma_1<\gamma<\gamma_{\max},\end{cases} \tag{16}$$

and $N_e^0 = 4\pi r^3 n_0/3$ for the assumed system. In the weak cooling regime, $s = p$, $\gamma_0 = \gamma_{\min}$ and $\gamma_1 = \gamma_c$, whereas in the strong cooling regime, $s = 2$, $\gamma_0 = \gamma_c$, and $\gamma_1 = \gamma_{\min}$.

As an example of the elementary theory, consider the temporal index in the strongly cooled regime for high energy electrons with $\gamma \geq \gamma_1 = \gamma_m \propto \Gamma$, $\gamma_0 = \gamma_c \propto 1/(\Gamma^3 t)$, and $s = 2$. From equation (14), $f_\epsilon \propto \Gamma^4 r^3\ \gamma_0^{s-1}\gamma_1^{p+1-s}\epsilon^{(2-p)/2}$ $/\Gamma^{2-p} \propto \Gamma^{2(p-1)}r^3\epsilon^{(2-p)/2}/t \propto t^\chi\epsilon^{\alpha_\nu}$, with $\chi = (2-3p)/4$ and $\alpha_\nu = (2-p)/2$. A few other such results include the decay of a strong and weak cooling νF_ν peak frequency $\epsilon_{pk} \propto t^{-3/2}$ and $\epsilon_{pk} \propto t^{-3p/2}$, respectively. A cooling index change by $\Delta(\alpha_\nu) = 1/2$ is accompanied by a change of temporal index from $\chi = 3(1-p)/4$ to $\chi = (2-3p)/4$ at late times, so that $\Delta\chi = 1/4$. Extension to power-law radial electron profiles is obvious, and beaming breaks introduce two additional factors of Γ from $\delta \approx \Gamma$ in equation (15), when the observer sees beyond the causally connected regions of the jetted blob, noting that N_e^0 must be renormalised appropriately.

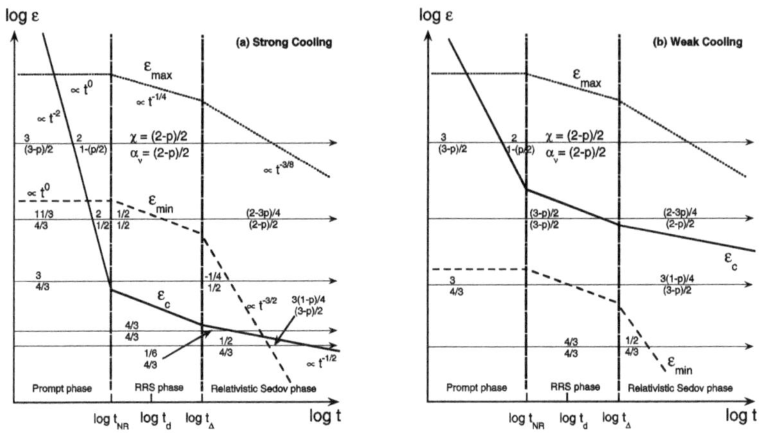

Fig. 11. Possible time profiles and spectral behaviour of the nonthermal synchrotron radiation that could arise for a system that is in the fast cooling regime (or slow cooling regime) during a portion of the evolution of a blast wave in a uniform surrounding medium. The temporal indices χ, and νF_ν spectral indices α_ν are shown, respectively above and below the lines that represent different families of possible emission trajectories.

Figure 11 shows spectral and temporal indices that are derived from the preceding analytic considerations of a spherical adiabatic blast wave that is in the strong cooling regime (or weak cooling regime) during a portion of its evolution. We use the notation that the νF_ν spectrum is described by

$$f_\epsilon \propto t^\chi \epsilon^{\alpha_\nu} \ . \tag{17}$$

Observations at a specific photon energy ϵ will detect the system evolving through regimes with different spectral and temporal behaviours. The relativistic Sedov phase corresponds to the afterglow regime, and we also consider a possible relativistic reverse shock (RRS) phase. If equation (2) is satisfied with Γ replaced by Γ_0, then the blast wave will not evolve through the RRS phase. Even for this simple system, a wide variety of behaviours is possible. Inclusion of additional effects, including a nonuniform external medium [70] or blast-wave evolution in a partially radiative phase [19] will introduce additional complications.

The heavy solid and dotted lines in Figure 11 represent the evolution of the cooling photon energies ϵ_c and ϵ_{min}, respectively. In the lower-right hand corner of the diagram, one sees that observations in the afterglow phase may detect a transition from the uncooled portion of the synchrotron spectrum where $\chi = (3-p)/2$ and $\alpha_\nu = (3-p)/2$ to a cooled portion of the synchrotron spectrum where $\chi = (2-3p)/4$ and $\alpha_\nu = (2-p)/2$. Thus a change in photon index by 0.5 units is accompanied by a change in temporal index by 1/4 unit, independent of p. Comparison with afterglow observations can be used to test the blast-wave model [78]. The recently launched *Swift* telescope will be able to monitor behaviours in the time interval between the prompt and afterglow phase, and to search for other regimes of blast wave evolution and evidence for evolution in the non-adiabatic regime.

3.2 Isotropic or Beamed Outflows?

An observer will receive most emission from those portions of a GRB blast wave that are within an angle $\sim 1/\Gamma$ to the direction to the observer. The afterglow is thus a probe for the geometry of the ejecta - at late stages, if the outflow is beamed, we expect a spherically-symmetric assumption to be inadequate; the deviations from the predictions of such a model would then tell us about the ejection in directions away from our line of sight [93]. As the blast wave decelerates by sweeping up material from the CM, a break in the light curve will occur when the jet opening half-angle θ_j becomes smaller than $1/\Gamma$. This is due to a change from a spherical blast wave geometry, given by equation (14), to a geometry defined by a localised emission region, as given by equation (15). Assuming that the blast wave decelerates adiabatically in a uniform surrounding medium, the condition $\theta_j \cong 1/\Gamma = \Gamma_0^{-1}(r_{br}/r_d)^{3/2} = \Gamma_0^{-1}(t_{br}/t_d)^{3/8}$ implies

$$t_{br} \approx 45(1+z)\left(\frac{E_{52}}{n_0}\right)^{1/3} \theta_j^{8/3} \text{ days}\,, \tag{18}$$

from which the jet angle

$$\theta_j \approx 0.1 \left[\frac{t_{br}(\mathrm{d})}{1+z}\right]^{3/8} \left(\frac{n_0}{E_{52}}\right)^{1/8} \tag{19}$$

can be derived [103]. Note that the beaming angle is only weakly dependent on n_0 and E_0. The appearance of achromatic breaks in the development of GRB afterglows has been interpreted as indicating that they are jet flows beamed towards us [33], though these observations may also be associated with the trans-relativistic evolution of spherical blast waves. Collimation factors of $\Omega_i/4\pi < 0.01$ have been derived from such steepenings [33, 78]. If GRBs are mostly jets, then this reduces the energy per burst by two or three orders of magnitude at the expense of increasing their overall frequency [33].

4 An Unsteady Relativistic Outflow

A widely recognised problem is that if the rest mass energy of entrained baryons in the outflow exceeded even 10^{-5} of the total energy, the associated opacity would trap the radiation so that it was degraded by adiabatic expansion (and thermalised) before escape. This problem arises if the *event* is approximated as an instantaneous fireball, or as an outflowing wind which is *steady* over the entire burst duration. In the previous chapter we have discussed how kinetic energy can be reconverted into radiation by relativistic shocks which form when the ejecta run into external matter. Alternatively, we postulate here an outflow persisting (typically) for a few seconds. But instead of assuming this to be a steady wind we suppose that it is irregular on much shorter timescales. For instance, if the Lorentz factor in an outflowing wind varied by a factor of more than 2, then the shocks that developed when fast material overtook slower material would be internally relativistic. Dissipation would then take place whenever internal shocks developed in the ejecta – it need not await the deceleration by swept-up ambient external matter.

Whenever part of the ejecta *catches up* with other material ejected earlier at a lower Lorentz factor, an internal shock forms, which dissipates the relative kinetic energy. To illustrate the basic idea, suppose that two shells of equal rest mass, but with different Lorentz factors Γ_i and Γ_j (with $\Gamma_i > \Gamma_j \gg 1$) are ejected at times t_1 and t_2, where $t_2 - t_1 = \delta t$. In the case of highly relativistic ejecta, the shock develops after a distance of order $c\delta t \Gamma_j \Gamma_i$. For high Lorentz factors, therefore, the shock takes a long time to develop, even if $\Gamma \gg 1$. This is, of course, because the distance that must be caught up is (in the stationary frame) of order $c\delta t$, but the speeds all differ from c by less than $1/\Gamma_j^2$.

For illustration, suppose that the Lorentz factor of the outflow is, on average, $\Gamma_* \sim 100$, but varies from $\Gamma_*/2$ to $2\Gamma_*$ on a timescale δt. The velocity differences are of order $10^{-4}c$, so the distance for the shock to develop in the lab frame is $10^4 c\delta t$. The reconversion of bulk energy can nevertheless be very efficient: when the two blobs share their momentum, they move with $\Gamma_{ij} = \sqrt{\Gamma_i \Gamma_j}$, so the fraction of the energy dissipated is

$$\varepsilon = \frac{\Gamma_i + \Gamma_j - 2\sqrt{\Gamma_i \Gamma_j}}{\Gamma_i + \Gamma_j}. \tag{20}$$

For the previous numerical example, the efficiency would be 0.2. High efficiency does not, therefore, require an impact on matter at rest; all that is needed is that the relative motions in the comoving frame be relativistic – i.e. $\Gamma_i/\Gamma_j > 2$.

Suppose that the mean outflow (over some time $t_w \sim t_{\rm grb}$) can be characterised by a steady wind with given values of L_w and $\eta = L_w/\dot{M}c^2$. We then assume that the actual value of η (or L_w) is unsteady.

The mean properties of the wind determine the average bulk Lorentz factor $\Gamma \sim (r/r_0)$ for $r < r_\eta$ or $\Gamma \sim \eta$ otherwise. Here r_0 is the central source dimension. The Lorentz factor saturates to $\Gamma \sim \eta$ at a saturation radius $r_\eta/r_0 \sim \eta$ where the wind energy density, in radiation or in magnetic energy, drops below the baryon rest mass density in the comoving frame [21]. The comoving density in the (continuous) wind regime is $n' = (L_o/4\pi r^2 m_p c^3 \eta \Gamma)$, and using the above behaviour of Γ below and above the coasting radius, as well as the definition of the Thomson optical depth in a continuous wind $\tau_T \simeq n'\sigma_T(r/\Gamma)$ we find that the baryonic photosphere, where $\tau_T = 1$, due to electrons associated with baryons, is

$$r_\tau = \frac{\dot{M}\sigma_T}{4\pi m_p c \Gamma^2} \approx 10^{13} L_{w,52} \eta_2^{-3} \text{ cm}, \tag{21}$$

where $\eta_2 = (\eta/10^2)$ and $L_{w,52} = (L_w/10^{52}\text{erg})$. The above equation holds provided that η is low enough that the wind has already reached its *terminal* Lorentz factor at r_τ. This requires $\eta < 10^3 L_{w,51}^{1/4} \delta t_0^{-1/4}$, if one takes $r_0 \sim c\delta t$, where $\delta t_0 = (\delta t/1\text{sec})$.

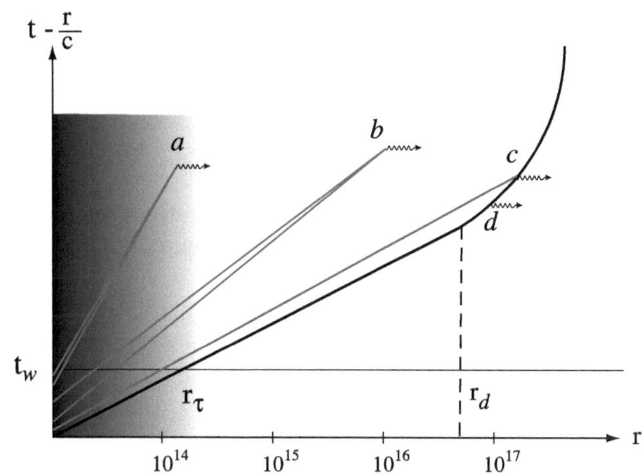

Fig. 12. Schematic spacetime diagram in source frame coordinates of a relativistic outflow, with a range of Lorentz factors, triggered by an explosion that can be approximated as instantaneous. The axes (logarithmic) are r versus $t-(r/c)$, where t is time measured by a distant observer, and is zero when the burst is observed to start. In this plot, light rays are horizontal lines. The primary gamma-ray emission is assumed to continue, with a quasi-steady luminosity L_w, for a time t_w. If Γ fluctuates by a factor of ~ 2 around its mean value, relative motions within the outflowing material give rise to internal shocks. Decreasing η values lead to world lines further to the left. In case (a), $\eta < \eta_\tau$ and the dissipation occurs when the wind is optically thick. In case (b), with $\eta_\mathrm{d} > \eta > \eta_\tau$, ejecta collide in an optically thin region before reaching the contact discontinuity. The contact discontinuity and the forward shock are being decelerated because of the increasing amount of external matter being swept up (giving rise to a long-term afterglow; case [d]), so that they lag behind the light cone by an increasing amount Δr , whose increase with r is steeper than linear. This deceleration allows ejecta to catch up and pass through a reverse shock just inside the contact discontinuity (case [c]).

If the value of η at the base increases by a factor > 2 over a timescale δt, then the later ejecta will catch up and dissipate a significant fraction of their energy at some radius $r_\iota > r_\eta$ given by

$$r_\iota \sim c\delta t \eta^2 \sim 3 \times 10^{14} \delta t_0 \eta_2^2 \text{ cm}. \tag{22}$$

Dissipation, to be most effective, must occur when the wind is optically thin. Otherwise it will suffer adiabatic cooling before escaping, and could be thermalised. Outside r_τ, where radiation has decoupled from the plasma, the relativistic internal motions in the comoving frame will lead to shocks in the gas. This implies the following lower limit on η:

$$\eta > \eta_\tau \approx 3 \times 10^1 L_{w,51}^{1/5} \delta t_0^{-1/5}. \tag{23}$$

The initial wind starts to decelerate when it has swept up $\sim \eta^{-1}$ of its initial mass. For sufficiently high η, the deceleration radius given in equation (4) can formally become smaller than the collisional radius of equation (22). This requires

$$\eta > \eta_\mathrm{d} \approx 8 \times 10^2 L_{w,52}^{1/8} t_{w,1}^{1/8} n_0^{-1/8} \delta t_0^{-3/8}. \tag{24}$$

This deceleration allows slower ejecta to catch up, replenishing and re-energising the reverse shock and boosting the momentum in the blast wave [88].

Figure 12 shows the schematic world-lines of a relativistic outflow with a range of Lorentz factors. We identify three types of contributions to the observed time history, each with a different character. For a relatively low Lorentz factor $\eta < \eta_\tau$, as in curve (a) of Figure 12, the radius $r_\iota < r_\tau$, and the dissipation occurs before the wind is optically thin. Below the baryonic photosphere, shocks would occur at high optical depths and their spectrum would suffer adiabatic cooling before escaping. For a larger Lorentz factor $\eta > \eta_\mathrm{d}$, corresponding to curve (c) of Figure 12, the ejecta would expand freely until the contact discontinuity had been decelerated by sweeping up external material. It would then crash into the reverse shock, thermalising its energy and boosting the power of the afterglow. In curve (b), with intermediate η, deceleration occurs at radii $r_\mathrm{d} > r_\iota$, and dissipation takes place when the wind is optically thin (i.e. when it is most effective).

If the Poynting flux provides a fraction α of the total luminosity L_w at the base of the wind (at r_0), the magnetic field there is $B_0 \sim 10^{10} \alpha^{1/2} L_{w,51}^{1/2} \delta t_0^{-1}$ Gauss. The comoving magnetic field at r_ι is

$$B_\iota = B_0 (r_0/r_\eta)^2 (r_\eta/r_\iota) \sim 10^4 \alpha^{1/2} L_{w,51}^{1/2} \delta t_0^{-1} \eta_2^{-3} \text{Gauss} \tag{25}$$

If the electrons are accelerated in the dissipation shocks to a Lorentz factor $\gamma = 10^3 \gamma_3$, the ratio of the synchrotron cooling time to the dynamic expansion time in the comoving frame is

$$(t_{\rm sy}/t_{\rm adi})_\iota \sim 5 \times 10^{-3}\alpha^{-1}L_{w,51}^{-1}\gamma_3^{-1}\delta t_0 \eta_2^5 \tag{26}$$

so a very high radiative efficiency is ensured even for δt as high as seconds.

If the shock dissipation leads to photons whose energy in the comoving frame exceeds 1 MeV, then there is the possibility of extra pair production from photon collisions. For the dissipation to occur outside any possible pair-dominated photosphere (a requirement that may actually be unnecessary if the pairs annihilate on a shock cooling timescale) r_τ would need to be higher than in equation (21).

It is clear from equation (26) that a magnetic field can ensure efficient cooling even if it is not strong enough to be dynamically significant (i.e. even for $\alpha \ll 1$). If, however, the field is dynamically significant in the wind, then its stresses will certainly dominate the (pre-shock) gas pressure. Indeed, in a wind with $\alpha = 1$ the magnetosonic and Alfvén speeds may remain marginally relativistic even beyond r_η if the field becomes predominantly transverse. In this extreme case, magnetic fields could inhibit shock formation unless η varied by much more than a factor of 2. On the other hand, the presence of a dynamically-significant and non-uniform field could actually drive internal motions leading to dissipation even in a constant-η wind [107].

5 General Considerations

In the preceding sections, we have endeavoured to outline some of the physical process that are believed to be most relevant to interpreting GRBs. Although some of the features now observed in GRB sources (especially afterglows) were anticipated by theoretical discussions, the recent burst of observational discovery has left theory lagging behind. There are, however, some topics on which we do believe that there will be steady work of direct relevance to interpreting observations.

Foremost amongst these topics is the development and use of sophisticated hydrodynamical codes for numerical simulation of GRB sources. Existing two dimensional codes have already uncovered some gas-dynamical properties of relativistic flows unanticipated by analytical models, but there are some key questions that they cannot yet tackle. In particular, high resolution is needed because even a tiny mass fraction of baryons loading down the outflow severely limits the attainable Lorentz factor. We must wait for useful and affordable three dimensional simulations before we can understand the nonlinear development of instabilities. Well-resolved three dimensional simulations are becoming increasingly common and they rarely fail to surprise us. The symmetry-breaking involved in transitioning from two to three dimensions is crucial and leads to qualitatively new phenomena. The key to using simulations productively is to isolate questions that can realistically be addressed and where we do not know what the outcome will be, and then to analyse the simulations so that we can learn what is the correct way to think about the problem and to describe it in terms of elementary principles.

Simulations in which the input physics is so circumscribed that they merely illustrate existing prejudice are of less value!

Another topic which seems ripe for a more sophisticated treatment concerns the intensity and shape of the intrinsic spectrum of the emitted radiation. Few would dispute the statement that the photons which bring us all our information about the nature of GRBs are the result of particle acceleration in relativistic shocks. Since charged particles radiate only when accelerated, one must attempt to deduce from the spectrum *how* the particles are being accelerated, *why* they are being accelerated, and to identify the macroscopic source driving the microphysical acceleration process.

Collisionless shocks are among the main agents for accelerating ions as well as electrons to high energies whenever sufficient time is available. Particles reflected from the shock and from scattering centres behind the shock in the turbulent compressed region when coming back across the shock into the turbulent upstream region have a good chance to experience multiple scattering and acceleration by first-order Fermi acceleration. Second-order or stochastic Fermi acceleration in the broadband turbulence downstream of collisionless shocks will also contribute to acceleration. In addition, ions may be trapped at perpendicular shocks. The trapping forces are provided by the electrostatic potential of the shock and the Lorentz force exerted on the particle by the magnetic and electric fields in the upstream region. With each reflection at the shock the particles gyrate parallel to the motional electric field, picking up energy and surfing along the shock surface. All these mechanisms are still under investigation, but there is evidence that shocks play a most important role in the acceleration of cosmic rays and other particles to very high energies.

There is no in situ information available from astrophysical plasmas. So one is forced to refer to indirect methods and analogies with accessible plasmas. Such plasmas are found only in near-Earth space. Actually, most of the ideas about and models of the behaviour of astrophysical plasmas have been borrowed from space physics and have been rescaled to astrophysical scales. However, the large spatial and long temporal scales in astrophysics and astrophysical observations do not allow for the resolution of the collisionless state of the plasmas. For instance, in the solar wind the collisional mean free path is of the order of a few AU. Looked at from the outside, the heliosphere, the region which is affected by the solar wind, will thus be considered collision dominated over time scales longer than a typical propagation time from the Sun to Jupiter. On any of the smaller and shorter scales this is wrong, because collisionless processes govern the solar wind here. Similar arguments apply to stellar winds, molecular clouds, pulsar magnetospheres and the hot gas in clusters of galaxies. One should thus be aware of the mere fact of a lack of small-scale observations. Collisionless processes generate anomalous transport coefficients. This helps in deriving a more macroscopic description. However, small-scale genuinely collisionless processes are thereby hidden. This implies that it will be difficult, if not impossible, to infer anything about the real structure, for instance, of collisionless astrophysical shock waves. The reader is refer to [6], and [109] for an excellent presentation of the basic kinetic collisionless (space) plasma theory.

When terrestrial plasma physics becomes an exact science, a change of scale by 10 or 20 orders of magnitude, and the incorporation of the effects of special relativity may suffice to provide us with a fully predictive theory!

The most interesting problem for a theorist remains, however, the nature of the central engine and the means of extracting power in a useful collimated form. In all observed cases of relativistic jets, the central object is compact, either a neutron star or black hole, and is accreting matter and angular momentum. In addition, in most systems there is direct or indirect evidence that magnetic fields are present – detected in the synchrotron radiation in galactic and extragalactic radio sources or inferred in collapsing supernova cores from the association of remnants with radio pulsars. This combination of magnetic field and rotation may be very relevant to the production of relativistic jets. Much of what we have summarised in this respect is conjecture and revolves largely around different prejudices as to how magnetic, three dimensional flows behave in strong gravitational fields. There are serious issues of theory that need to be settled independent of what guidance we get from observations of astrophysical black holes. The best prospects probably lie with performing numerical magneto-hydrodynamical simulations.

References

1. Akerlof K. *et al.*, 1999, Nature, 398, 400
2. Amati L. *et al.*, 2000, Science, 290, 953
3. Antonelli L. A. *et al.*, 2000, ApJ, 545, L39
4. Atkins R. *et al.*, 2000, ApJ, 533, L119
5. Band D. *et al.*, 1993, ApJ, 413, 281
6. Baumjohann W., Treumann R. A. 1996, *Basic Space Plasma Physics* (London: Imperial College Press)
7. Berger E. *et al.*, 2002, ApJ, 581, 981
8. Berger E. *et al.*, 2003, ApJ, 588, 99
9. Berger E. *et al.*, 2005, Nature, 438, 988
10. Bhat P. N. *et al.*, 1992, Nature, 359, 217
11. Blain A. W., Natarajan P., 2000, MNRAS, 312, L35
12. Blandford R. D., 1999, in Astrophysical Disks, ed. Sellwood D. and Goodman J. (ASP, New York), 265
13. Blandford R. D., McKee C. F., 1976, Phys. Fluids, 19, 1130
14. Blandford R. D., Znajek R. L., 1977, MNRAS, 179, 433
15. Bloom J. S. *et al.*, 1999, Nature, 401, 453
16. Bloom J. S. *et al.*, 2001, AJ, 121, 2879
17. Bloom J. S. *et al.*, 2002, AJ, 123, 1111
18. Bloom J. S., *et al.*, 2006, ApJ, 638, 354
19. Böttcher M., Dermer C. D., 2000, ApJ., 532, 281
20. Brunner R., Connolly A., Szalay A., 1999, ApJ, 516, 563
21. Cavallo G., Rees M. J., 1978, MNRAS, 183, 359

22. Costa E. *et al.*, 1997, Nature, 387, 783
23. Costa E. *et al.*, 1999, A&AS, 138, 425
24. Derishev E. V., Kocharovsky V. V., Kocharovsky Vl. V., 1999, ApJ, 521, 640
25. Dermer C. D., Humi M., 2001, ApJ, 536, 479
26. Djorgovski S. G. *et al.*, 1997, Nature, 387, 876
27. Djorgovski S. G. *et al.*, 2001a, in Gamma Ray Bursts in the Afterglow Era, ed. E. Costa, Frontera F. and Hjorth J. (Springer: Berlin), 218
28. Djorgovski S. G. *et al.*, 2001b, ApJ, 562, 654
29. Eichler D., Livio M., Piran T., Schramm D. N., 1989, Nature, 340, 126
30. Fishman G., Meegan C., 1995, ARA&A, 33, 415
31. Fox D. B. et al, 2005, Nature, 437, 845
32. Frail D. A. *et al.*, 1997, Nature, 389, 261
33. Frail D. A. *et al.*, 2001, ApJ, 562, L55
34. Frail D.A. *et al.*, 2002, ApJ, 565, 829
35. Frontera F. *et al.*, 2000, ApJS, 127, 59
36. Fryer C. L., Woosley S. E., Hartmann D. H., 1999, ApJ, 526, 152
37. Fryer C. L., Woosley S. E., 1998, ApJ, 502, L9
38. Galama T. *et al.*, 1998, Nature, 387, 479
39. Galama T., Wijers R. A. M. J., 2000, ApJ, 549, L209
40. Garnavich P. M. *et al.*, 2003, ApJ, 582, 924
41. Gehrels N. *et al.*, 2005, Nature, 437, 851
42. Goodman J. 1986, ApJ, 308, L47
43. Greiner J. *et al.*, 1994, A&A, 302, 1216
44. Groot P. J. *et al.*, 1998, ApJ, 493, L27
45. Heise J. *et al.*, 2001, in Gamma Ray Bursts in the Afterglow Era, ed. E. Costa, F. Frontera, and J. Hjorth (Springer: Berlin), 16
46. Hjorth J. *et al.*, 2003, Nature, 423, 847
47. Hjorth J. *et al.*, 2005, ApJ, 630, L117
48. Hogg D. *et al.*, 1998, ApJ, 504, 622
49. Hurley K. C. *et al.*, 1994, Nature, 372, 652
50. Iwamoto K. *et al.*, 1998, Nature, 395, 672
51. Katz J., Piran T., 1997, ApJ, 490, 772
52. Klebesadel R. W., Strong I. B., Olson R. A., 1973, ApJ, 182, L85
53. Klebesadel R. W., Laros J., Fenimore E. E., 1984, BAAS, 16, 1016
54. Kluźniak W., Lee W. H., 1998, ApJ, 494, L53
55. Kommers J. M. *et al.*, 2000, ApJ, 533, 696
56. Königl A., Granot J., 2001, ApJ, 560, 145
57. Koshut T. M. *et al.*, 1995, ApJ, 452, 145
58. Kouveliotou C. *et al.*, 1993, ApJ, 413, L101
59. Kulkarni S. *et al.*, 1998, Nature, 395, 663
60. Lamb D. Q., Reichart D., 2000, ApJ, 536, 1
61. Lattimer J. M., Schramm D. N., 1976, ApJ, 210, 549
62. Lee W. H., Ramirez-Ruiz E., 2002, ApJ, 577, 893
63. MacFadyen A. I., Woosley S. E., 1999, ApJ, 524, 262
64. Mallozzi R. S. *et al.*, 1995, ApJ, 454, 597
65. Meegan C. A. *et al.*, 1992, Nature, 355, 143
66. Mészáros P., 2002, ARA&A, 40, 137
67. Mészáros P., Rees M. J., 1993, ApJ, 405, 278
68. Mészáros P., Laguna P., Rees M. J., 1993, ApJ, 415, 181

69. Mészáros P., Rees M. J., 1997, ApJ, 482, L29
70. Mészáros P., Rees M. J., Wijers R., 1998, ApJ, 499, 301
71. Metzger M. *et al.*, 1997, Nature, 387, 878
72. Mirabal N. *et al.*, 2002, ApJ, 578, 818
73. Murakami T. *et al.*, 1991, Nature, 350, 592
74. Narayan R., Paczyński B., Piran T., 1992, ApJ, 395, L83
75. Paczyński B., 1986, ApJ, 308, L43
76. Paczyński B., 1991, Acta Astron. 41, 257
77. Paczyński B., 1998, ApJ, 494, L45
78. Panaitescu A., Kumar P., 2001, ApJ, 554, 667
79. Patat F. *et al.*, 2001, ApJ, 555, 900
80. Phinney E. S., 1991, ApJ, 380, L17
81. Piran T., 1999, Phys. Rep., 314, 575
82. Piro L. *et al.*, 1999, ApJ, 514, L73
83. Piro L. *et al.*, 2000, Science, 290, 955
84. Pohl M., Schlickeiser R., 2000, A&A, 354, 395
85. Popham R., Woosley S. E., Fryer C., 1999, ApJ, 518, 356
86. Prochaska J. X. et al., 2005, astro-ph/0510022
87. Ramirez-Ruiz E., Merloni A., 2001, MNRAS, 320, L25
88. Ramirez-Ruiz E., Lloyd-Ronning N. M., 2002, NewA, 7, 197
89. Ramirez-Ruiz E., Trentham N., Blain A. W., 2002, MNRAS, 329, 465
90. Rees M.J., Mészáros P., 1992, MNRAS, 258, 41
91. Rees M.J., Mészáros P., 1994, ApJ, 430, L93
92. Reeves G. D. *et al.*, 2002, Nature, 416, 512
93. Rhoads J. E., 1999, ApJ, 525, 737
94. Rosswog S. *et al.*, 1999, A&A, 341, 499
95. Rosswog S., Ramirez-Ruiz E., 2002, MNRAS, 336, L7
96. Rosswog S., Ramirez-Ruiz E., Davies M. B., 2003, MNRAS, 345, 1077
97. Ruffert M. *et al.*, 1997, A&A, 319, 122
98. Ruffert M., Janka H.-T., 1999, A&A, 344, 573
99. Salpeter E. E., 1964, ApJ, 140, 796
100. Sari R., Piran T., 1995, ApJ, 455, L143
101. Sari R., Narayan R., Piran T., 1996, ApJ, 473, 204
102. Sari R., Piran T., Narayan R., 1998, ApJ, 497, L17
103. Sari R., Piran T., Halpern J., 1999, ApJ, 519, L17
104. Schmidt M., 2001, ApJ, 523, L117
105. Shemi A., Piran T., 1990, ApJ, 365, L55
106. Stanek K. Z. *et al.*, 2003, ApJ, 591, L17
107. Thompson C., 1994, MNRAS, 270, 480
108. Trentham N., Ramirez-Ruiz E. Blain A. W., 2002, MNRAS, 334, 983
109. Treumann R. A., Baumjohann W., 1997, *Advanced Space Plasma Physics* (London: Imperial College Press)
110. Van Paradijs J. *et al.*, 1997, Nature, 386, 686
111. van Paradijs J., Kouveliotou C., Wijers R. A. M. J., 2000, ARA&A, 38, 379
112. van den Bergh S., Tammann G. A., 1991, ARA&A, 29, 363
113. Vietri M., 1997, ApJ, 478, L9
114. Waxman E., 1997, ApJ, 485, L5
115. Waxman E., Frail D., Kulkarni S., 1998, ApJ, 497, 288
116. Waxman E., Draine B., 2000, ApJ, 537, 796
117. Wijers R. A. M. J. *et al.*, 1998, MNRAS, 294, L13
118. Woosley S. E., 1993, ApJ, 405, 273
119. Zel'dovich Ya. B., Novikov I. D., 1964, Sov. Phys. Dok., 158, 811
120. Zhang W., Fryer C. L., 2001, ApJ, 550, 357

Understanding Galaxy Formation and Evolution

Vladimir Avila-Reese[1]

Instituto de Astronomía, Universidad Nacional Autónoma de México, A.P. 70-264, 04510, México, D.F. avila@astroscu.unam.mx

The old dream of integrating into one the study of micro and macrocosmos is now a reality. Cosmology, astrophysics and particle physics converge within a scenario (but still not a theory) of cosmic structure formation and evolution called Λ Cold Dark Matter (ΛCDM). This scenario emerged mainly to explain the origin of galaxies. In these lecture notes, I first present a review of the main galaxy properties, highlighting the questions that any theory of galaxy formation should explain. Then, the cosmological background and the main aspects of primordial perturbation generation and evolution are pedagogically detached. Next, I focus on the "dark side" of galaxy formation, presenting a review on ΛCDM halo assembling and properties, and on the main candidates for non–baryonic dark matter. Finally, the complex processes of baryon dissipation inside the non–linearly evolving CDM halos, formation of disks and spheroids, and transformation of gas into stars are briefly described, remarking on the possibility of a few driving factors and parameters able to explain the main body of galaxy properties. A summary and discussion of some of the issues and open problems of the ΛCDM paradigm are given in the final part of these notes.

1 Introduction

Our vision of the cosmic world and in particular of the whole Universe has been changing dramatically in the last century. As we will see, galaxies were repeatedly the main protagonist in the scene of these changes. It is about 80 years since Edwin Hubble established the nature of galaxies as gigantic self-bound stellar systems and used their kinematics to show that the Universe as a whole is expanding uniformly at the present time. Galaxies, as the building blocks of the Universe, are also tracers of its large–scale structure and of its evolution in the last 13 Gyrs or more. By looking inside galaxies we find that they are the arena where stars form, evolve and collapse in constant interaction with the interstellar medium (ISM), a complex mix of gas and

A. Carramiñana et al. (eds.), Solar, Stellar and Galactic Connections between Particle Physics and Astrophysics, 115–164.

plasma, dust, radiation, cosmic rays, and magnetics fields. The center of a significant fraction of galaxies harbor supermassive black holes. When these "monsters" are fed with infalling material, the accretion disks around them release, mainly through powerful plasma jets, the largest amounts of energy known in astronomical objects. This phenomenon of Active Galactic Nuclei (AGN) was much more frequent in the past than in the present, being the high–redshift quasars (QSO's) the most powerful incarnation of the AGN phenomenon. But the most astonishing surprise of galaxies comes from the fact that luminous matter (stars, gas, AGN's, etc.) is only a tiny fraction ($\sim 1-5\%$) of all the mass measured in galaxies and the giant halos around them. What this dark component of galaxies is made of? This is one of the most acute enigmas of modern science.

Thus, exploring and understanding galaxies is of paramount interest to cosmology, high–energy and particle physics, gravitation theories, and, of course, astronomy and astrophysics. As astronomical objects, among other questions, we would like to know how do they take shape and evolve, what is the origin of their diversity and scaling laws, why they cluster in space as observed, following a sponge–like structure, what is the dark component that predominates in their masses. By answering to these questions we would able also to use galaxies as a true link between the observed universe and the properties of the early universe, and as physical laboratories for testing fundamental theories.

The content of these notes is as follows. In §2 a review on main galaxy properties and correlations is given. By following an analogy with biology, the taxonomical, anatomical, ecological and genetical study of galaxies is presented. The observational inference of dark matter existence, and the baryon budget in galaxies and in the Universe is highlighted. Section 3 is dedicated to a pedagogical presentation of the basis of cosmic structure formation theory in the context of the Λ Cold Dark Matter (ΛCDM) paradigm. The main questions to be answered are: why CDM is invoked to explain the formation of galaxies? How is explained the origin of the seeds of present–day cosmic structures? How these seeds evolve?. In §4 an updated review of the main results on properties and evolution of CDM halos is given, with emphasis on the aspects that influence the propertied of the galaxies expected to be formed inside the halos. A short discussion on dark matter candidates is also presented (§§4.2). The main ingredients of disk and spheroid galaxy formation are reviewed and discussed in §5. An attempt to highlight the main drivers of the Hubble and color sequences of galaxies is given in §§5.3. Finally, some selected issues and open problems in the field are resumed and discussed in §6.

2 Galaxy Properties and Correlations

During several decades galaxies were considered basically as self–gravitating stellar systems so that the study of their physics was a domain of Galactic Dynamics. Galaxies in the local Universe are indeed mainly conglomerates of hundreds of millions to trillions of stars *supported against gravity either by rotation or by random motions.* In the former case, the system has the shape of a *flattened disk*, where most of the material is on circular orbits at radii that are the minimal ones allowed by the specific angular momentum of the material. Besides, disks are dynamically fragile systems, unstable to perturbations. Thus, the mass distribution along the disks is the result of the specific angular momentum distribution of the material from which the disks form, and of the posterior dynamical (internal and external) processes. In the latter case, the shape of the galactic system is a concentrated *spheroid/ellipsoid*, with mostly (disordered) radial orbits. The spheroid is dynamically hot, stable to perturbations. Are the properties of the stellar populations in the disk and spheroid systems different?

Stellar Populations

Already in the 40's, W. Baade discovered that according to the ages, metallicities, kinematics and spatial distribution of the stars in our Galaxy, they separate in two groups: 1) Population I stars, which populate the plane of the disk; their ages do not go beyond 10 Gyr –a fraction of them in fact are young ($\lesssim 10^6$ yr) luminous O,B stars mostly in the spiral arms, and their metallicities are close to the solar one, $Z \approx 2\%$; 2) Population II stars, which are located in the spheroidal component of the Galaxy (stellar halo and partially in the bulge), where velocity dispersion (random motion) is higher than rotation velocity (ordered motion); they are old stars (> 10 Gyr) with very low metallicities, on the average lower by two orders of magnitude than Population I stars. In between Pop's I and II there are several stellar subsystems[1].

Stellar populations are true fossils of the galaxy assembling process. The differences between them evidence differences in the formation and evolution of the galaxy components. The Pop II stars, being old, of low metallicity, and dominated by random motions (dynamically hot), had to form early in the assembling history of galaxies and through violent processes. In the meantime, the large range of ages of Pop I stars, but on average younger than the Pop II stars, indicates a slow star formation process that continues even today in the disk plane. Thus, the common wisdom says that *spheroids form early in a violent collapse (monolithic or major merger), while disks assemble by continuous infall of gas rich in angular momentum, keeping a self–regulated SF process.*

[1] Astronomers suspect also the existence of non–observable Population III of pristine stars with zero metallicities, formed in the first molecular clouds $\sim 4 \times 10^8$ yrs ($z \sim 20$) after the Big Bang. These stars are thought to be very massive, so that in scaletimes of 1Myr they exploded, injected a big amount of energy to the primordial gas and started to reionize it through expanding cosmological HII regions (see e.g., [20, 21] for recent reviews on the subject).

Interstellar Medium (ISM)

Galaxies are not only conglomerates of stars. The study of galaxies is incomplete if it does not take into account the ISM, which for late–type galaxies accounts for more mass than that of stars. Besides, it is expected that in the deep past, galaxies were gas–dominated and with the passing of time the cold gas was being transformed into stars. The ISM is a turbulent, non–isothermal, multi–phase flow. Most of the gas mass is contained in neutral instable HI clouds ($10^2 < T < 10^4$K) and in dense, cold molecular clouds ($T < 10^2$K), where stars form. Most of the volume of the ISM is occupied by diffuse ($n \approx 0.1\mathrm{cm}^{-3}$), warm–hot ($T \approx 10^4 - 10^5$K) turbulent gas that confines clouds by pressure. The complex structure of the ISM is related to (i) its peculiar thermodynamical properties (in particular the heating and cooling processes), (ii) its hydrodynamical and magnetic properties which imply development of turbulence, and (iii) the different energy input sources. The star formation unities (molecular clouds) appear to form during large–scale compression of the diffuse ISM driven by supernovae (SN), magnetorotational instability, or disk gravitational instability (e.g., [7]). At the same time, the energy input by stars influences the hydrodynamical conditions of the ISM: the star formation results self–regulated by a delicate energy (turbulent) balance.

Galaxies are true "ecosystems" where stars form, evolve and collapse in constant interaction with the complex ISM. Following a pedagogical analogy with biological sciences, we may say that the study of galaxies proceeded through taxonomical, anatomical, ecological and genetical approaches.

2.1 Taxonomy

As it happens in any science, as soon as galaxies were discovered, the next step was to attempt to classify these news objects. This endeavor was taken on by Edwin Hubble. The showiest characteristics of galaxies are the bright shapes produced by their stars, in particular those most luminous. Hubble noticed that by their external look (morphology), galaxies can be divided into three principal types: Ellipticals (E, from round to flattened elliptical shapes), Spirals (S, characterized by spiral arms emanating from their central regions where an spheroidal structure called bulge is present), and Irregulars (Irr, clumpy without any defined shape). In fact, the last two classes of galaxies are disk–dominated, rotating structures. Spirals are subdivided into Sa, Sb, Sc types according to the size of the bulge in relation to the disk, the openness of the winding of the spiral arms, and the degree of resolution of the arms into stars (in between the arms there are also stars but less luminous than in the arms). Roughly 40% of S galaxies present an extended rectangular structure (called bar) further from the bulge; these are the barred Spirals (SB), where the bar is evidence of disk gravitational instability.

From the physical point of view, the most remarkable aspect of the morphological Hubble sequence is the ratio of spheroid (bulge) to total luminosity.

This ratio decreases from 1 for the Es, to ~ 0.5 for the so–called lenticulars (S0), to $\sim 0.5 - 0.1$ for the Ss, to almost 0 for the Irrs. What is the origin of this sequence? Is it given by nature or nurture? Can the morphological types change from one to another and how frequently they do it? It is interesting enough that roughly half of the stars at present are in galaxy spheroids (Es and the bulges of S0s and Ss), while the other half is in disks (e.g., [11]), where some fraction of stars is still forming.

2.2 Anatomy

The morphological classification of galaxies is based on their external aspect and it implies somewhat subjective criteria. Besides, the "showy" features that characterize this classification may change with the color band: in blue bands, which trace young luminous stellar populations, the arms, bar and other features may look different to what it is seen in infrared bands, which trace less massive, older stellar populations. We would like to explore deeper the internal physical properties of galaxies and see whether these properties correlate along the Hubble sequence. Fortunately, this seems to be the case in general so that, in spite of the complexity of galaxies, some clear and sequential trends in their properties encourage us to think about regularity and the possibility to find driving parameters and factors beyond this complexity.

Figure 1 below resumes the main trends of the "anatomical" properties of galaxies along the Hubble sequence.

The advent of extremely large galaxy surveys made possible massive and uniform determinations of global galaxy properties. Among others, the Sloan Digital Sky Survey (SDSS[2]) and the Two–degree Field Galaxy Redshift Survey (2dFGRS[3]) currently provide uniform data already for around 10^5 galaxies in limited volumes. The numbers will continue growing in the coming years. The results from these surveys confirmed the well known trends shown in Fig. 1; moreover, it allowed to determine the distributions of different properties. Most of these properties present a *bimodal* distribution with two main sequences: the red, passive galaxies and the blue, active galaxies, with a fraction of intermediate types (see for recent results [68, 6, 114, 34, 127] and more references therein). The most distinct segregation in two peaks is for the specific star formation rate ($\dot{M}_s/M_s$); there is a narrow and high peak of passive galaxies, and a broad and low peak of star forming galaxies. The two sequences are also segregated in the luminosity function: the faint end is dominated by the blue, active sequence, while the bright end is dominated by the red, passive sequence. It seems that the transition from one sequence to the other happens at the galaxy stellar mass of $\sim 3 \times 10^{10} M_\odot$.

---

[2] *www.sdss.org/sdss.html*

[3] *www.aao.gov.au/2df/*

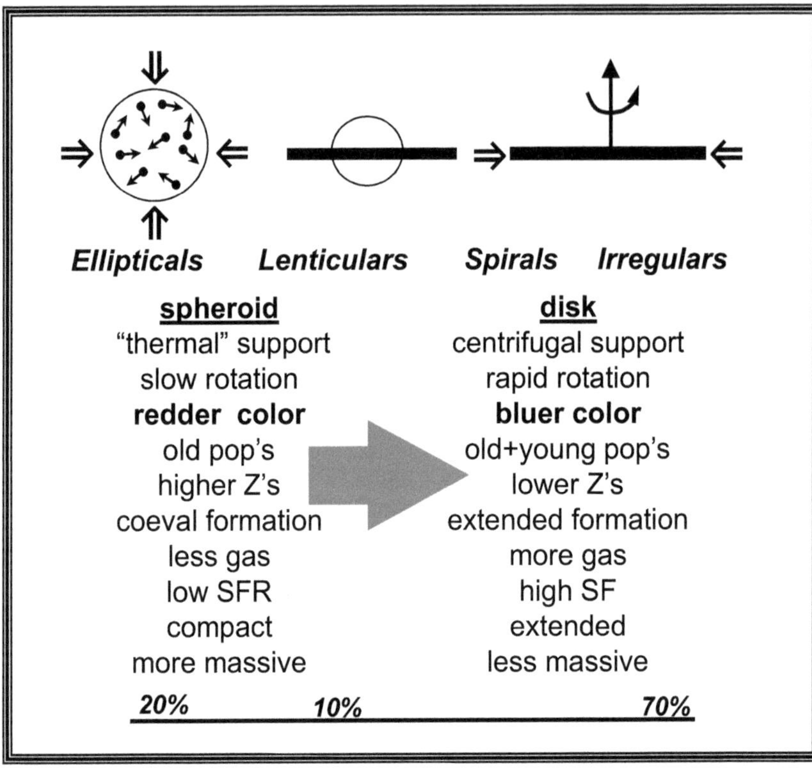

Fig. 1. Main trends of physical properties of galaxies along the Hubble morphological sequence. The latter is basically a sequence of change of the spheroid–to–disk ratio. Spheroids are supported against gravity by velocity dispersion, while disks by rotation.

The Hidden Component

Under the assumption of Newtonian gravity, the observed dynamics of galaxies points out to the presence of enormous amounts of mass not seen as stars or gas. Assuming that disks are in centrifugal equilibrium and that the orbits are circular (both are reasonable assumptions for non–central regions), the measured rotation curves are good tracers of the total (dynamical) mass distribution (Fig. 2). The mass distribution associated with the luminous galaxy (stars+gas) can be inferred directly from the surface brightness (density) profiles. For an exponential disk of scalelength R_d (=3 kpc for our Galaxy), the rotation curve beyond the optical radius ($R_{opt} \approx 3.2R_d$) decreases as in the Keplerian case. The observed HI rotation curves at radii around and beyond R_{opt} are far from the Keplerian fall–off, implying the existence of hidden mass called *dark matter (DM)* [99, 18]. The fraction of DM increases with radius.

It is important to remark the following observational facts:

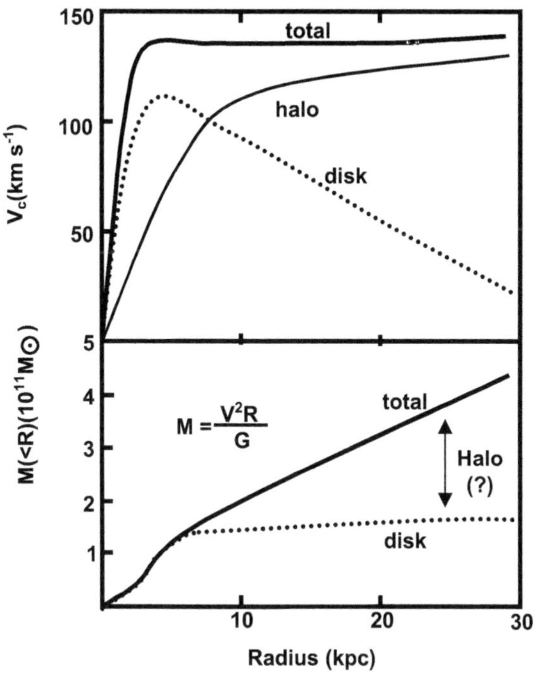

Fig. 2. Under the assumption of circular orbits, the observed rotation curve of disk galaxies traces the dynamical (total) mass distribution. The outer rotation curve of a nearly exponential disk decreases as in the Keplerian case. The observed rotation curves are nearly flat, suggesting the existence of massive dark halos.

- the *outer rotation curves are not universally flat* as it is assumed in hundreds of papers. Following, Salucci & Gentile [101], let us define the average value of the rotation curve logarithmic slope, $\bigtriangledown \equiv (dlogV/dlogR)$ between two and three R_d. A flat curve means $\bigtriangledown = 0$; for an exponential disk without DM, $\bigtriangledown = -0.27$ at $3R_s$. Observations show a large range of values for the slope: $-0.2 \leq \bigtriangledown \leq 1$
- the rotation curve shape ($\bigtriangledown$) correlates with the luminosity and surface brightness of galaxies [95, 123, 132]: it increases according the galaxy is fainter and of lower surface brightness
- at the optical radius R_{opt}, the DM–to–baryon ratio varies from ≈ 1 to 7 for luminous high–surface brightness to faint low–surface brightness galaxies, respectively
- the roughly smooth shape of the rotation curves implies a fine coupling between disk and DM halo mass distributions [25]

The HI rotation curves extend typically to $2 - 5R_{opt}$. The dynamics at larger radii can be traced with satellite galaxies if the satellite statistics allows for that. More recently, the technique of (statistical) *weak lensing* around galaxies began to emerge as the most direct way to trace the masses of galaxy halos. The results show that a typical L_* galaxy (early or late) with a stellar mass of $M_s \approx 6 \times 10^{10} M_\odot$ is surrounded by a halo of $\approx 2 \times 10^{12} M_\odot$ ([80] and more references therein). The extension of the halo is typically $\approx 200-250$kpc. These numbers are very close to the determinations for our own Galaxy.

The picture has been confirmed definitively: luminous galaxies are just the top of the iceberg (Fig. 3). The baryonic mass of (normal) galaxies is only $\sim 3-5\%$ of the DM mass in the halo! This fraction could be even lower for dwarf galaxies (because of feedback) and for very luminous galaxies (because the gas cooling time > Hubble time). On the other hand, the universal baryon–to–DM fraction ($\Omega_B/\Omega_{DM} \approx 0.04/0.022$, see below) is $f_{B,Un} \approx 18\%$. Thus, galaxies are not only dominated by DM, but are much more so than the average in the Universe! This begs the next question: if the majority of baryons is not in galaxies, where it is? Recent observations, based on highly ionized absorption lines towards low redshift luminous AGNs, seem to have found a fraction of the missing baryons in the interfilamentary warm–hot intergalactic medium at $T \lesssim 10^5 - 10^7$ K [89].

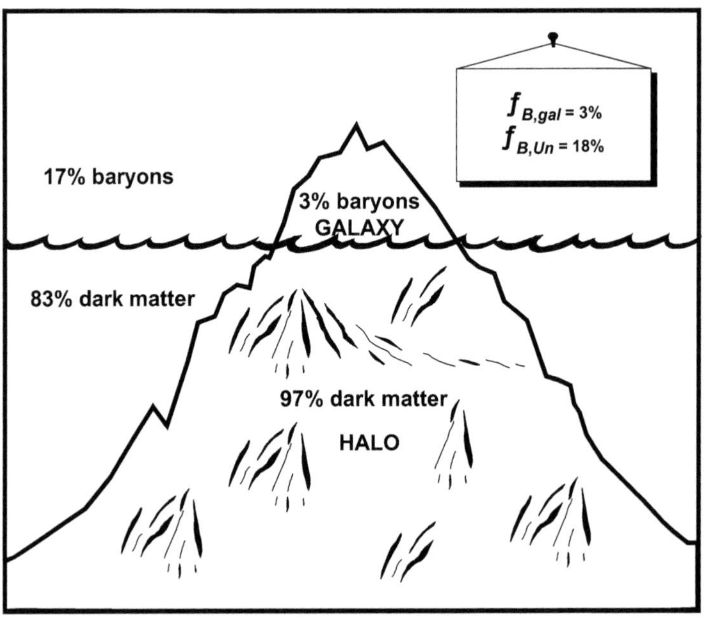

Fig. 3. Galaxies are just the top of the iceberg. They are surrounded by enormous DM halos extending 10–20 times their sizes, where baryon matter is only less than 5% of the total mass. Moreover, galaxies are much more DM–dominated than the average content of the Universe. The corresponding typical baryon–to–DM mass ratios are given in the inset.

Global baryon inventory: The different probes of baryon abundance in the Universe (primordial nucleosynthesis of light elements, the ratios of odd and even CMBR acoustic peaks heights, absorption lines in the Lyα forest) have been converging in the last years towards the same value of the baryon density: $\Omega_b \approx 0.042 \pm 0.005$. In Table 1 below, the densities (Ω's) of different baryon components at low redshifts and at $z > 2$ are given (from [48] and [89]).

Table 1. Abundances of the different baryon components ($h = 0.7$)

Component	Contribution to Ω
Low redshifts	
Galaxies: stars	0.0027 ± 0.0005
Galaxies: HI	$(4.2 \pm 0.7) \times 10^{-4}$
Galaxies: H_2	$(1.6 \pm 0.6) \times 10^{-4}$
Galaxies: others	$(\approx 2.0) \times 10^{-4}$
Intracluster gas	0.0018 ± 0.0007
IGM: (cold-warm)	0.013 ± 0.0023
IGM: (warm-hot)	≈ 0.016
$z > 2$	
Lyα forest clouds	> 0.035

The present–day abundance of baryons in virialized objects (normal stars, gas, white dwarfs, black holes, etc. in galaxies, and hot gas in clusters) is therefore $\Omega_B \approx 0.0037$, which accounts for $\approx 9\%$ of all the baryons at low redshifts. The gas in not virialized structures in the Intergalactic Medium (cold-warm Lyα/β gas clouds and the warm–hot phase) accounts for $\approx 73\%$ of all baryons. Instead, at $z > 2$ more than 88% of the universal baryonic fraction is in the Lyα forest composed of cold HI clouds. The baryonic budget's outstanding question is: *Why only $\approx 9\%$ of baryons are in virialized structures at the present epoch?*

2.3 Ecology

The properties of galaxies vary systematically as a function of environment. The environment can be relatively local (measured through the number of nearest neighborhoods) or of large scale (measured through counting in defined volumes around the galaxy). The morphological type of galaxies is earlier in the locally denser regions (morphology–density relation),the fraction of ellipticals being maximal in cluster cores [40] and enhanced in rich [96] and poor groups. The extension of the morphology–density relation to low local–density environment (cluster outskirts, low mass groups, field) has been a matter of debate. From an analysis of SDSS data, it was found that [54] (i) in the sparsest regions both relations flatten out, (ii) in the intermediate density regions (e.g., cluster outskirts) the intermediate–type galaxy (mostly

S0s) fraction increases towards denser regions whereas the late–type galaxy fraction decreases, and (iii) in the densest regions intermediate–type fraction decreases radically and early–type fraction increases. In a similar way, a study based on 2dFGRS data of the luminosity functions in clusters and voids shows that the population of faint late–type galaxies dominates in the latter, while, in contrast, very bright early–late galaxies are relatively overabundant in the former [34]. This and other studies suggest that the origin of the morphology–density (or morphology-radius) relation could be a combination of (i) *initial (cosmological) conditions* and (ii) of *external mechanisms* (ram-pressure and tidal stripping, thermal evaporation of the disk gas, strangulation, galaxy harassment, truncated star formation, etc.) that operate mostly in dense environments, where precisely the relation steepens significantly.

The morphology–environment relation evolves. It systematically flattens with z in the sense that the grow of the early-type (E+S0) galaxy fraction with density becomes less rapid ([97] and more references therein) the main change being in the high–density population fraction. Postman et al. conclude that the observed flattening of the relation up to $z \sim 1$ is due mainly to a deficit of S0 galaxies and an excess of Sp+Irr galaxies relative to the local galaxy population; the E fraction-density relation does not appear to evolve over the range $0 < z < 1.3$! Observational studies show that other properties besides morphology vary with environment. The galaxy properties most sensitive to environment are the integral color and specific star formation rate (e.g. [68, 114, 127]. The dependences of both properties on environment extend typically to lower densities than the dependence for morphology. These properties are tightly related to the galaxy star formation history, which in turn depends on internal formation/evolution processes related directly to initial cosmological conditions as well as to external astrophysical mechanisms able to inhibit or induce star formation activity.

2.4 Genetics

Galaxies definitively evolve. We can reconstruct the past of a given galaxy by matching the observational properties of its stellar populations and ISM with (parametric) spectro–photo–chemical models (inductive approach). These are well–established models specialized in following the spectral, photometrical and chemical evolution of stellar populations formed with different gas infall rates and star formation laws (e.g. [16] and the references therein). The inductive approach allowed to determine that spiral galaxies as our Galaxy can not be explained with closed–box models (a single burst of star formation); continuous infall of low–metallicity gas is required to reproduce the local and global colors, metal abundances, star formation rates, and gas fractions. On the other hand, the properties of massive ellipticals (specially their high α-elements/Fe ratios) are well explained by a single early fast burst of star formation and subsequent passive evolution.

A different approach to the genetical study of galaxies emerged after cosmology provided a reliable theoretical background. Within such a background it is possible to "handle" galaxies as physical objects that evolve according to the initial and boundary conditions given by cosmology. The deductive construction of galaxies can be confronted with observations corresponding to different stages of the proto-galaxy and galaxy evolution. The breakthrough for the deductive approach was the success of the inflationary theory and the consistency of the so–called Cold Dark Matter (CDM) scenario with particle physics and observational cosmology. The main goal of these notes is to describe the ingredients, predictions, and tests of this scenario.

Galaxy Evolution in Action

The dramatic development of observational astronomy in the last 15 years or so opened a new window for the study of galaxy genesis: the follow up of galaxy/protogalaxy populations and their environment at different redshifts. The Deep and Ultra Deep Fields of the Hubble Spatial Telescope and other facilities allowed to discover new populations of galaxies at high redshifts, as well as to measure the evolution of global (per unit of comoving volume) quantities associated with galaxies: the cosmic star formation rate density (SFRD), the cosmic density of neutral gas, the cosmic density of metals, etc. Overall, these global quantities change significantly with z, in particular the SFRD as traced by the UV–luminosity at rest of galaxies [79]: since $z \sim 1.5-2$ to the present it decreased by a factor close to ten (the Universe is literally lightening off), and for higher redshifts the SFRD remains roughly constant or slightly decreases ([51, 61] and the references therein). There exists indications that the SFRD at redshifts 2–4 could be approximately two times higher if considering Far Infrared/submilimetric sources (SCUBA galaxies), where intense bursts of star formation take place in a dust–obscured phase.

Concerning populations of individual galaxies, the Deep Fields evidence a significant increase in the fraction of blue galaxies at $z \sim 1$ for the blue sequence that at these epochs look more distorted and with higher SFRs than their local counterparts. Instead, the changes observed in the red sequence are small; it seems that most red elliptical galaxies were in place long ago. At higher redshifts ($z \gtrsim 2$), galaxy objects with high SFRs become more and more common. The most abundant populations are:

Lyman Break Galaxies (LBG) , selected via the Lyman break at 912Å in the rest–frame. These are star–bursting galaxies (SFRs of $10 - 1000 M_\odot/yr$) with stellar masses of $10^9 - 10^{11} M_\odot$ and moderately clustered.

Sub-millimeter (SCUBA) Galaxies, detected with sub–millimeter bolometer arrays. These are strongly star–bursting galaxies (SFRs of $\sim 1000 M_\odot/yr$) obscured by dust; they are strongly clustered and seem to be merging galaxies, probably precursors of ellipticals.

Lyman α emitters (LAEs), selected in narrow–band studies centered in the Lyman α line at rest at $z > 3$; strong emission Lyman α lines evidence phases of rapid star formation or strong gas cooling. LAEs could be young (disk?) galaxies in the early phases of rapid star formation or even before, when the gas in the halo was cooling and infalling to form the gaseous disk.

Quasars (QSOs), easily discovered by their powerful energetics; they are associated to intense activity in the nuclei of galaxies that apparently will end as spheroids; QSOs are strongly clustered and are observed up to $z \approx 6.5$.

There are many other populations of galaxies and protogalaxies at high redshifts (Luminous Red Galaxies, Damped Lyα disks, Radiogalaxies, etc.). A major challenge now is to put together all the pieces of the high–redshift puzzle to come up with a coherent picture of galaxy formation and evolution.

3 Cosmic Structure Formation

In the previous section we have learn that galaxy formation and evolution are definitively related to cosmological conditions. Cosmology provides the theoretical framework for the initial and boundary conditions of the cosmic structure formation models. At the same time, the confrontation of model predictions with astronomical observations became the most powerful testbed for cosmology. As a result of this fruitful convergence between cosmology and astronomy, there emerged the current paradigmatic scenario of cosmic structure formation and evolution of the Universe called Λ Cold Dark Matter (ΛCDM). The ΛCDM scenario integrates nicely: (1) cosmological theories (Big Bang and Inflation), (2) physical models (standard and extensions of the particle physics models), (3) astrophysical models (gravitational cosmic structure growth, hierarchical clustering, gastrophysics), and (4) phenomenology (CMBR anisotropies, non-baryonic DM, repulsive dark energy, flat geometry, galaxy properties).

Nowadays, cosmology passed from being the Cinderella of astronomy to be one of the highest precision sciences. Let us consider only the Inflation/Big Bang cosmological models with the F-R-W metric and adiabatic perturbations. The number of parameters that characterize these models is high, around 15 to be more precise. No single cosmological probe constrain all of these parameters. By using multiple data sets and probes it is possible to constrain with precision several of these parameters, many of which correlate among them (degeneracy). The main cosmological probes used for precision cosmology are the CMBR anisotropies, the type–Ia SNe and long Gamma–Ray Bursts, the Lyα power spectrum, the large–scale power spectrum from galaxy surveys, the cluster of galaxies dynamics and abundances, the peculiar velocity surveys, the weak and strong lensing, the baryonic acoustic oscillation in the large–scale galaxy distribution. There is a model that is systematically consistent with most of these probes and one of the goals in the last years has been to improve the error bars of the parameters for this 'concordance' model. The geometry in the concordance model is flat with an energy composition dominated in $\sim 2/3$ by the cosmological constant Λ (generically called Dark Energy), responsible for the current accelerated expansion of the Universe. The other $\sim 1/3$ is matter, but $\sim 85\%$ of this 1/3 is in form of non–baryonic DM. Table 2 presents the central values of different parameters of the ΛCDM cosmology from combined model fittings to the recent 3–year *WMAP* CMBR and several other cosmological probes [109] (see the WMAP website).

Table 2. Constraints to the parameters of the ΛCDM model

Parameter	Constraint
Total density	$\Omega = 1$
Dark Energy density	$\Omega_\Lambda = 0.74$
Dark Matter density	$\Omega_{DM} = 0.216$
Baryon Matter dens.	$\Omega_B = 0.044$
Hubble constant	$h = 0.71$
Age	13.8 Gyr
Power spectrum norm.	$\sigma_8 = 0.75$
Power spectrum index	$n_s(0.002) = 0.94$

In the following, I will describe some of the ingredients of the ΛCDM scenario, emphasizing that most of these ingredients are well established aspects that any alternative scenario to ΛCDM should be able to explain.

3.1 Origin of Fluctuations

The Big Bang[4] is now a mature theory, based on well established observational pieces of evidence. However, the Big Bang theory has limitations. One of them is namely the origin of fluctuations that should give rise to the highly inhomogeneous structure observed today in the Universe, at scales of less than $\sim$ 200Mpc. The smaller the scales, the more clustered is the matter. For example, the densities inside the central regions of galaxies, within the galaxies, cluster of galaxies, and superclusters are about 10^{11}, 10^6, 10^3 and few times the average density of the Universe, respectively.

The General Relativity equations that describe the Universe dynamics in the Big Bang theory are for an homogeneous and isotropic fluid (Cosmological Principle); inhomogeneities are not taken into account in this theory "by definition". Instead, the concept of fluctuations is inherent to the Inflationary theory introduced in the early 80's by A. Guth and A. Linde namely to

[4] It is well known that the name of 'Big Bang' is not appropriate for this theory. The key physical conditions required for an explosion are temperature and pressure gradients. These conditions contradict the Cosmological Principle of homogeneity and isotropy on which is based the 'Big Bang' theory.

overcome the Big Bang limitations. According to this theory, at the energies of Grand Unification ($\gtrsim 10^{14}$GeV or $T \gtrsim 10^{27}$K!), the matter was in the state known in quantum field theory as vacuum. Vacuum is characterized by quantum fluctuations –temporary changes in the amount of energy in a point in space, arising from Heisenberg uncertainty principle. For a small time interval Δt, a virtual particle–antiparticle pair of energy ΔE is created (in the GU theory, the field particles are supposed to be the X- and Y-bosons), but then the pair disappears so that there is no violation of energy conservation. Time and energy are related by $\Delta E \Delta t \approx \frac{h}{2\pi}$. The vacuum quantum fluctuations are proposed to be the seeds of present–day structures in the Universe.

How is that quantum fluctuations become density inhomogeneities? During the inflationary period, the expansion is described approximately by the de Sitter cosmology, $a \propto e^{Ht}$, $H \equiv \dot{a}/a$ is the Hubble parameter and it is constant in this cosmology. Therefore, the proper length of any fluctuation grows as $\lambda_p \propto e^{Ht}$. On the other hand, the proper radius of the horizon for de Sitter metric is equal to c/H =const, so that initially causally connected (quantum) fluctuations become suddenly supra–horizon (classical) perturbations to the spacetime metric. After inflation, the Hubble radius grows proportional to ct, and at some time a given curvature perturbation cross again the horizon (becomes causally connected, $\lambda_p < L_H$). It becomes now a true density perturbation. The interesting aspect of the perturbation 'trip' outside the horizon is that its amplitude remains roughly constant, so that if the amplitude of the fluctuations at the time of exiting the horizon during inflation is constant (scale invariant), then their amplitude at the time of entering the horizon should be also scale invariant. In fact, the computation of classical perturbations generated by a quantum field during inflation demonstrates that the amplitude of the scalar fluctuations at the time of crossing the horizon is nearly constant, $\delta\phi_H \propto$const. This can be understood on dimensional grounds: due to the Heisenberg principle $\delta\phi/\delta t \propto$ const, where $\delta t \propto H^{-1}$. Therefore, $\delta\phi_H \propto H$, but H is roughly constant during inflation, so that $\delta\phi_H \propto$const.

3.2 Gravitational Evolution of Fluctuations

The ΛCDM scenario assumes the gravitational instability paradigm: the cosmic structures in the Universe were formed as a consequence of the growth of primordial tiny fluctuations (for example seeded in the inflationary epochs) by gravitational instability in an expanding frame. The fluctuation or perturbation is characterized by its density contrast,

$$\delta \equiv \frac{\delta\rho}{\overline{\rho}} = \frac{\rho - \overline{\rho}}{\overline{\rho}}, \tag{1}$$

where $\overline{\rho}$ is the average density of the Universe and ρ is the perturbation density. At early epochs, $\delta << 1$ for perturbation of all scales, otherwise the homogeneity condition in the Big Bang theory is not anymore obeyed. When $\delta << 1$, the perturbation is in the *linear* regime and its physical size grows with the expansion proportional to $a(t)$. The perturbation analysis in the linear approximation shows whether a given perturbation is stable ($\delta \sim$ const or even $\rightarrow 0$) or unstable (δ grows). In the latter case, when $\delta \rightarrow 1$, the linear approximation is not anymore valid, and the perturbation "separates" from the expansion, collapses, and becomes a self–gravitating structure. The gravitational evolution in the *non–linear regime* is complex for realistic cases and is studied with numerical N–body simulations. Next, a pedagogical review of the linear evolution of perturbations is presented. More detailed explanations on this subject can be found in the books [72, 94, 90, 30, 77, 92].

Relevant Times and Scales

The important times in the problem of linear gravitational evolution of perturbations are: (a) the epoch when inflation finished ($t_{inf} \approx 10^{-34}$s, at this time the primordial fluctuation field is established); (b) the epoch of matter–radiation equality t_{eq} (corresponding to $a_{eq} \approx 1/3.9{\times}10^4(\Omega_0 h^2)$, before t_{eq} the dynamics of the universe is dominated by radiation density, after t_{eq} dominates matter density); (c) the epoch of recombination t_{rec}, when radiation decouples from baryonic matter (corresponding to $a_{rec} = 1/1080$, or $t_{rec} \approx 3.8 \times 10^5$yr for the concordance cosmology).

Scales: first of all, we need to characterize the size of the perturbation. In the linear regime, its physical size expands with the Universe: $\lambda_p = a(t)\lambda_0$, where λ_0 is the comoving size, by convention fixed (extrapolated) to the present epoch, $a(t_0) = 1$. In a given (early) epoch, the size of the perturbation can be larger than the so–called *Hubble radius*, the typical radius over which physical processes operate coherently (there is causal connection): $L_H \equiv (a/\dot{a})^{-1} = H^{-1} = n^{-1}ct$. For the radiation or matter dominated cases, $a(t) \propto t^n$, with $n = 1/2$ and $n = 2/3$, respectively, that is $n < 1$. Therefore, L_H grows faster than λ_p and at a given "crossing" time t_{cross}, $\lambda_p < L_H$. Thus, the perturbation is supra–horizon sized at epochs $t < t_{cross}$ and sub–horizon sized at $t > t_{cross}$. Notice that if $n > 1$, then at some time the perturbation "exits" the Hubble radius. This is what happens in the inflationary epoch, when $a(t) \propto e^t$: causally–connected fluctuations of any size are are suddenly "taken out" outside the Hubble radius becoming causally disconnected.

For convenience, in some cases it is better to use masses instead of sizes. Since in the linear regime $\delta << 1$ ($\rho \approx \overline{\rho}$), then $M \approx \rho_M(a)\ell^3$, where ℓ is the size of a given region of the Universe with average matter density ρ_M. The mass of the perturbation, M_p, is invariant.

Supra–horizon Sized Perturbations

In this case, causal, microphysical processes are not possible, so that it does not matter what perturbations are made of (baryons, radiation, dark matter, etc.); they are in general just perturbations to the metric. To study the gravitational growth of metric perturbations, a General Relativistic analysis is necessary. A major issue in carrying out this program is that the metric perturbation is not a gauge invariant quantity. See e.g., [72] for an outline of how E. Lifshitz resolved brilliantly this difficult problem in 1946. The result is quite simple and it shows that the amplitude of metric perturbations outside the horizon grows *kinematically* at different rates, depending on the dominant component in the expansion dynamics. For the critical cosmological model (at early epochs all models approach this case), the growing modes of metric perturbations according to what dominates the background are:

$$\delta_{m,+} \propto a(t) \propto t^{2/3}, \ldots\ldots\ldots\ldots\ldots\ldots matter \quad (2)$$

$$\delta_{m,+} \propto a(t)^2 \propto t, \ldots\ldots\ldots\ldots\ldots\ldots radiation$$

$$\delta_{m,+} \propto a(t)^{-2} \propto e^{-2Ht}, ..\Lambda \ (deSitter) \quad (3)$$

Sub–horizon Sized Perturbations

Once perturbations are causally connected, microphysical processes are switched on (pressure, viscosity, radiative transport, etc.) and the gravitational evolution of the perturbation depends on what it is made of. Now, we deal with true *density* perturbations. For them applies the classical perturbation analysis for a fluid, originally introduced by J. Jeans in 1902, in the context of the problem of star formation in the ISM. But unlike in the ISM, in the cosmological context the fluid is expanding. What can prevent the perturbation amplitude from growing gravitationally? The answer is pressure support. If the fluid pressure gradient can re–adjust itself in a timescale t_{press} smaller than the gravitational collapse timescale, t_{grav}, then pressure prevents the gravitational growth of δ. Thus, the condition for gravitational instability is:

$$t_{grav} \approx \frac{1}{(G\rho)^{1/2}} < t_{press} \approx \frac{\lambda_p}{v}, \quad (4)$$

where ρ is the density of the component that is most gravitationally dominant in the Universe, and v is the sound speed (collisional fluid) or velocity dispersion (collisionless fluid) of the perturbed component. In other words, if the perturbation scale is larger than a critical scale $\lambda_J \sim v(G\rho)^{-1/2}$, then pressure loses, gravity wins.

The perturbation analysis applied to the hydrodynamical equations of a fluid at rest shows that δ grows *exponentially* with time for perturbations obeying the Jeans instability criterion $\lambda_p > \lambda_J$, where the exact value of λ_J is $v(\pi/G\rho)^{1/2}$. If $\lambda_p < \lambda_J$, then the perturbations are described by stable *gravito–acustic oscillations*. The situation is conceptually similar for perturbations in an expanding cosmological fluid, but the growth of δ in the unstable regime is *algebraical* instead of exponential. Thus, the cosmic structure formation process is relatively slow. Indeed, the typical epochs of galaxy and cluster of galaxies formation are at redshifts $z \sim 1 - 5$ (ages of $\sim 1.2 - 6$ Gyrs) and $z < 1$ (ages larger than 6 Gyrs), respectively.

Baryonic matter. The Jeans instability analysis for a relativistic (plasma) fluid of baryons *ideally* coupled to radiation and expanding at the rate $H = \dot{a}/a$ shows that there is an instability critical scale $\lambda_J = v(3\pi/8G\rho)^{1/2}$, where the sound speed for adiabatic perturbations is $v = p/\rho = c/\sqrt{3}$; the latter equality is due to pressure radiation. At the epoch when *radiation dominates*, $\rho = \rho_r \propto a^{-4}$ and then $\lambda_J \propto a^2 \propto ct$. It is not surprising that at this epoch λ_J approximates the Hubble scale $L_H \propto ct$ (it is in fact $\sim$ 3 times larger). Thus, perturbations that might collapse gravitationally are in fact outside the horizon, and those that already entered the horizon, have scales smaller than λ_J: they are stable gravito–acoustic oscillations. When *matter dominates*, $\rho = \rho_M \propto a^{-3}$, and $a \propto t^{2/3}$. Therefore, $\lambda_J \propto a \propto t^{2/3} \lesssim L_H$, but still radiation is coupled to baryons, so that radiation pressure is dominant and λ_J remains large. However, when radiation decouples from baryons at t_{rec}, the pressure support drops dramatically by a factor of $P_r/P_b \propto n_r T/n_b T \approx 10^8$! Now, the Jeans analysis for a gas mix of H and He at temperature $T_{\rm rec} \approx 4000$ K shows that baryonic clouds with masses $\gtrsim 10^6 \rm M_\odot$ can collapse gravitationally, i.e. all masses of cosmological interest. But this is literally too "ideal" to be true.

The problem is that as the Universe expands, radiation cools ($T_r = T_0 a^{-1}$) and the photon–baryon fluid becomes less and less perfect: the mean free path for scattering of photons by electrons (which at the same time are coupled electrostatically to the protons) increases. Therefore, photons can diffuse out of the bigger and bigger density perturbations as the photon mean free path increases. If perturbations are in the gravito–acoustic oscillatory regime, then the oscillations are damped out and the perturbations disappear. The "ironing out" of perturbations continues until the epoch of recombination. In a pioneering work, J. Silk [104] carried out a perturbation analysis of a relativistic cosmological fluid taking into account radiative transfer in the diffusion approximation. He showed that all photon–baryon perturbations of masses smaller than M_S are "ironed out" until t_{rec} by the (Silk) damping process. The first crisis in galaxy formation theory emerged: calculations showed that M_S is of the order of $10^{13} - 10^{14} \rm M_\odot h^{-1}$! If somebody (god, inflation, ...) seeded primordial fluctuations in the Universe, by Silk damping all galaxy–sized perturbation are "ironed out". [5]

Non–baryonic matter. The gravito–acoustic oscillations and their damping by photon diffusion refer to baryons. What happens for a fluid of non–baryonic DM? After all, astronomers, since Zwicky in the 1930s, find routinely pieces

[5] In the 1970s Y. Zel'dovich and collaborators worked out a scenario of galaxy formation starting from very large perturbations, those that were not affected by Silk damping. In this elegant scenario, the large–scale perturbations, considered in a first approximation as ellipsoids, collapse most rapidly along their shortest axis, forming flattened structures ("pancakes"), which then fragment into galaxies by gravitational or thermal instabilities. In this 'top-down' scenario, to obtain galaxies in place at $z \sim 1$, the amplitude of the large perturbations at recombination should be $\geq 3 \times 10^{-3}$. Observations of the CMBR anisotropies showed that the amplitudes are 1–2 order of magnitudes smaller than those required.

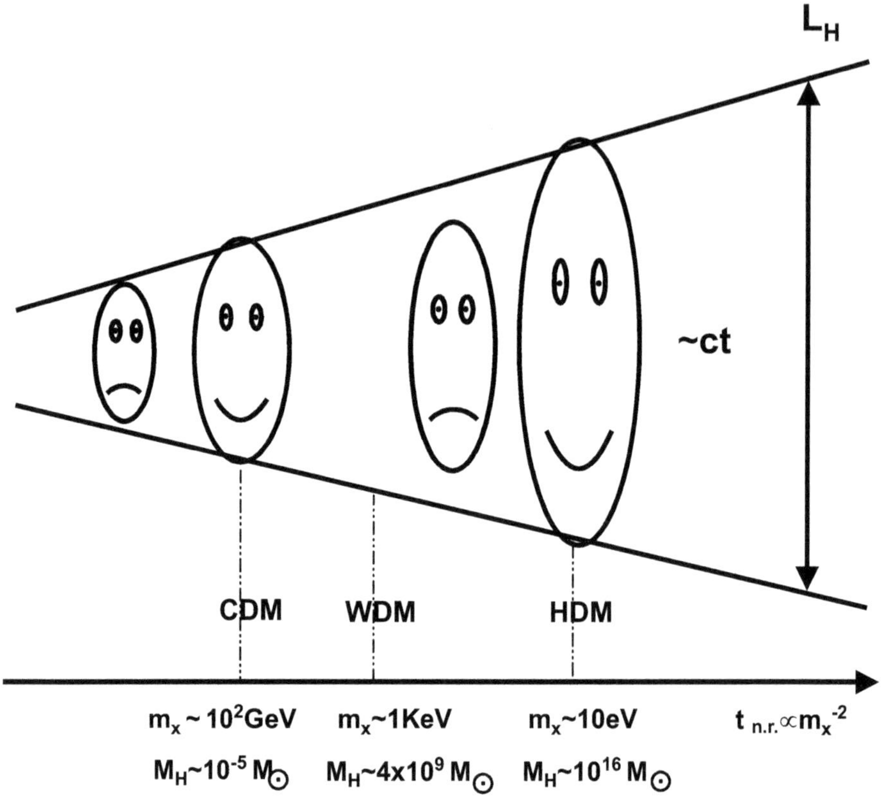

Fig. 4. Free–streaming damping kills perturbations of sizes roughly smaller than the horizon length if they are made of relativistic particles. The epoch $t_{n.r.}$ when thermal–coupled particles become non–relativistic is inverse proportional to the square of the particle mass m_X. Typical particle masses of CDM, WDM and HDM are given together with the corresponding horizon (filtering) masses.

of evidence for the presence of large amounts of DM in the Universe. As DM is assumed to be collisionless and not interacting electromagnetically, then the radiative or thermal pressure supports are not important for linear DM perturbations. However, DM perturbations can be damped out by *free streaming* if the particles are relativistic: the geodesic motion of the particles at the speed of light will iron out any perturbation smaller than a scale close to the particle horizon radius, because the particles can freely propagate from an overdense region to an underdense region. Once the particles cool and become non relativistic, free streaming is not anymore important. A particle of mass m_X and temperature T_X becomes non relativistic when $k_B T_X \sim m_X c^2$. Since $T_X \propto a^{-1}$, and $a \propto t^{1/2}$ when radiation dominates, one then finds that the epoch when a thermal–relic particle becomes non relativistic is $t_{nr} \propto m_X^{-2}$.

The more massive the DM particle, the earlier it becomes non relativistic, and the smaller are therefore the perturbations damped out by free streaming (those smaller than $\sim ct$; see Fig. 4). According to the epoch when a given thermal DM particle species becomes non relativistic, DM is called Cold Dark Matter (CDM, very early), Warm Dark Matter (WDM, early) and Hot Dark Matter (HDM, late)[6].

The only non–baryonic particles confirmed experimentally are (light) neutrinos (HDM). For neutrinos of masses $\sim 1 - 10$eV, free streaming attains to iron out perturbations of scales as large as massive clusters and superclusters of galaxies (see Fig. 4). Thus, HDM suffers the same problem of baryonic matter concerning galaxy formation[7]. At the other extreme is CDM, in which case survive free streaming practically all scales of cosmological interest. This makes CDM appealing to galaxy formation theory. In the minimal CDM model, it is assumed that perturbations of all scales survive, and that CDM particles are collisionless (they do not self–interact). Thus, if CDM dominates, then the first step in galaxy formation study is reduced to the calculation of the linear and non–linear gravitational evolution of collisionless CDM perturbations. Galaxies are expected to form in the centers of collapsed CDM structures, called *halos*, from the baryonic gas, first trapped in the gravitational potential of these halos, and second, cooled by radiative (and turbulence) processes (see §5).

The CDM perturbations are free of any physical damping processes and in principle their amplitudes may grow by gravitational instability. However, when radiation dominates, the perturbation growth is stagnated by expansion. The gravitational instability timescale for sub–horizon linear CDM perturbations is $t_{grav} \sim (G\rho_{DM})^{-2}$, while the expansion (Hubble) timescale is given by $t_{exp} \sim (G\overline{\rho})^{-2}$. When radiation dominates, $\overline{\rho} \approx \rho_r$ and $\rho_r >> \rho_M$. Therefore $t_{exp} << t_{grav}$, that is, expansion is faster than the gravitational shrinking.

Fig. 5 resumes the evolution of primordial perturbations. Instead of spatial scales, in Fig. 5 are shown masses, which are invariant for the perturbations. We highlight the following conclusions from this plot: (1) Photon–baryon perturbations of masses $< M_S$ are washed out ($\delta_B \to 0$) as long as baryon matter is coupled to radiation. (2) The amplitude of CDM perturbations that enter the horizon before t_{eq} is "freezed-out" ($\delta_{DM} \propto$const) as long as radiation dominates; these are perturbations of masses smaller than $M_{H,eq} \approx 10^{13}(\Omega_M h^2)^{-2}M_\odot$, namely galaxy scales. (3) The baryons are trapped gravitationally by CDM perturbations, and within a factor of two in z, baryon perturbations attain amplitudes half that of δ_{DM}. For WDM

---

[6] The reference to "early" and "late" is given by the epoch and the corresponding radiation temperature when the largest galaxy–sized perturbations ($M \sim 10^{13}$M$_\odot$) enter the horizon: $a_{gal} \sim a_{eq} \approx 1/3.9\times10^4(\Omega_0 h^2)$ and $T_r \sim 1$KeV.

[7] Neutrinos exist and have masses larger than 0.05 eV according to determinations based on solar neutrino oscillations. Therefore, neutrinos contribute to the matter density in the Universe. Cosmological observations set a limit: $\Omega_\nu h^2 < 0.0076$, otherwise too much structure is erased.

or HDM perturbations, the free–streaming damping introduces a mass scale $M_{fs} \approx M_{H,n.r.}$ in Fig. 5, below which $\delta \rightarrow 0$; M_{fs} increases as the DM mass particle decreases (Fig. 4).

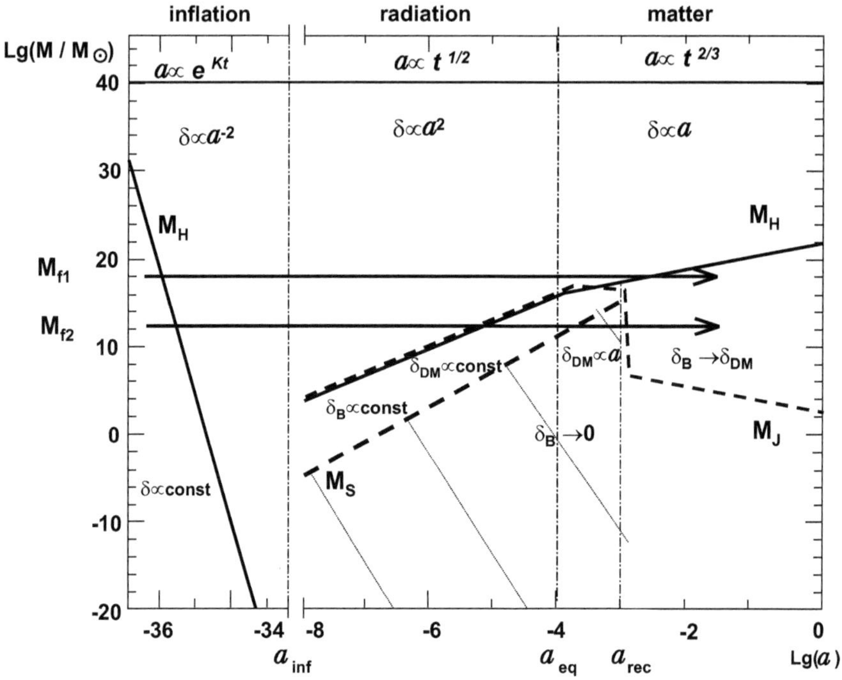

Fig. 5. Different evolutive regimes of perturbations δ. The suffixes "B" and "DM" are for baryon–photon and DM perturbations, respectively. The evolution of the horizon, Jeans and Silk masses (M_H, M_J, and M_S) are showed. M_{f1} and M_{f2} are the masses of two perturbations. See text for explanations.

The processed power spectrum of perturbations. The exact solution to the problem of linear evolution of cosmological perturbations is much more complex than the conceptual aspects described above. Starting from a primordial fluctuation field, the perturbation analysis should be applied to a cosmological mix of baryons, radiation, neutrinos, and other non–baryonic dark matter components (e.g., CDM), at sub– and supra–horizon scales (the fluid assumption is relaxed). Then, coupled relativistic hydrodynamic and Boltzmann equations in a general relativity context have to be solved taking into account radiative and dissipative processes. The outcome of these complex calculations is the full description of the processed fluctuation field at the recombination epoch (when fluctuations at almost all scales are still in the linear regime). The goal is double and of crucial relevance in cosmology and astrophysics:

1) to predict the physical and statistical properties of CMBR anisotropies, which can be then compared with observations, and 2) to provide the initial conditions for calculating the non–linear regime of cosmic structure formation and evolution. Fortunately, there are now several public friendly-to-use codes that numerically solve the cosmological linear perturbation equations (e.g., CMBFast and CAMB [8]).

The description of the density fluctuation field is statistical. As any random field, it is convenient to study perturbations in the Fourier space. The Fourier expansion of $\delta(\mathbf{x})$ is:

$$\delta(\mathbf{x}) = \frac{V}{(2\pi)^3} \int \delta_k e^{-i\mathbf{kx}} d^3k, \tag{5}$$

$$\delta_k = V^{-1} \int \delta(\mathbf{x}) e^{i\mathbf{kx}} d^3x \tag{6}$$

The Fourier modes δ_k evolve independently while the perturbations are in the linear regime, so that the perturbation analysis can be applied to this quantity. For a Gaussian random field, any statistical quantity of interest can be specified in terms of the power spectrum $P(k) \equiv |\delta_k|^2$, which measures the amplitude of the fluctuations at a given scale k[9]. Thus, from the linear perturbation analysis we may follow the evolution of $P(k)$. A more intuitive quantity than $P(k)$ is the mass variance $\sigma_M^2 \equiv \langle(\delta M/M)_R^2\rangle$ of the fluctuation field. The physical meaning of σ_M is that of an *rms* density contrast on a given sphere of radius R associated to the mass $M = \rho V_W(R)$, where $W(R)$ is a window (smoothing) function. The mass variance is related to $P(k)$. By assuming a power law power spectrum, $P(k) \propto k^n$, it is easy to show that

$$\sigma_M \propto R^{-(3+n)} \propto M^{-(3+n)/3} = M^{-2\alpha} \tag{7}$$

$$\alpha = \frac{3+n}{6},$$

for $4 < n < -3$ using a Gaussian window function. The question is: How the scaling law of perturbations, σ_M, evolves starting from an initial $(\sigma_M)_i$?

In the early 1970s, Harrison and Zel'dovich independently asked themselves about the functionality of σ_M (or the density contrast) at the time adiabatic perturbations cross the horizon, that is, if $(\sigma_M)_H \propto M^{\alpha_H}$, then what is the value of α_H? These authors concluded that $-0.1 \leq \alpha_H \leq 0.2$, i.e.

---

8 *http://www.cmbfast.org* and *http://camb.info/*

9 The phases of the Fourier modes in the Gaussian case are *uncorrelated.* Gaussianity is the simplest assumption for the primordial fluctuation field statistics and it seems to be consistent with some variants of inflation. However, there are other variants that predict non–Gaussian fluctuations (for a recent review on this subject see e.g. [8]), and the observational determination of the primordial fluctuation statistics is currently an active field of investigation. The properties of cosmic structures depend on the assumption about the primordial statistics, not only at large scales but also at galaxy scales; see for a review and new results [4].

$\alpha_H \approx 0$ ($n_H \approx -3$). If $\alpha_H >> 0$ ($n_H >> -3$), then $\sigma_M \to \infty$ for $M \to 0$; this means that for a given small mass scale M, the mass density of the perturbation at the time of becoming causally connected can correspond to the one of a (primordial) black hole. Hawking evaporation of black holes put a constraint on $M_{BH,prim} \lesssim 10^{15}$g, which corresponds to $\alpha_H \leq 0.2$, otherwise the γ–ray background radiation would be more intense than that observed. If $\alpha_H << 0$ ($n_H << -3$), then larger scales would be denser than the small ones, contrary to what is observed. The scale–invariant *Harrison–Zel'dovich power spectrum,* $P_H(k) \propto k^{-3}$, is for perturbations at the time of entering the horizon. How should the primordial power spectrum $P_i(k) = Ak_i^n$ or $(\sigma_M)_i = BM^{-\alpha_i}$ (defined at some fixed initial time) be to produce such scale invariance? Since t_i until the horizon crossing time $t_{cross}(M)$ for a given perturbation of mass M, $\sigma_M(t)$ evolves as $a(t)^2$ (supra–horizon regime in the radiation era). At t_{cross}, the horizon mass M_H is equal by definition to M. We have seen that $M_H \propto a^3$ (radiation dominion), so that $a_{cross} \propto M_H^{1/3} = M^{1/3}$. Therefore,

$$\sigma_M(t_{cross}) \propto (\sigma_M)_i (a_{cross}/a_i)^2 \propto M^{-\alpha_i} M^{2/3}, \tag{8}$$

i.e. $\alpha_H = 2/3 - \alpha_i$ or $n_H = n_i - 4$. A similar result is obtained if the perturbation enters the horizon during the matter dominion era. From this analysis one concludes that for the perturbations to be scale invariant at horizon crossing ($\alpha_H = 0$ or $n_H = -3$), the primordial (initial) power spectrum should be $P_i(k) = Ak^1$ or $(\sigma_M)_i \propto M^{-2/3} \propto \lambda_0^{-2}$ (i.e. $n_i = 1$ and $\alpha = 2/3$; A is a normalization constant). Does inflation predict such power spectrum? We have seen that, according to the quantum field theory and assuming that H =const during inflation, the fluctuation amplitude is scale invariant at the time to exit the horizon, δ_H ~const. On the other hand, we have seen that supra–horizon curvature perturbations during a de Sitter period evolve as $\delta \propto a^{-2}$ (eq. 4). Therefore, at the end of inflation we have that $\delta_{inf} = \delta_H(\lambda_0)(a_{inf}/a_H)^{-2}$. The proper size of the fluctuation when crossing the horizon is $\lambda_p = a_H \lambda_0 \approx H^{-1}$; therefore, $a_H \approx 1/(\lambda_0 H)$. Replacing now this expression in the equation for δ_{inf} we get that:

$$\delta_{inf} \approx \delta_H(\lambda_0)(a_{inf}\lambda_0 H)^{-2} \propto \lambda_0^{-2} \propto M^{-2/3}, \tag{9}$$

if δ_H ~const. Thus, inflation predicts α_i nearly equal to 2/3 ($n_i \approx 1$)! Recent results from the analysis of CMBR anisotropies by the *WMAP* satellite [109] seem to show that n_i is slightly smaller than 1 or that n_i changes with the scale (running power–spectrum index). This is in more agreement with several inflationary models, where H actually slightly vary with time introducing some scale dependence in δ_H.

The perturbation analysis, whose bases were presented in §3.2 and resumed in Fig. 5, show us that σ_M grows (kinematically) while perturbations are in the supra–horizon regime. Once perturbations enter the horizon (first the smaller ones), if they are made of CDM, then the gravitational growth is "freezed out" whilst radiation dominates (stangexpantion). As shown schematically

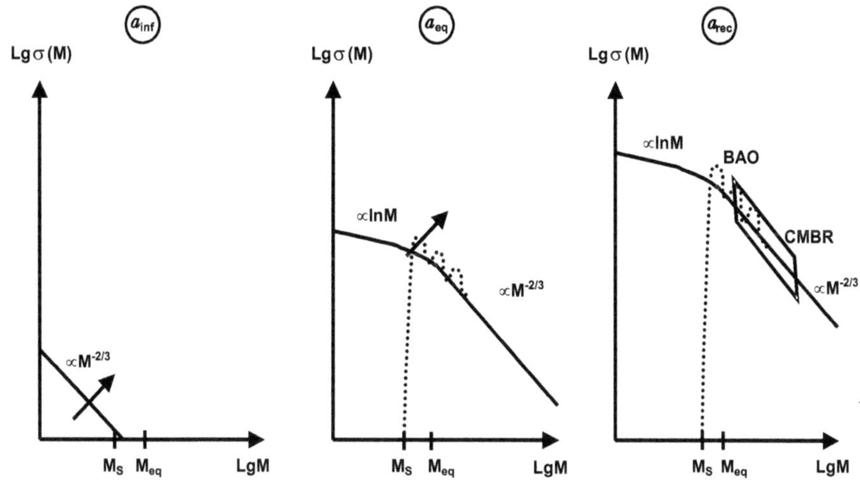

Fig. 6. Linear evolution of the perturbation mass variance σ_M. The perturbation amplitude in the supra–horizon regime grow kinematically. DM perturbations (solid curve) that cross the horizon during the radiation dominion, freeze–out their grow due to stangexpantion, producing a flattening in the scaling law σ_M for all scales smaller than the corresponding to the horizon at the equality epoch (galaxy scales). Baryon–photon perturbations smaller than the Silk mass M_S are damped out (dotted curve) and those larger than M_S but smaller than the horizon mass at recombination are oscillating (Baryonic Acoustic Oscillation, BAO).

in Fig. 6, this "flattens" the variance σ_M at scales smaller than $M_{H,eq}$; in fact, $\sigma_M \propto ln(M)$ at these scales, corresponding to galaxies! After t_{eq} the CDM variance (or power spectrum) grows at the same rate at all scales. If perturbations are made out of baryons, then for scales smaller than M_S, the gravito–acoustic oscillations are damped out, while for scales close to the Hubble radius at recombination, these oscillations are present. The "final" processed mass variance or power spectrum is defined at the recombination epoch. For example, the power spectrum is expressed as:

$$P_{rec}(k) = Ak^{n_i} \times (D(t_{rec})/D(t_i))^2 \times T^2(k), \quad (10)$$

where the first term is the initial power spectrum $P_i(k)$; the second one is how much the fluctuation amplitude has grown in the linear regime ($D(t)$ is the so–called linear growth factor), and the third one is a transfer function that encapsulates the different damping and freezing out processes able to deform the initial power spectrum shape. At large scales, $T^2(k) = 1$, i.e. the primordial shape is conserved (see Fig. 6).

Besides the mass power spectrum, it is possible to calculate the *angular power spectrum of temperature fluctuation in the CMBR.* This spectrum consists basically of 2 ranges divided by a critical angular scales: the angle θ_h corresponding to the horizon scale at the epoch of recombination

($(L_H)_{rec} \approx 200(\Omega h^2)^{-1/2}$ Mpc, comoving). For scales grander than θ_h the spectrum is featureless and corresponds to the scale–invariant supra–horizon Sachs-Wolfe fluctuations. For scales smaller than θ_h, the sub–horizon fluctuations are dominated by the Doppler scattering (produced by the gravito–acoustic oscillations) with a series of decreasing in amplitude peaks; the position (angle) of the first Doppler peak depends strongly on Ω, i.e. on the geometry of the Universe. In the last 15 years, high–technology experiments as *COBE*, *Boomerang*, *WMAP* provided valuable information (in particular the latter one) on CMBR anisotropies. The results of this exciting branch of astronomy (called sometimes anisotronomy) were of paramount importance for astronomy and cosmology (see for a review [62] and the W. Hu website[10]).

Just to highlight some of the key results of CMBR studies, let us mention the next ones: 1) detailed predictions of the ΛCDM scenario concerning the linear evolution of perturbations were accurately proved, 2) several cosmological parameters as the geometry of the Universe, the baryonic fraction Ω_B, and the index of the primordial power spectrum, were determined with high precision (see the actualized, recently delivered results from the 3 year analysis of *WMAP* in [109]), 3) by studying the polarization maps of the CMBR it was possible to infer the epoch when the Universe started to be significantly reionized by the formation of first stars, 4) the amplitude (normalization) of the primordial fluctuation power spectrum was accurately measured. The latter is crucial for further calculating the non–linear regime of cosmic structure formation. I should emphasize that while the shape of the power spectrum is predicted and well understood within the context of the ΛCDM model, the situation is fuzzy concerning the power spectrum normalization. We have a phenomenological value for A but not a theoretical prediction.

4 The Dark Side of Galaxy Formation and Evolution

A great triumph of the ΛCDM scenario was the overall consistency found between predicted and observed CMBR anisotropies generated at the recombination epoch. In this scenario, the gravitational evolution of CDM perturbations is the driver of cosmic structure formation. At scales much larger than galaxies, (i) mass density perturbations are still in the (quasi)linear regime, following the scaling law of primordial fluctuations, and (ii) the dissipative physics of baryons does not affect significantly the matter distribution. Thus, the large–scale structure (LSS) of the Universe is determined basically by DM perturbations yet in their (quasi)linear regime. At smaller scales, non–linearity strongly affects the primordial scaling law and, moreover, the dissipative physics of baryons "distorts" the original DM distribution, particularly inside galaxy–sized DM halos. However, DM in any case provides the original "mold" where gas dynamics processes take place.

[10] *http://background.uchicago.edu/∼whu/physics/physics.html*

The ΛCDM scenario describes successfully the observed LSS of the Universe (for reviews see e.g., [49, 58], and for some recent observational results see e.g. [115, 102, 109]). The observed filamentary structure can be explained as a natural consequence of the CDM gravitational instability occurring preferentially in the shortest axis of 3D and 2D protostructures (the Zel'dovich panckakes). The clustering of matter in space, traced mainly by galaxies, is also well explained by the clustering properties of CDM. At scales r much larger than typical galaxy sizes, the galaxy 2-point correlation function $\xi_{gal}(r)$ (a measure of the average clustering strength on spheres of radius r) agrees rather well with $\xi_{CDM}(r)$. Current large statistical galaxy surveys as SDSS and 2dFGRS, allow now to measure the redshift–space 2-point correlation function at large scales with unprecedented accuracy, to the point that weak "bumps" associated with the baryon acoustic oscillations at the recombination epoch begin to be detected [41]. At small scales ($\lesssim 3$Mpch^{-1}), $\xi_{gal}(r)$ departs from the predicted pure $\xi_{CDM}(r)$ due to the emergence of two processes: (i) the strong non–linear evolution that small scales underwent, and (ii) the complexity of the baryon processes related to galaxy formation. The difference between $\xi_{gal}(r)$ and $\xi_{CDM}(r)$ is parametrized through one "ignorance" parameter, b, called bias, $\xi_{gal}(r) = b\xi_{CDM}(r)$. Below, some basic ideas and results related to the former processes will be described. The baryonic process will be sketched in the next Section.

4.1 Nonlinear Clustering Evolution

The scaling law of the processed ΛCDM perturbations, is such that σ_M at galaxy–halo scales decreases slightly with mass (logarithmically) and for larger scales, decreases as a power law (see Fig. 6). Because the perturbations of higher amplitudes collapse first, the first structures to form in the ΛCDM scenario are typically the smallest ones. Larger structures assemble from the smaller ones in a process called *hierarchical clustering* or bottom–up mass assembling. It is interesting to note that the concept of hierarchical clustering was introduced several years before the CDM paradigm emerged. Two seminal papers settled the basis for the current theory of galaxy formation: Press & Schechter 1974 [98] and White & Rees 1979 [131]. In the latter it was proposed that "the smaller–scale virialized [dark] systems merge into an amorphous whole when they are incorporated in a larger bound cluster. Residual gas in the resulting potential wells cools and acquires sufficient concentration to self–gravitate, forming luminous galaxies up to a limiting size".

The Press & Schechter (P-S) formalism was developed to calculate the mass function (per unit of comoving volume) of halos at a given epoch, $n(M, z)$. The starting point is a Gaussian density field filtered (smoothed) at different scales corresponding to different masses, the mass variance σ_M being the characterization of this filtering process. A collapsed halo is identified when the evolving density contrast of the region of mass M, $\delta_M(z)$, attains a critical value, δ_c, given by the spherical top–hat collapse model[11]. This way, the Gaussian probability distribution for δ_M is used to calculate the mass distribution of objects collapsed at the epoch z. The P-S formalism assumes implicitly that the only objects to be counted as collapsed halos at a given epoch are those with $\delta_M(z) = \delta_c$. For a mass variance decreasing with mass, as is the case for CDM models, this implies a "hierarchical" evolution of $n(M, z)$: as z decreases, less massive collapsed objects disappear in favor of more massive ones (see Fig. 8). The original P-S formalism had an error of 2 in the sense that integrating $n(M, z)$ half of the mass is lost. The authors multiplied $n(M, z)$ by 2, argumenting that the objects duplicate their masses by accretion from the sub–dense regions. The problem of the factor of 2 in the P-S analysis was partially solved using an excursion set statistical approach [17, 73].

To get an idea of the typical formation epochs of CDM halos, the spherical collapse model can be used. According to this model, the density contrast of given overdense region, δ, grows with z proportional to the growing factor, $D(z)$, until it reaches a critical value, δ_c, after which the perturbation is supposed to collapse and virialize[12]. at redshift $z_{\rm col}$ (for example see [90]):

$$\delta(z_{\rm col}) \equiv \delta_0 D(z_{\rm col}) = \delta_{c,0}. \qquad (11)$$

The convention is to fix all the quantities to their linearly extrapolated values at the present epoch (indicated by the subscript "0") in such a way that $D(z = 0) \equiv D_0 = 1$. Within this convention, for an Einstein–de Sitter cosmology, $\delta_{c,0} = 1.686$, while for the ΛCDM cosmology, $\delta_{c,0} = 1.686\Omega_{M,0}^{0.0055}$, and the growing factor is given by

$$D(z) = \frac{g(z)}{g(z_0)(1+z)}, \qquad (12)$$

[11] The spherical top–hat model refers to the exact calculation of the collapse of a uniform spherical density perturbation in an otherwise uniform Universe; the dynamics of such a region is the same of a closed Universe. The solution of the equations of motion shows that the perturbation at the beginning expands as the background Universe (proportional to a), then it reaches a maximum expansion (size) in a time t_{max}, and since that moment the perturbation separates of the expanding background, collapsing in a time $t_{col} = 2t_{max}$.

[12] The mathematical solution gives that the spherical perturbed region collapses into a point (a black hole) after reaching its maximum expansion. However, real perturbations are lumpy and the particle orbits are not perfectly radial. In this situation, during the collapse the structure comes to a dynamical equilibrium under the influence of large scale gravitational potential gradients, a process named by the oxymoron "violent relaxation" (see e.g. [14]); this is a typical collective phenomenon. The end result is a system that satisfies the virial theorem: for a self–gravitating system this means that the internal kinetic energy is half the (negative) gravitational potential energy. Gravity is supported by the velocity dispersion of particles or lumps. The collapse factor is roughly 1/2, i.e. the typical virial radius R_v of the collapsed structure is ≈ 0.5 the radius of the perturbation at its maximum expansion.

where a good approximation for $g(z)$ is [24]:

$$g(z) \simeq \frac{5}{2}\left[\Omega_M^{\frac{4}{7}} - \Omega_\Lambda + \left(1+\frac{\Omega_M}{2}\right)\left(1+\frac{\Omega_\Lambda}{70}\right)\right]^{-1}, \qquad (13)$$

and where $\Omega_M = \Omega_{M,0}(1+z)^3/E^2(z)$, $\Omega_\Lambda = \Omega_\Lambda/E^2(z)$, with $E^2(z) = \Omega_\Lambda + \Omega_{M,0}\,(1+z)^3$. For the Einstein–de Sitter model, $D(z) = (1+z)$. We need now to connect the top–hat sphere results to a perturbation of mass M. The processed perturbation field, fixed at the present epoch, is characterized by the mass variance σ_M and we may assume that $\delta_0 = \nu\sigma_M$, where δ_0 is δ linearly extrapolated to $z=0$, and ν is the peak height. For average perturbations, $\nu = 1$, while for rare, high–density perturbations, from which the first structures arose, $\nu >> 1$. By introducing $\delta_0 = \nu\sigma_M$ into eq. (11) one may infer z_{col} for a given mass. Fig. 7 shows the typical z_{col} of 1σ, 2σ, and 3σ halos. The collapse of galaxy–sized 1σ halos occurs within a relatively small range of redshifts. This is a direct consequence of the "flattening" suffered by σ_M during radiation–dominated era due to stangexpansion (see §3.2). Therefore, in a ΛCDM Universe it is not expected to observe a significant population of galaxies at $z \gtrsim 5$.

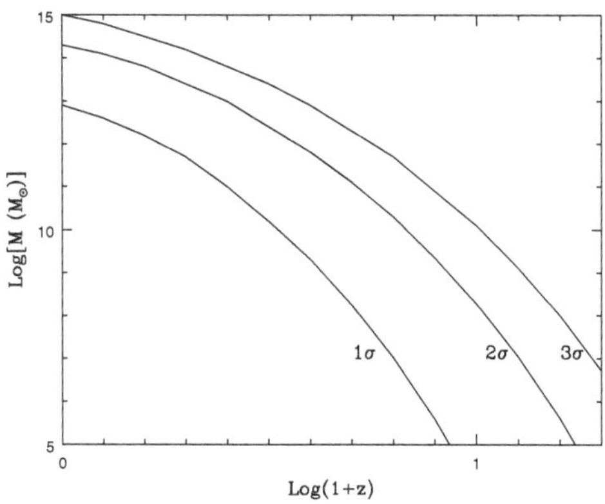

Fig. 7. Collapse redshifts of spherical top–hat 1σ, 2σ and 3σ perturbations in a ΛCDM cosmology with $\sigma_8 = 0.9$. Note that galaxy–sized ($M \sim 10^8 - 10^{13}\mathrm{M}_\odot$) 1σ halos collapse in a redshift range, from $z \sim 3.5$ to $z = 0$, respectively; the corresponding ages are from ~ 1.9 to 13.8 Gyr, respectively.

The problem of cosmological gravitational clustering is very complex due to non–linearity, lack of symmetry and large dynamical range. Analytical and semi–analytical approaches provide illuminating results but numerical N–body simulations are necessary to tackle all the aspects of this problem. In the last 20 years, the "industry" of numerical simulations had an impressive development. The first cosmological simulations in the middle 80s used a few 10^4 particles (e.g., [36]). The currently largest simulation (called the Millenium simulation [111]) uses $\sim 10^{10}$ particles! A main effort is done to reach larger and larger dynamic ranges in order to simulate encompassing volumes large enough to contain representative populations of all kinds of halos (low mass and massive ones, in low– and high–density environments, high–peak rare halos), yet resolving the inner structure of individual halos.

Halo Mass Function

The CDM halo mass function (comoving number density of halos of different masses at a given epoch z, $n(M,z)$) obtained in the N–body simulations is consistent with the P-S function in general, which is amazing given the approximate character of the P-S analysis. However, in more detail, the results of large N–body simulations are better fitted by modified P-S analytical functions, as the one derived in [103] and showed in Fig. 8. Using the Millennium simulation, the halo mass function has been accurately measured in the range that is well sampled by this run ($z \leq 12, M \geq 1.7 \times 10^{10}\mathrm{M}_{\odot}\mathrm{h}^{-1}$). The mass function is described by a power law at low masses and an exponential cut–off at larger masses. The "cut-off", most typical mass, increases with time and is related to the hierarchical evolution of the 1σ halos shown in Fig. 7. The halo mass function is the starting point for modeling the luminosity function of galaxies. From Fig. 8 we see that the evolution of the abundances of massive halos is much more pronounced than the evolution of less massive halos. This is why observational studies of abundance of massive galaxies or cluster of galaxies at high redshifts provide a sharp test to theories of cosmic structure formation. The abundance of massive rare halos at high redshifts are for example a strong function of the fluctuation field primordial statistics (Gaussianity or non-Gaussianity).

Subhalos. An important result of N–body simulations is the existence of subhalos, i.e. halos inside the virial radius of larger halos, which survived as self–bound entities the gravitational collapse of the higher level of the hierarchy. Of course, subhalos suffer strong mass loss due to tidal stripping, but this is probably not relevant for the luminous galaxies formed in the innermost regions of (sub)halos. This is why in the case of subhalos, the maximum circular velocity V_m (attained at radii much smaller than the virial radius) is used instead of the virial mass. The V_m distribution of subhalos inside cluster–sized and galaxy–sized halos is similar [83]. This distribution agrees with the distribution of galaxies seen in clusters, but for galaxy–sized halos the number of subhalos overwhelms by 1–2 orders of magnitude the observed number of satellite galaxies around galaxies like Milky Way and Andromeda [70, 83].

Fig. 9 (right side) shows the subhalo cumulative V_m–distribution for a CDM Milky Way–like halo compared to the observed satellite V_m–distribution. In this Fig. are also shown the V_m–distributions obtained for the same Milky–Way halo but using the power spectrum of three WDM models with particle masses $m_X \approx 0.6, 1$, and 1.7 KeV. The smaller m_X, the larger is the free–streaming (filtering) scale, R_f, and the more substructure is washed out (see

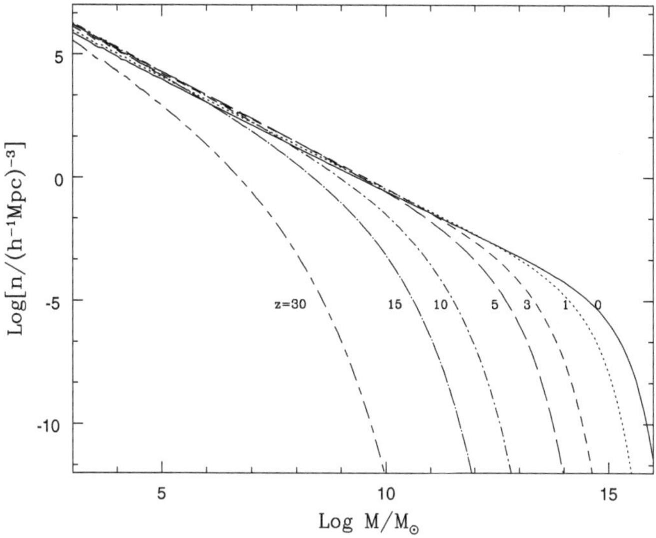

Fig. 8. Evolution of the comoving number density of collapsed halos (P–S mass function) according to the ellipsoidal modification by [103]. Note that the "cut–off" mass grows with time. Most of the mass fraction in collapsed halos at a given epoch are contained in halos with masses around the "cut–off" mass.

§3.2). In the left side of Fig. 9 is shown the DM distribution inside the Milky–Way halo simulated by using a CDM power spectrum (top) and a WDM power spectrum with $m_X \approx 1$KeV (sterile neutrino, bottom). For a student it should be exciting to see with her(his) own eyes this tight connection between micro– and macro–cosmos: the mass of the elemental particle determines the structure and substructure properties of galaxy halos!

Halo Density Profiles

High–resolution N–body simulations [87] and semi–analytical techniques (e.g., [3]) allowed to answer the following questions: How is the inner mass distribution in CDM halos? Does this distribution depend on mass? How universal is it? The two–parameter density profile established in [87] (the Navarro-Frenk-White, NFW profile) departs from a single power law, and it was proposed to be universal and not depending on mass. In fact the slope $\beta(r) \equiv -\mathrm{dlog}\rho(r)/\mathrm{dlog}r$ of the NFW profile changes from -1 in the center to -3 in the periphery. The two parameters, a normalization factor, ρ_s and a shape factor, r_s, were found to be related in a such a way that the profile depends only on one shape parameter that could be expressed as the concentration, $c_{NFW} \equiv r_s/R_v$. The more massive the halo, the less concentrated on the average. For the ΛCDM model, $c \approx 20-5$ for $M \sim 2\times10^8-2\times10^{15}\mathrm{M}_\odot\mathrm{h}^{-1}$, respectively [42]. However, for a given M, the scatter of c_{NFW} is large ($\approx 30-40\%$), and it is related to the halo formation history [3, 22, 125] (see

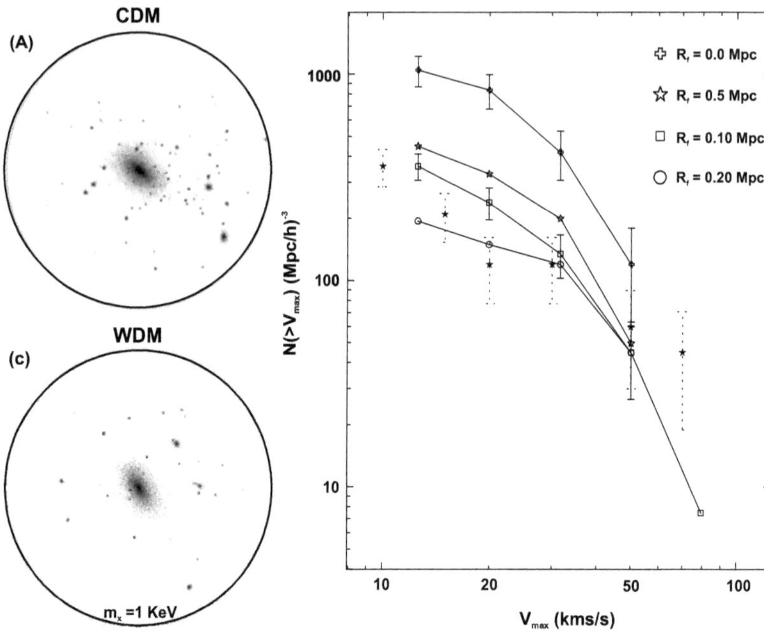

Fig. 9. Dark matter distribution in a sphere of 400Mpch^{-1} of a simulated Galaxy–sized halo with CDM (a) and WDM (m_X = 1KeV, b). The substructure in the latter case is significantly erased. Right panel shows the cumulative maximum V_c distribution for both cases (open crosses and squares, respectively) as well as for an average of observations of satellite galaxies in our Galaxy and in Andromeda (dotted error bars). *Adapted from [31].*

below). A significant fraction of halos depart from the NFW profile. These are typically not relaxed or disturbed by companions or external tidal forces.

Is there a "cusp" crisis? More recently, it was found that the inner density profile of halos can be steeper than $\beta = -1$ (e.g. [84]). However, it was shown that in the limit of resolution, β never is as steep a -1.5 [88]. The inner structure of CDM halos can be tested in principle with observations of (i) the inner rotation curves of DM dominated galaxies (Irr dwarf and LSB galaxies; the inner velocity dispersion of dSph galaxies is also being used as a test), and (ii) strong gravitational lensing and hot gas distribution in the inner regions of clusters of galaxies. Observations suggest that the DM distribution in dwarf and LSB galaxies has a roughly constant density core, in contrast to the cuspy cores of CDM halos (the literature on this subject is extensive; see for recent results [37, 50, 107, 128] and more references therein). If the observational studies confirm that halos have constant–density cores, then either astrophysical mechanisms able to expand the halo cores should work efficiently or the ΛCDM scenario should be modified. In the latter case, one of the possibilities is to introduce weakly self–interacting DM particles. For small cross sections, the interaction is effective only in the more dense inner regions of galaxies, where heat inflow may expand the core. However, the gravo–thermal catastrophe can also be triggered. In [32] it was shown that in order to avoid the gravo–thermal instability and to produce shallow cores with densities approximately constant for all masses, as suggested by observations, the DM cross section per unit of particle mass should be $\sigma_{DM}/m_X = 0.5 - 1.0v_{100}^{-1}$ cm^2/gr, where v_{100} is the relative velocity of the colliding particles in unities of 100 km/s; v_{100} is close to the halo maximum circular velocity, V_m.

The DM mass distribution was inferred from the rotation curves of dwarf and LSB galaxies under the assumptions of circular motion, halo spherical symmetry, the lack of asymmetrical drift, etc. In recent studies it was discussed that these assumptions work typically in the sense of lowering the observed inner rotation velocity [59, 100, 118]. For example, in [118] it is demonstrated that non-circular motions (due to a bar) combined with gas pressure support and projection effects systematically underestimate by up to 50% the rotation velocity of cold gas in the central 1 kpc region of their simulated dwarf galaxies, creating the illusion of a constant density core.

Mass–velocity relation. In a very simplistic analysis, it is easy to find that $M \propto V_c^3$ if the average halo density ρ_h does not depend on mass. On one hand, $V_c \propto (GM/R)^{1/2}$, and on the other hand, $\rho_h \propto M/R^3$, so that $V_c \propto M^{1/3}\rho_h^{1/6}$. Therefore, for ρ_h =const, $M \propto V_c^3$. We have seen in §3.2 that the CDM perturbations at galaxy scales have similar amplitudes (actually $\sigma_M \propto \ln M$) due to the stangexpansion effect in the radiation–dominated era. This implies that galaxy–sized perturbations collapse within a small range of epochs attaining more or less similar average densities. The CDM halos actually have a mass distribution that translates into a circular velocity profile $V_c(r)$. The maximum of this profile, V_m, is typically the circular velocity that characterizes a given halo of virial mass M. Numerical and semi–numerical results show that (ΛCDM model):

$$M \approx 5.2 \times 10^4 \left(\frac{V_m}{kms^{-1}}\right)^{3.2} \mathrm{M_\odot h^{-1}}, \tag{14}$$

Assuming that the disk infrared luminosity $L_{IR} \propto M$, and that the disk maximum rotation velocity $V_{rot,m} \propto V_m$, one obtains that $L_{IR} \propto V_{rot,m}^{3.2}$, amazingly similar to the observed infrared Tully–Fisher relation [116], one of the most robust and intriguingly correlations in the galaxy world! I conclude that this relation is a *clear imprint of the CDM power spectrum of fluctuations.*

Mass Assembling Histories

One of the key concepts of the hierarchical clustering scenario is that cosmic structures form by a process of continuous mass aggregation, opposite to the monolithic collapse scenario. The mass assembly of CDM halos is characterized by the mass aggregation history (MAH), which can alternate *smooth mass accretion* with *violent major mergers*. The MAH can be calculated by using semi–analytical approaches based on extensions of the P-S formalism. The main idea lies in the estimate of the *conditional* probability that given a collapsed region of mass M_0 at z_0, a region of mass M_1 embedded within the volume containing M_0, had collapsed at an earlier epoch z_1. This probability is calculated based on the excursion set formalism starting from a Gaussian density field characterized by an evolving mass variance σ_M[17, 73]. By using the conditional probability and random trials at each temporal step, the "backward" MAHs corresponding to a fixed mass M_0 (defined for instance at $z = 0$) can be traced. The MAHs of isolated halos by definition decrease toward the past, following different tracks (Fig. 10), sometimes with abrupt big jumps that can be identified as major mergers in the halo assembly history.

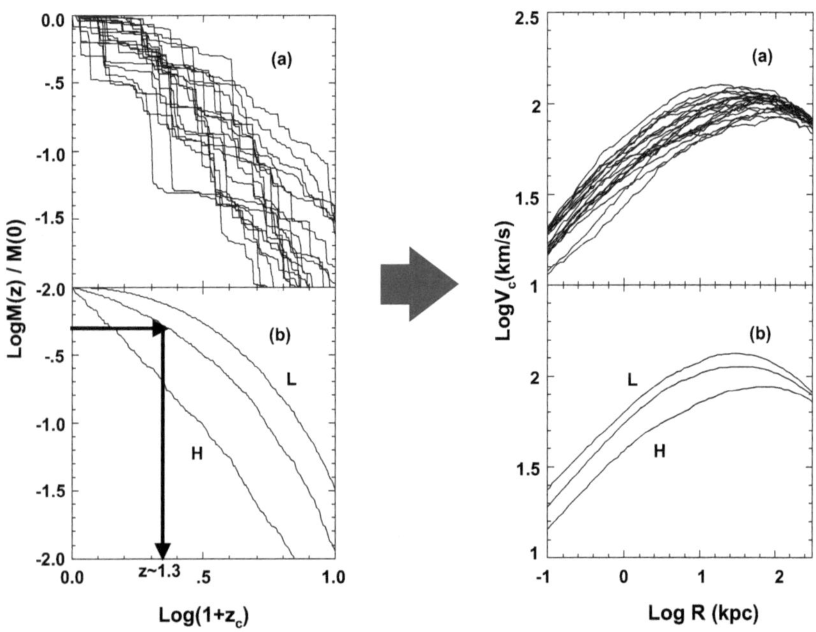

Fig. 10. *Upper panels (a).* A score of random halo MAHs for a present–day virial mass of $3.5 \times 10^{11} M_\odot$ and the corresponding circular velocity profiles of the virialized halos. *Lower panels (b).* The average MAH and two extreme deviations from 10^4 random MAHs for the same mass as in (a), and the corresponding halo circular velocity profiles. The MAHs are diverse for a given mass and the V_c (mass) distribution of the halos depend on the MAH. *Adapted from [45].*

To characterize typical behaviors of the halo MAHs, one may calculate the average MAH for a given virial mass M_0, for a given "population" of halos selected by its environment, etc. In the left panels of Fig. 10 are shown 20 individual MAHs randomly selected from 10^4 trials for $M_0 = 3.5 \times 10^{11} M_\odot$ in a ΛCDM cosmology [45]. In the bottom panel are plotted the average MAH from these 10^4 trials as well as two extreme deviations from the average. The average MAHs depend on mass: more massive halos have a more extended average MAH, i.e. they aggregate a given fraction of M_0 latter than less massive halos. It is a convention to define the typical halo formation redshift, z_f, when half of the current halo mass M_0 has been aggregated. For instance, for the ΛCDM cosmology the average MAHs show that $z_f \approx 2.2, 1.2$ and 0.7 for $M_0 = 10^{10} M_\odot, 10^{12} M_\odot$ and $10^{14} M_\odot$, respectively. A more physical definition of halo formation time is when the halo maximum circular velocity V_m attains its maximum value. After this epoch, the mass can continue growing, but the inner gravitational potential of the system is already set.

Right panels of Fig. 10 show the present–day halo circular velocity profiles, $V_c(r)$, corresponding to the MAHs plotted in the left panels. The average $V_c(r)$ is well described by the NFW profile. There is a direct relation between the MAH and the halo structure as described by $V_c(r)$ or the concentration parameter. The later the MAH, the more extended is $V_c(r)$ and the less concentrated is the halo [3, 125]. Using high–resolution simulations some authors have shown that the halo MAH presents two regimes: an early phase of fast mass aggregation (mainly by major mergers) and a late phase of slow aggregation (mainly by smooth mass accretion) [133, 75]. The potential well of a present–day halo is set mainly at the end of the fast, major–merging driven, growth phase.

From the MAHs we may infer: (i) the mass aggregation rate evolution of halos (halo mass aggregated per unit of time at different z's), and (ii) the major merging rates of halos (number of major mergers per unit of time per halo at different z's). These quantities should be closely related to the star formation rates of the galaxies formed within the halos as well as to the merging of luminous galaxies and pair galaxy statistics. By using the ΛCDM model, several studies showed that most of the mass of the present–day halos has been aggregated by accretion rather than major mergers (e.g., [85]). Major merging was more frequent in the past [55], and it is important for understanding the formation of massive galaxy spheroids and the phenomena related to this process like QSOs, supermassive black hole growth, obscured star formation bursts, etc. Both the mass aggregation rate and major merging rate histories depend strongly on environment: the denser the environment, the higher is the merging rate in the past. However, in the dense environments (group and clusters) form typically structures more massive than in the less dense regions (field and voids). Once a large structure virializes, the smaller, galaxy–sized halos become subhalos with high velocity dispersions: the mass growth of the subhalos is truncated, or even reversed due to tidal stripping, and the merging probability strongly decreases. Halo assembling (and therefore, galaxy assembling) definitively depends on environment. Overall, by integrating the MAHs of the whole galaxy–sized ΛCDM halo population in a given volume, the general result is that the peak in halo assembling activity was at $z \approx 1-2$. After these redshifts, the global mass aggregation rate strongly decreases (e.g., [121].

To illustrate the driving role of DM processes in galaxy evolution, I mention briefly here two concrete examples:

1). Distributions of present–day specific mass aggregation rate, $(\dot{M}/M)_0$*, and halo lookback formation time,* $T_{1/2}$. For a ΛCDM model, these distributions are bimodal, in particular the former. We have found that roughly 40% of halos (masses larger than $\approx 10^{11}\mathrm{M}_\odot \mathrm{h}^{-1}$) have $(\dot{M}/M)_0 \leq 0$; they are basically subhalos. The remaining 60% present a broad distribution of $(\dot{M}/M)_0 > 0$ peaked at $\approx 0.04\mathrm{Gyr}^{-1}$. Moreover, this bimodality strongly changes with large–scale environment: the denser is the environment the, higher is the fraction of halos with $(\dot{M}/M)_0 \leq 0$. It is interesting enough that similar fractions and dependences on environment are found for the specific star formation rates of galaxies in large statistical surveys (§§2.3); the situation is similar when confronting the distributions of $T_{1/2}$ and observed colors. Therefore, it seems that the *the main driver of the observed bimodalities in $z = 0$ specific star formation rate and color of galaxies is the nature of the CDM halo mass aggregation process.* Astrophysical processes of course are important but the main body of the bimodalities can be explained just at the level of DM processes.

2. Major merging rates. The observational inference of galaxy major merging rates is not an easy task. The two commonly used methods are based on the statistics of galaxy pairs (pre–mergers) and in the morphological distortions of ellipticals (post–mergers). The results show that the merging rate increases as $(1+z)^x$, with $x \sim 0-4$. The predicted major merging rates in the ΛCDM scenario agree roughly with those inferred from statistics of galaxy pairs. From the fraction of normal galaxies in close companions (with separations less than 50 kpch^{-1}) inferred from observations at $z = 0$ and $z = 0.3$ [91], and assuming an average merging time of ~ 1 Gyr for these separations, we estimate that the major merging rate at the present epoch is $\sim 0.01\ \mathrm{Gyr}^{-1}$ for halos in the range of $0.1-2.0\ 10^{12}\mathrm{M}_\odot$, while at $z = 0.3$ the rate increased to $\sim 0.018\ \mathrm{Gyr}^{-1}$. These values are only slightly lower than predictions for the ΛCDM model.

Angular Momentum

The origin of the angular momentum (AM) is a key ingredient in theories of galaxy formation. Two mechanisms of AM acquirement were proposed for the CDM halos (e.g., [93, 23, 78]): 1. tidal torques of the surrounding shear field when the perturbation is still in the linear regime, and 2. transfer of orbital AM to internal AM in major and minor mergers of collapsed halos. The angular momentum of DM halos is parametrized in terms of the dimensionless spin parameter $\lambda \equiv J\sqrt{E}/(GM^{5/2}$, where J is the modulus of the total angular momentum and E is the total (kinetic plus potential). It is easy to show that λ can be interpreted as the level of rotational support of a gravitational system, $\lambda = \omega/\omega_{sup}$, where ω is the angular velocity of the system and ω_{sup} is the angular velocity needed for the system to be rotationally supported against gravity (see [90]).

For disk and elliptical galaxies, $\lambda \sim 0.4-0.8$ and $\sim 0.01-0.05$, respectively. Cosmological N–body simulations showed that the CDM halo spin parameter is log–normal distributed, with a median value $\lambda \approx 0.04$ and a standard deviation $\sigma_\lambda \approx 0.5$; this distribution is almost independent from cosmology. A related quantity, but more straightforward to compute is $\lambda' \equiv \frac{J}{\sqrt{2}MV_vR_v}$ [23], where R_v is the virial radius and V_v the circular velocity at this radius. Recent simulations show that $(\lambda', \sigma_{\lambda'}) \approx (0.035, 0.6)$, though some variations with environment and mass are measured [5]. The evolution of the spin parameter depends on the AM acquirement mechanism. In general, a significant systematical change of λ with time is not expected, but relatively strong changes are measured in short time steps, mainly after merging of halos, when λ increases.

How is the internal AM distribution in CDM halos? Bullock et al. [23] found that in most of cases this distribution can be described by a simple (universal) two–parameter function that departs significantly from the solid–body rotation distribution. In addition, the spatial distribution of AM in CDM halos tends to be cylindrical, being well aligned for 80% of the halos, and misaligned at different levels for the rest. The mass distribution of the galaxies formed within CDM halos, under the assumption of specific AM conservation, is established by λ, the halo AM distribution, and its alignment.

4.2 Non–baryonic Dark Matter Candidates

The non–baryonic DM required in cosmology to explain observations and cosmic structure formation should be in form of elemental or scalar field particles or early formed quark nuggets. Modifications to fundamental physical theories (modified Newtonian Dynamics, extra–dimensions, etc.) are also plausible if DM is not discovered.

There are several docens of predicted elemental particles as DM candidates. The list is reduced if we focus only on well–motivated exotic particles from the point of view of particle physics theory alone (see for a recent review [53]). The most popular particles beyond the standard model are the *supersymmetric (SUSY)* particles in supersymmetric extensions of the Standard Model of particle physics. Supersymmetry is a new symmetry of space–time introduced in the process of unifying the fundamental forces of nature (including gravity). An excellent CDM candidate is the lightest stable SUSY particle under the requirement that superpartners are only produced or destroyed in pairs (called R-parity conservation). This particle called *neutralino* is weakly interacting and massive (WIMP). Other SUSY particles are the gravitino and the sneutrino; they are of WDM type. The predicted masses for neutralino range from $\sim$ 30 to 5000 GeV. The cosmological density of neutralino (and of other thermal WIMPs) is naturally as required when their interaction cross section is of the order of a weak cross section. The latter gives the possibility to detect neutralinos in laboratory.

The possible discovery of WIMPs relies on two main techniques:

(i) Direct detections. The WIMP interactions with nuclei (elastic scattering) in ultra–low–background terrestrial targets may deposit a tiny amount of energy (< 50 keV) in the target material; this kinetic energy of the recoiling nucleus is converted partly into scintillation light or ionization energy and partly into thermal energy. Dozens of experiments worldwide -of cryogenic or scintillator type, placed in mines or underground laboratories, attempt to measure these energies. Predicted event rates for neutralinos range from 10^{-6} to 10 events per kilogram detector material and day. The nuclear recoil spectrum is featureless, but depends on the WIMP and target nucleus mass. To convincingly detect a WIMP signal, a specific signature from the galactic halo particles is important. The Earth's motion through the galaxy induces both a seasonal variation of the total event rate and a forward–backward asymmetry in a directional signal. The detection of structures in the dark velocity space, as those predicted to be produced by the Sagittarius stream, is also an specific signature from the Galactic halo; directional detectors are needed to measure this kind of signatures.

The DAMA collaboration reported a possible detection of WIMP particles obeying the seasonal variation; the most probable value of the WIMP mass was ~ 60 GeV. However, the interpretation of the detected signal as WIMP particles is controversial. The sensitivity of current experiments (e.g., CDMS and EDEL-WEISS) limit already the WIMP–proton spin–independent cross sections to values $\lesssim 2\ 10^{-42} - 10^{-40}\mathrm{cm}^{-2}$ for the range of masses $\sim 50 - 10^4$ GeV, respectively; for smaller masses, the cross–section sensitivities are larger, and WIMP signals were not detected. Future experiments will be able to test the regions in the cross-section–WIMP mass diagram, where most of models make certain predictions.

(ii) Indirect detections. We can search for WIMPS by looking for the products of their annihilation. The flux of annihilation products is proportional to the square of the WIMP density, thus regions of interest are those where the WIMP concentration is relatively high. There are three types of searches according to the place where WIMP annihilation occur: (i) in the Sun or the Earth, which gives rise to a signal in high-energy neutrinos; (ii) in the galactic halo, or in the halo of external galaxies, which generates $\gamma-$rays and other cosmic rays such as positrons and antiprotons; (iii) around black holes, specially around the black hole at the Galactic Center. The predicted radiation fluxes depend on the particle physics model used to predict the WIMP candidate and on astrophysical quantities such as the dark matter halo structure, the presence of sub–structure, and the galactic cosmic ray diffusion model.

Most of WIMPS were in thermal equilibrium in the early Universe (thermal relics). Particles which were produced by a non-thermal mechanism and that never had the chance of reaching thermal equilibrium are called non-thermal relics (e.g., axions, solitons produced in phase transitions, WIMPZILLAs produced gravitationally at the end of inflation). From the side of WDM, the most popular candidate are the $\sim$ 1KeV sterile neutrinos. A sterile neutrino is a fermion that has no standard model interactions other than a coupling to the standard neutrinos through their mass generation mechanism. Cosmological probes, mainly the power spectrum of Lyα forest at high redshifts, constrain the mass of the sterile neutrino to values larger than $\sim$ 2KeV.

5 The Bright Side of Galaxy Formation and Evolution

The ΛCDM scenario of cosmic structure formation has been well tested for perturbations that are still in the linear or quasilinear phase of evolution. These tests are based, among other cosmological probes, on accurate measurements of:

- the CMBR temperature fluctuations at large and small angular scales
- the large–scale mass power spectrum as traced by the spatial distribution of galaxies and cluster of galaxies, by the Lyα forest clouds, by maps of gravitational weak and strong lensing, etc.
- the peculiar large–scale motions of galaxies[13].
- the statistics of strong gravitational lensing (multiple–lensed arcs).

Although these cosmological probes are based on observations of luminous (baryonic) objects, the physics of baryons plays a minor or indirect role in the properties of the linear mass perturbations. The situation is different at small (galaxy) scales, where perturbations went into the non–linear regime and the dissipative physics of baryons becomes relevant. The interplay of DM and baryonic processes is crucial for understanding galaxy formation and evolution. The progress in this field was mostly heuristic; the ΛCDM scenario provides the initial and boundary conditions for modeling galaxy evolution, but the complex physics of the baryonic processes, in the absence of fundamental theories, requires a model adjustment through confrontation with the observations.

Following, I will outline some key concepts, ingredients, and results of the galaxy evolution study based on the ΛCDM scenario. Some of the pioneer papers in this field are those of Gunn [57], White & Reese [131], Fall & Efstathiou [43], Blumental et al. [15], Davis et al. [36], Katz & Gunn [65], White & Frenk [130], Kauffmann et al. [66]. For useful lecture notes and recent reviews see e.g., Longair [76, 77], White [129], Steinmetz [113], Firmani & Avila-Reese [46].

[13] Recall that linear theory relates the peculiar velocity, that is the velocity deviation from the Hubble flow, to the density contrast. It is said that the cosmological velocity field is *potential*; any primordial rotational motion able to give rise to a density perturbation decays as the Universe expands due to angular momentum conservation.

The main methods of studying galaxy formation and evolution in the ΛCDM context are:

• Semi-analytical Models (e.g., [130, 66, 28, 9, 108, 29, 12, 10]), where the halo mass assembling histories are calculated with the extended Press–Schechter formalism and galaxies are seeded within the halos by means of phenomenological recipes. This method is very useful for producing whole populations of galaxies at a given epoch and predicting statistical properties as the luminosity function and the morphological mix.

• Semi-numerical Models (e.g, [45, 2, 119, 16]), where the internal physics of the galaxies, including those of the halos, are modeled numerically but under simplifying assumptions; the initial and boundary conditions are taken from the ΛCDM scenario by using the extended Press–Schechter formalism and halo AM distributions from simulations. This method is useful to predict the local properties of galaxies and correlations among the global properties, as well as to follow the overall evolution of individual galaxies.

• Numerical N–body+hydrodynamical simulations (e.g., [65, 27, 64, 86, 112, 126, 1, 110, 56]), where the DM and baryonic processes are followed in cosmological simulations. This is the most advanced and complete approach to galaxy evolution. However, current limitations in the computational capabilities and the lack of fundamental theories for several of the physical processes involved, do not allow yet to exploit optimally this method. A great advance is being made currently with an hybrid approach: in the high–resolution cosmological N–body simulations of only DM, galaxies are grafted by using the semi–analytical models (e.g., [67, 60, 38, 13, 111, 63]).

5.1 Disks

The formation of galaxy disks deep inside the CDM halos is a generic process in the ΛCDM scenario. Let us outline the (simplified) steps of disk galaxy formation in this scenario:

1. DM halo growth. The "mold" for disk formation is provided by the mass and AM distributions of the virialized halo, which grows hierarchically. A description of these aspects were presented in the previous Section.

2. Gas cooling and infall, and the maximum mass of galaxies. It is common to assume that the gas in a halo is shock–heated during collapse to the virial temperature [131]. The gas then cools radiatively and falls in a free–fall time to the center. The cooling function $\Lambda(n, T_k; Z)$ depends on the gas density, temperature, and composition[14]. Since the seminal work by White & Frenk (1990) [130], the rate infall of gas available to form the galaxy is assumed to

[14] The main cooling processes for the intrahalo gas are collisional excitation and ionization, recombination, and bremsstrahlung. The former is the most efficient for kinetic temperatures $T_k \approx 10^4 - 10^5$K and for neutral hydrogen and single ionized helium; for a meta–enriched gas, cooling is efficient at temperatures between $10^5 - 10^7$K. At higher temperatures, where the gas is completely ionized, the

be driven either by the free–fall time, t_{ff}, if $t_{ff} > t_{cool}$ or by the cooling time t_{cool} if $t_{ff} < t_{cool}$. The former case applies to halos of masses smaller than approximately $5\times10^{11}M_{\odot}$, whilst the latter applies to more massive halos. The cooling flow from the quasistatic hot atmosphere *is the process that basically limits the baryonic mass of galaxies* [105], and therefore the bright end of the galaxy luminosity function; for the outer, dilute hot gas in large halos, t_{cool} becomes larger than the Hubble time. However, detailed calculations show that even so, in massive halos too much gas cools, and the bright end of the predicted luminosity function results with a decrease slower than the observed one [12]. Below we will see some solutions proposed to this problem.

More recently it was shown that the cooling of gas trapped in filaments during the halo collapse may be so rapid that the gas flows along the filaments to the center, thus avoiding shock heating [69]. However, this process is efficient only for halos less massive than $2.5\times10^{11}M_{\odot}$, which in any case (even if shock–heating happens), cool their gas very rapidly [19]. Thus, for modeling the formation of disks, and for masses smaller than $\sim 5\times10^{11}M_{\odot}$, we may assume that gas infalls in a dynamical time since the halo has virialized, or in two dynamical times since the protostructure was at its maximum expansion.

3. Disk formation, the origin of exponentially, and rotation curves. The gas, originally distributed in mass and AM as the DM, cools and collapses until it reaches centrifugal balance in a disk. Therefore, assuming detailed AM conservation, the radial mass distribution of the disk can be calculated by equating its specific AM to the AM of its final circular orbit in centrifugal equilibrium. The typical collapse factor of the gas within a DM halo is $\sim 10-15$[15], depending on the initial halo spin parameter λ; the higher the λ, the more extended (lower surface density) is the resulting disk. The surface density profile of the disks formed within CDM halos is nearly exponential, which provides an explanation to the long–standing question of why galaxy disks are exponential. This is a direct consequence of the AM distribution acquired by the halos by tidal torques and mergers. In more detail, however, the profiles are more concentrated in the center and with a slight excess in the periphery than the exponential law [45, 23]. The cusp in the central disk could give rise to either a photometrical bulge [120] or to a real kinematical bulge due to disk gravitational instability enhanced by the higher central surface density [2] (bulge secular formation). In a few cases (high–λ, low–concentrated halos), purely exponential disks can be formed.

Baryons are a small mass fraction in the CDM halos, however, the disk formed in the center is very dense (recall the high collapse factors), so that

---

dominant cooling process is bremsstrahlung. At temperatures lower than 10^4K (small halos) and in absence of metals, the main cooling process is by H_2 and HD molecule line emission.

15 It is interesting to note that in the absence of a massive halo around galaxies, the collapse factor would be larger by $\sim M/M_d \approx 20$, where M and M_d are the total halo and disk masses, respectively [90].

the contribution of the baryonic disk to the inner gravitational potential is important or even dominant. The formed disk will drag gravitationally DM, producing an inner halo contraction that is important to calculate for obtaining the rotation curve decomposition. The method commonly used to calculate it is based on the approximation of radial adiabatic invariance, where spherical symmetry and circular orbits are assumed (e.g., [47, 82]). However, the orbits in CDM halos obtained in N–body simulations are elliptical rather than circular; by generalizing the adiabatic invariance to elliptical orbits, the halo contraction becomes less efficient [132, 52].

The rotation curve decomposition of disks within contracted ΛCDM halos are in general consistent with observations [82, 45, 132] (nearly–flat total rotation curves; maximum disk for high–surface brightness disks; submaximum disk for the LSB disks; in more detail, the outer rotation curve shape depends on surface density, going from decreasing to increasing at the disk radius for higher to lower densities, respectively). However, there are important non–solved issues. For example, from a large sample of observed rotation curves, Persic et al. [95] inferred that the rotation curve shapes are described by an "universal" profile that (i) depends on the galaxy luminosity and (ii) implies a halo profile different from the CDM (NFW) profile. Other studies confirm only partially these claims [123, 132, 26]. Statistical studies of rotation curves are very important for testing the ΛCDM scenario.

In general, the structure and dynamics of disks formed within ΛCDM halos under the assumption of detailed AM conservation seem to be consistent with observations. An important result to remark is the successful prediction of the infrared Tully–Fisher relation and its scatter[16]. The core problem mentioned in §4.2 is the most serious potential difficulty. Other potential difficulties are: (i) the predicted disk size (surface brightness) distribution implies a $P(\lambda)$ distribution narrower than that corresponding to ΛCDM halos by almost a factor of two [74]; (ii) the internal AM distribution inferred from observations of dwarf galaxies seems not to be in agreement with the ΛCDM halo AM distribution [122]; (iii) the inference of the halo profile from the statistical study of rotation curve shapes seems not to be agreement with CMD halos. In N–body+hydrodynamical simulations of disk galaxy formation there was common another difficulty called the 'angular momentum catastrophe': the simulated disks ended too much concentrated, apparently due to AM transference of baryons to DM during the gas collapse. The formation of highly concentrated disks also affects the shape of the rotation curve (strongly decreasing), as well as the zero–point of the Tully–Fisher relation. Recent numerical

[16] In §4.1 we have shown that the basis of the Tully–Fisher relation is the CDM halo $M - V_m$ relation. From the pure halo to the disk+halo system there are several intermediate processes that could distort the original $M - V_m$ relation. However, it was shown that the way in which the CDM halo couples with the disk and the way galaxies transform their gas into stars "conspire" to keep the relation. Due to this conspiring, the Tully–Fisher relation is robust to variations in the baryon fraction f_B (or mass–to–luminosity ratios) and in the spin parameter λ [45].

simulations are showing that the 'angular momentum catastrophe', rather than a physical problem, is a problem related to the resolution of the simulations and the correct inclusion of feedback effects.

4. Star formation and feedback. We are coming to the less understood and most complicated aspects of the models of galaxy evolution, which deserve separate notes. The star formation (SF) process is studied at two levels (each one by two separated communities!): (i) the small–scale physics, related to the complex processes by which the cold gas inside molecular clouds fragments and collapses into stars, and (ii) the large–scale physics, related to the disk global instabilities that give rise to the largest unities of SF, the molecular clouds. The SF physics incorporated to galaxy evolution models is still oversimplified, phenomenological and refers to the latter item. The large-scale SF cycle in normal galaxies is believed to be self–regulated by a balance between the energy injection due to SF (mainly SNe) and dissipation (radiative or turbulent). Two main approaches have been used to describe the SF self–regulation in models of galaxy evolution: **(a)** the halo cooling-feedback approach [130]), **(b)** the disk turbulent ISM approach [44, 124].

According to the former, the cool gas is reheated by the "galaxy" SF feedback and driven back to the *intrahalo medium* until it again cools radiatively and collapses into the galaxy. This approach has been used in semi–analytical models of galaxy formation where the internal structure and hydrodynamics of the disks are not treated in detail. The reheating rate is assumed to depend on the halo circular velocity V_c: $\dot{M}_{rh} \propto \dot{M}_s/V_c^{\alpha}$, where $\dot{M}_s$ is the SF rate (SFR) and $\alpha \geq 2$. Thus, the galaxy SFR, gas fraction and luminosity depend on V_c. In these models, the disk ISM is virtually ignored and the SN–energy injection is assumed to be as efficient as to reheat the cold gas up to the virial temperature of the halo. A drawback of the model is that it predicts hot X-ray halos around disk galaxies much more luminous than those observed.

Approach (b) is more appropriate for models where the internal processes of the disk are considered. In this approach, the SF at a given radius r is assumed to be triggered by disk gravitational instabilities (Toomre criterion) and self–regulated by a balance between energy injection (mainly by SNe) and dissipation in the turbulent ISM in the direction perpendicular to the disk plane:

$$Q_g(r) \equiv \frac{v_g(r)\kappa(r)}{\pi G \Sigma_g(r)} < Q_{crit} \tag{15}$$

$$\gamma_{SN}\epsilon_{SN}\dot{\Sigma}_*(r) + \dot{\Sigma}_{E,accr}(r) = \frac{\Sigma_g(r)v_g^2(r)}{2t_d(r)}, \tag{16}$$

where v_g and Σ_g are the gas velocity dispersion and surface density, κ is the epicyclic frequency, Q_{crit} is a critical value for instability, γ_{SN} and ϵ_{SN} are the kinetic energy injection efficiency of the SN into the gas and the SN energy generated per gram of gas transformed into stars, respectively, $\dot{\Sigma}_*$ is the surface SFR, and $\dot{\Sigma}_{E,accr}$ is the kinetic energy input due to mass accretion rate (or eventually any other energy source as AGN feedback). The key parameter in the self–regulating process is the dissipation time t_d. The disk ISM is a turbulent, non-isothermal, multi-temperature flow. Turbulent dissipation in the ISM is typically efficient ($t_d \sim 10^7 - 10^8$yr) in such a way that self–regulation happens at the characteristic vertical scales of the disk. Thus, there is not too much room for strong feedback with the gas at heights larger than the vertical scaleheigth of normal present–day disks: self–regulation is at the level of the disk, but not at the level of the gas corona around. With this approach the predicted SFR is proportional to Σ_g^n (Schmidt law), with $n \approx 1.4 - 2$ varying along the disk, in good agreement with observational inferences. The typical SF timescales are not longer than $3 - 4$Gyr. Therefore, to keep active SFRs in the disks, gas infall is necessary, a condition perfectly fulfilled in the ΛCDM scenario.

Given the SFR radius by radius and time by time, and assuming an IMF, the corresponding luminosities in different color bands can be calculated with stellar population synthesis models. The final result is then an evolving inside–out luminous disk with defined global and local colors.

5. Secular Evolution The "quiet" evolution of galaxy disks as described above can be disturbed by minor mergers (satellite accretion) and interactions with close galaxy companions. However, as several studies have shown, the disk may suffer even intrinsic instabilities which lead to secular changes in its structure, dynamics, and SFR. The main effects of secular evolution, i.e. dynamical processes that act in a timescale longer than the disk dynamical time, are the vertical thickening and "heating" of the disk, the formation of bars, which are efficient mechanisms of radial AM and mass redistribution, and the possible formation of (pseudo)bulges (see for recent reviews [71, 33]). Models of disk galaxy evolution should include these processes, which also can affect disk properties, for example increasing the disk scale radii [117].

5.2 Spheroids

As mentioned in §2, the simple appearance, the dominant old stellar populations, the α–elements enhancement, and the dynamically hot structure of spheroids suggest that they were formed by an early ($z \gtrsim 4$) single violent event with a strong burst of star formation, followed by passive evolution of their stellar population (*monolithic* mechanism). Nevertheless, both observations and theory point out to a more complex situation. There are two ways to define the formation epoch of a spheroid: when most of its stars formed or when the stellar spheroid acquired its dynamical properties in violent or secular processes. For the monolithic collapse mechanism both epochs coincide.

In the context of the ΛCDM scenario, spheroids are expected to be formed basically as the result of major mergers of disks. However,

- if the major mergers occur at high redshifts, when the disks are mostly gaseous, then the situation is close to the monolithic collapse;
- if the major mergers occur at low redshifts, when the galaxies have already transformed a large fraction of their gas into stars, then the spheroids assemble by the "classical" dissipationless collision.

Besides, stellar disks may develop spheroids in their centers (bulges) by secular evolution mechanisms, both intrinsic or enhanced by minor mergers and interactions; this channel of spheroid formation should work for late–type galaxies and it is supported by a large body of observations [71]. But the picture is even more complex in the hierarchical cosmogony as galaxy morphology may be continuously changing, depending on the MAH (smooth accretion and violent mergers) and environment. An spheroid formed early should continue accreting gas so that a new, younger disk grows around. A naive expectation in the context of the ΛCDM scenario is that massive elliptical galaxies should be assembled mainly by late major mergers of the smaller galaxies in the hierarchy. It is also expected that the disks in galaxies with small bulge–to–disk ratios should be on average redder than those in galaxies with large bulge–to–disk ratios, contrary to observations.

Although it is currently subject of debate, a more elaborate picture of spheroid formation is emerging now in the context of the ΛCDM hierarchical scenario (see [106, 46, 39] and the references therein). The basic ideas are that massive ellipticals formed early ($z \gtrsim 3$) and in a short timescale by the merging of gas–rich disks in rare high–peak, clustered regions of the Universe. The complex physics of the merging implies (i) an ultraluminous burst of SF obscured by dust (cool ULIRG phase) and the establishment of a spheroidal structure, (ii) gas collapse to the center, a situation that favors the growth of the preexisting massive black hole(s) through an Eddington or even super–Eddington regime (warm ULIRG phase), (iii) the switch on of the AGN activity associated to the supermassive black hole when reaching a critical mass, reverting then the gas inflow to gas outflow (QSO phase), (iv) the switch off of the AGN activity leaving a giant stellar spheroid with a supermassive black hole in the center and a hot gas corona around (passive elliptical evolution). In principle, the hot corona may cool by cooling flows and increase the mass of the galaxy, likely renewing a disk around the spheroid. However, it seems that recurrent AGN phases (less energetic than the initial QSO phase) are possible during the life of the spheroid. Therefore, the energy injected from AGN in the form of radio jets (feedback) can be responsible for avoiding the cooling flow. This way is solved the problem of disk formation around the elliptical, as well as the problem of the extended bright end in the luminosity function. It is also important to note that as soon as the halo hosting the elliptical becomes a subhalo of the group or cluster, the MAH is truncated (§4). According to the model just described, massive elliptical galaxies were in place at high redshifts, while less massive galaxies (collapsing from more common density peaks) assembled later. This model was called *downsizing* or anti-hierarchical. In spite of the name, it fits perfectly within the hierarchical ΛCDM scenario.

5.3 Drivers of the Hubble Sequence

• Disks are generic objects formed by gas dissipation and collapse inside the growing CDM halos. Three (cosmological) initial and boundary conditions related to the halos define the main properties of disks in isolated halos:

1. The virial mass, which determines extensive properties
2. The spin parameter λ, which determines mainly the disk surface brightness (SB; it gives rise to the sequence from high SB to low SB disks) and strongly influences the rotation curve shape and the bulge–to–disk ratio (within the secular scenario). λ also plays some role in the SFR history.
3. The MAH, which drives the gas infall rate and, therefore, the disk SFR and color; the MAH determines also the halo concentration, and its scatter is reflected in the scatter of the Tully–Fisher relation.

The two latter determine the intensive properties of disks, suggesting a biparametrical sequence in SB and color. There is a fourth important parameter, the galaxy baryon fraction f_B, which influences the disk SB and rotation curve shape. We have seen that f_B in galaxies is 3–5 times lower than the universal Ω_B/Ω_{DM} fraction. This parameter is related probably to astrophysical processes as gas dissipation and feedback.

• The clustering of CDM halos follows an spatial distribution with very different large–scale environments. In low–density environments, halos live mostly isolated, favoring the formation of disks, whose properties are driven by the factors mentioned above. However, as we move to higher–density environments, halos form from more and more clustered high–peak perturbations that assemble early by violent major mergers: this is the necessary condition to form massive ellipticals. At some time, the larger scale in the hierarchy collapses and the halo becomes a subhalo: the mass aggregation is then truncated and the probability of merging decreases dramatically. Elliptical galaxies are settled and continue evolving passively. Thus, the environment of CDM halos is another important driver of the Hubble sequence, able to establish the main body of the observed blue–red and early–type morphology sequences and their dependences on density.

• Although the initial, boundary and environmental conditions provided by the ΛCDM scenario are drivers of several of the main properties and correlations of galaxies, astrophysical processes should also play an important role. The driving astrophysical processes are global SF and feedback. They should come in two modes that drive the disk and elliptical sequences: (i) the quiescent disk mode, where disk instabilities trigger SF and local (negative) feedback self–regulates the SFR, and (ii) the bursting mode of violent mergers of gaseous galaxies, where local shocks and gravothermal catastrophe trigger SF, and presumably a positive feedback increases its efficiency. Other important astrophysical drivers of galaxy properties are: (i) the SN–induced wind–driven outflows, which are important to shape the properties of dwarf galaxies ($M \lesssim 10^{10} M_\odot$, $V_m \lesssim 80$km/s), (ii) the AGN–induced hydrodynamical outflows, which are important to prevent cooling flows in massive ellipticals, (iii) several processes typical of high–density environments such as ram pressure, harassment, strangulation, etc., presumably important to shape some properties of galaxies in clusters.

6 Issues and Outlook

Our understanding of galaxy formation and evolution is in its infancy. So far, only the first steps were given in the direction of consolidating a theory in this field. The process is apparently so complex and non–linear that several specialists do not expect the emergence of a theory in the sense that a few driving parameters and factors might explain the main body of observations. Instead, the most popular trend now is to attain some description of galaxy evolution by simulating it in expensive computational runs. I believe that simulations are a valuable tool to extend a bridge between reality and the distorted (biased) information given by observations. However, the search of basic theories for explaining galaxy formation and evolution should not be replaced by the only effort of simulating in detail what in fact we want to get. The power of science lies in its predictive capability. Besides, if galaxy theory becomes predictive, then its potential to test fundamental and cosmological theories will be enormous.

Along this notes, potential difficulties or unsolved problems of the ΛCDM scenario were discussed. Now I summarize and complement them:

Physics

• What is non–baryonic DM? From the structure formation side, the preferred (and necessary!) type is CDM, though WDM with filtering masses below $\sim 10^9 M_\odot$ is also acceptable. So far none of the well–motivated cold or warm non–baryonic particles have been detected in Earth experiments. The situation is even worth for proposals not based on elemental particles as DM from extra–dimensions.

• What is Dark Energy? Dark Energy does not play apparently a significant role in the internal evolution of perturbations but it crucially defines the cosmic timescale and expansion rate, which are important for the growing factor of perturbations. The simplest interpretation of Dark Energy is the homogeneous and inert cosmological constant Λ, with equation of state parameter $w = -1$ and $\rho_\Lambda =$const. The combinations of different cosmological probes tend to favor the flat-geometry Λ models with $(\Omega_M, \Omega_\Lambda) \approx (0.26, 0.74)$. However, the cosmological constant explanation of Dark Energy faces serious theoretical problems. Several alternatives to Λ were proposed to ameliorate partially these problems (e.g. quintaessence, k–essence, Chaplygin gas, etc.). Also have been proposed unifying schemes of DM and Dark Energy through scalar fields (e.g, [81]).

Cosmology

• Inflation provides a natural mechanism for the generation of primordial fluctuations. The nearly scale–invariance of the primordial power spectrum is well predicted by several inflation models, but its amplitude, rather than being predicted, is empirically inferred from observations of CMBR anisotropies. Another aspect of primordial fluctuations not well understood is related to their statistics, i.e., whether they are Gaussian–distributed or not. And this is crucial for cosmic structure formation.

• Indirect pieces of evidence are consistent with the main predictions of inflation regarding primordial fluctuations. However, more direct tests of this theory are highly desirable. Hopefully, CMBR anisotropy observations will allow for some more direct tests (e.g., effects from primordial gravitational waves).

Astrophysics

• Issues at small scales. The excess of substructure (satellite galaxies) can be apparently solved by inhibition of galaxy formation in small halos due to UV–radiation produced by reionization and due to feedback, rather than to modifications to the scenario (e.g., the introduction of WDM). Observational inferences of the inner volume and phase–space densities of dwarf satellite galaxies are crucial to explore this question. The direct detection (with gravitational lensing) of the numerous subhalo (dark galaxy) population predicted by CDM for the Galaxy halo is a decisive test on the problem of substructure. The CDM prediction of cuspy halos is a more involved problem when confronting it with observational inferences. If the disagreement persists, then either the ΛCDM scenario will need a modification (e.g., introduction of self–interaction or annihilation), or astrophysical processes involving gas baryon physics should be in action. However, there are still unsolved issues at the intermediate level: for example, the central halo density profile of galaxies is inferred from observations of inner rotation curves *under* several assumptions that could be incorrect. An interesting technique to overcome this problem is being currently developed: to simulate as realistically as possible a given galaxy, "observe" its rotation curve and then compare with that of the real galaxy (see §§4.1).

• The early formation of massive red elliptical galaxies can be accommodated in the hierarchical ΛCDM scenario (§§5.2) *if* spheroids are produced by the major merger of gaseous disks, and if the cold gas is transformed rapidly into stars during the merger in a dynamical time or so. Both conditions should be demonstrated, in particular the latter. A kind of positive feedback seems to be necessary for such an efficient star formation rate (ISM shocks produced by the jets generated in the vicinity of supermassive black holes?).

• Once the elliptical has formed early, the next difficulty is how to avoid further (disk) growth around it. The problem can be partially solved by considering that ellipticals form typically in dense, clustered environments, and at some time they become substructures of larger virialized groups or clusters, truncating any possible accretion to the halo/galaxy. However, (i) galaxy halos, even in clusters, are filled with a reservoir of gas, and (ii) there are some ellipticals in the field. Therefore, negative feedback mechanisms are needed to stop gas cooling and accretion. AGN–triggered radio jets have been proposed as a possible mechanism, but further investigation is necessary.

• The merging mechanism of bulge formation within the hierarchical model implies roughly bluer (later formed) disks as the bulge–to–disk ratio is larger, contrary to the observed trend. The secular scenario could solve this problem but it is not still clear whether bars disolve or not in favor of pseudobulges. It is not clear also if the secular scenario could predict the central supermassive black hole mass–velocity dispersion relation.

• We lack a fundamental theory of star formation. So far, simple models, or even just phenomenological recipes, have been used in galaxy formation studies. The two proposed modes of star formation (the quiescent, inefficient, disk self–regulated regime, and the violent efficient star–bursting regime in mergers) are oversimplifications of a much more complex problem with more physical mechanisms (shocks, turbulence, etc.). Closely related to star formation is the problem of feedback. The feedback mechanisms are different in the ISM of disks, in the gaseous medium of merging galaxies with a powerful energy source (the AGN) other than stars, and in the diluted and hot intrahalo medium around galaxies.

• We have seen in §§2.2 that at the present epoch only $\approx 9\%$ of baryons are within virialized structures. Where are the remaining 91% of the baryons? The fraction of particles in halos measured in ΛCDM N–body cosmological simulations is $\sim 50\%$. This sounds good but still we have to explain, within the ΛCDM scenario, the $\sim 40\%$ of missing baryons. The question is were these baryons never trapped by collapsed halos or were they trapped but later expelled due to galaxy feedback. Large–scale N–body+hydrodynamical simulations have shown that the gravitational collapse of filaments may heat the gas and keep a big fraction of baryons outside the collapsed halos [35]. Nevertheless, feedback mechanisms, especially at high redshifts, are also predicted to be strong enough as to expel enriched gas back to the Intergalactic Medium. The problem is open.

The field has plenty of open and exciting problems. The ΛCDM scenario has survived many observational tests but it still faces the difficulties typical of a theory constructed phenomenologically and heuristically. Even if in the future it is demonstrated that CDM does not exist (which is little probable), the ΛCDM scenario would serve as an excellent "fitting" model to reality, which would strongly help researchers in developing new theories.

Acknowledgments

I am in debt with Dr. I. Alcántara-Ayala and R. Núñez-López for their help in the preparation of the figures. I am also grateful to J. Brenda for grammar corrections, and to the Editors for their infinite patience.

References

1. Abadi, M. G. et al. 2003, ApJ, 591, 499
2. Avila-Reese, V., & Firmani, C. 2000, RevMexAA, 36, 23
3. Avila-Reese, V., Firmani, C. & Hernández, X. 1998, ApJ, 505, 37
4. Avila-Reese, V. et al. 2003, ApJ, 598, 36
5. Avila-Reese, V. et al. 2005, ApJ, 634, 51
6. Balogh, M. L. et al. 2004, ApJ, 615, L101
7. Ballesteros–Paredes, J. et al. 2006, in "Protostars and Planets V", in press (astro-ph/0603357)
8. Bartolo, N. et al. 2004, Phys.Rep., 402, 103
9. Baugh, C.M., Cole, S., & Frenk, C.S. 1996, MNRAS, 283, 136
10. Baugh, C.M. et al. 2005, MNRAS, 356, 1191
11. Bell E. F. et al. 2003, ApJSS, 149, 289
12. Benson, A.J. et al. 2002, ApJ, 599, 38
13. Berlind, A. A. et al. 2005, ApJ, 629, 625
14. Binney, J. & Tremaine S. 1987, *Galactic Dynamics*, Princeton Univ. Press, Princeton
15. Blumenthal, G. R. et al. 1984, Nature, 311, 517
16. Boissier, S. & Prantzos, N. 2001, MNRAS, 325, 321
17. Bond, J. R. et al. 1991, ApJ, 379, 440
18. Bosma, A. 1981, AJ, 86, 1791
19. Bower, R.G. et al. 2005, preprint, astro-ph/0511338
20. Bromm, V., & Larson, R. B. 2004, ARA&A, 42, 79
21. Ciardi, B. & Ferrara, A. 2005, Space Science Reviews, 116, 625
22. Bullock, J. S. et al. 2001, MNRAS, 321, 559
23. Bullock, J.S. et al. 2001, ApJ, 555, 240
24. Carroll S.M., Press W.H., Turner E.L., 1992, ARA&A, 30, 499
25. Casertano, S., & van Gorkom, J.H. 1991, AJ, 101, 1231
26. Catinella, B., Giovanelli, R., & Haynes, M. P. 2005, astro-ph/0512051
27. Cen, R., & Ostriker, J. 1992, ApJ, 393, 22
28. Cole, S. et al. 1994, MNRAS, 271, 781
29. Cole, S. et al. 2000, MNRAS, 319, 168
30. Coles, P., & Lucchin, F. 1995, *Cosmology. The origin and evolution of cosmic structure*, Chichester: Wiley, —c1995,
31. Colín, P., Avila-Reese, V., & Valenzuela, O. 2000, ApJ, 542, 622
32. Colín, P., Avila-Reese, V., Valenzuela, O., & Firmani, C. 2002, ApJ, 581, 777
33. Combes, F., preprint, astro-ph/0506265
34. Croton, D. J. et al. 2005, MNRAS, 356, 1155
35. Davé, R. et al. 2001, ApJ, 552, 473
36. Davis, M. et al. 1985, ApJ, 292, 371
37. de Blok, W. J. G. 2005, ApJ, 634, 227

38. De Lucia, G., Kauffmann, G., & White, S. D. M. 2004b, MNRAS, 349, 1101
39. De Lucia, G. et al. 2006, MNRAS, 366, 499
40. Dressler, A. 1980, ApJ, 236, 351
41. Eisenstein D. J. et al., 2005, ApJ, 633, 560
42. Eke, V.R., Navarro, J.F. & Steinmetz, M., 2001, ApJ, 554, 114
43. Fall, S.M. & Efstathiou, G. (1980), MNRAS, 193, 189
44. Firmani, C., & Tutukov, A.V. 1994, A&A, 288, 713
45. Firmani, C., & Avila-Reese, V. 2000, MNRAS, 315, 457
46. Firmani, C., & Avila-Reese, V. 20003, RevMexAA (SC), v. 17, 106
47. Flores R.A. et al. 1993, ApJ, 412, 443
48. Fukugita, M., & Peebles, P. J. E. 2004, ApJ, 616, 643
49. Frenk, C.S. 2002, preprint, astro-ph/0208219
50. Gentile, G. et al. 2004, MNRAS, 351, 903
51. Giavalisco, M. et al. 2004, ApJ, 600, L103
52. Gnedin, O. Y. et al. 2004, ApJ, 616, 16
53. Gondolo, P. 2004, preprint, astro-ph/0403064
54. Goto, T. et al. 2003, MNRAS, 346, 601
55. Gottlöber, S., Klypin, A., & Kravtsov, A.V. 2001, ApJ, 546, 223
56. Governato, F. et al. 2004, ApJ, 607, 688
57. Gunn, J. E. 1977, ApJ, 218, 592
58. Guzzo, L. 2002, in *Modern Cosmology*, eds. S. Bonometto, V. Gorini & U. Moschella, Bristol, UK, 344
59. Hayashi, E. et al. 2004, MNRAS, 355, 794
60. Helly, J. C. et al. 2003, MNRAS, 338, 903
61. Hopkins, A.M. & Beacom, J. F. 2006, preprint, astro-ph:0601463
62. Hu, W., & Dodelson, S. 2002, ARA&A, 40, 171
63. Kang, X. et al. 2005, ApJ, 631, 21
64. Katz, N. 1992, ApJ, 391, 502
65. Katz, N., & Gunn, J. E. 1991, ApJ, 377, 365
66. Kauffmann, G., White, S.D.M., & Guiderdoni, B. 1993, MNRAS, 264, 201
67. Kauffmann, G. et al. 1999, MNRAS, 303, 188
68. Kauffmann, G. et al. 2004, MNRAS, 353, 713
69. Kereš, D., Katz, N., Weinberg, D. H., & Davé, R. 2005, MNRAS, 363, 2
70. Klypin, A.A. et al. 1999, ApJ, 522, 82
71. Kormendy, J., & Kennicutt, R. C. 2004, ARA&A, 42, 603
72. Kolb, E.W. & Turner, M. S. 1990, *The Early Universe*, Redwood City, California-Addison-Wesley Publishing Company
73. Lacey, C.G., & Cole, S. 1993, MNRAS, 262, 627
74. de Jong, R. S., & Lacey, C. 2000, ApJ, 545, 781
75. Lin, Y., Mo, H.J. & van den Bosch F.C., 2005, astro-ph/0510372
76. Longair, M. S. 1989, LNP Vol. 333: *Evolution of Galaxies: Astronomical Observations*, 333, 1
77. Longair, M. S. 1998, *Galaxy formation*. Springer-Verlag (Germany)
78. Maller, A.H., Dekel, A., & Somerville, R. 2002, MNRAS, 329, 423
79. Madau, P., Pozzetti, L., & Dickinson, M. 1998, ApJ, 498, 106
80. Mandelbaum, R. et al. 2005, preprint, astro-ph/0511164
81. Matos, T., & Ureña-López, L. 2001, Phys.Rev.D, 63, 063506
82. Mo, H.J., Mao, S., White, S.D.M. 1998, MNRAS, 295,319
83. Moore, B. et al. 1999, ApJ, 524, L19

84. Moore, B. et al. 1998, ApJ, 499, L5
85. Murali, C. et al. 2002, ApJ, 571, 1
86. Navarro, J. F., & White, S. D. M. 1993, MNRAS, 265, 271
87. Navarro, J.F., Frenk, C.S., & White, S.D.M. 1997, ApJ, 490, 493
88. Navarro, J. F. et al. 2004, MNRAS, 349, 1039
89. Nicastro, F. et al. 2005, Nature, 433, 495
90. Padmanabhan, T. 1993, *Cosmic structure formation in the universe*, Cambridge Univ. Press
91. Patton, D.R. et al. 2002, ApJ, 565, 208
92. Peacock, J. A. 1999, *Cosmological Physics*, Cambridge University Press
93. Peebles, P.J.E. 1969, ApJ, 155, 393
94. Peebles, P.J.E. 1993, *Principles of Physical Cosmology*, Princeton: Princeton University Press
95. Persic, M., Salucci, P. & Stel, F. 1996, MNRAS, 281, 27
96. Postman, M. & Geller, M. J. 1984, ApJ, 281, 95
97. Postman, M. et al. 2005, ApJ, 623, 721
98. Press, W. H., & Schechter, P. 1974, ApJ, 187, 425
99. Rubin, V.C., Thonnard, N., & Ford, W.K. 1980, ApJ, 238, 471
100. Rhee, G. et al. 2004, ApJ, 617, 1059
101. Salucci, P., & Gentile, G. 2005, preprint, astro-ph/0510716
102. Seljak U. et al., 2005, Phys. Rev. D, 71, 103515
103. Sheth, R. K., Mo, H. J., & Tormen, G. 2001, MNRAS, 323, 1
104. Silk, J. 1968, ApJ, 151, 459
105. Silk, J. 1977, ApJ, 211, 638
106. Silk, J., & Rees, M. J. 1998, A&A, 331, L1
107. Simon, J. et al. 2005, ApJ, 621, 757
108. Somerville, R.S., & Primack, J.R. 1999, MNRAS, 310, 1087
109. Spergel, D. N. et al. 2006, prepint, astro-ph/0603449
110. Springel, V., & Hernquist, L. 2003, MNRAS, 339, 289
111. Springel, V. et al. 2005, Nature, 435, 629
112. Steinmetz, M., & Müller, E. 1994, A&A, 281, L97
113. Steinmetz, M. 1996, in *Dark Matter in the Universe*, Eds. S. Bonometto, J.R. Primack, and A. Provenzale Oxford, GB: IOS Press, 1996, p.479.
114. Tanaka, M. et al. 2004, AJ, 128, 2677
115. Tegmark M. et al., 2004, Phys.Rev. D, 69, 103501
116. Tully, R.B., & Pierce, M.J. 2000, ApJ, 533, 744
117. Valenzuela, O.& Klypin, A. 2003, MNRAS, 345, 406
118. Valenzuela, O. et al. 2005, preprint, astro-ph/0509644
119. van den Bosch, F. C. 2000, ApJ, 530, 177
120. van den Bosch, F. C. 2001, MNRAS, 327, 1334
121. van den Bosch, F. C. 2002, MNRAS, 331, 98
122. van den Bosch 2001, F. C., Burkert, A., & Swaters, R. A. MNRAS, 326, 1205
123. Verheijen, M.A.W. 1997, PhD. Thesis, Groningen University
124. Wang, B., & Silk, J. 1994, ApJ, 427, 759
125. Wechsler, R.H. et al. 2002, ApJ, 568, 52
126. Weil, M., Eke, V. R., & Efstathiou, G. P. 1998, MNRAS, 300, 773
127. Weinmann, S. M. et al. 2006, MNRAS, 366, 2
128. Weldrake, D. T. F., de Blok, W. J. G., & Walter, F. 2003, MNRAS, 340, 12
129. White, S. D. M. 1996, *Cosmology and Large Scale Structure*, 349
130. White, S.D.M., & Frenk, C.S. 1991, ApJ, 379, 52
131. White, S. D. M. & Rees, M. J. 1978, MNRAS, 183, 341
132. Zavala, J. et al. 2003, A&A, 412, 633
133. Zhao, D. H. et al. 2003, ApJ, 597, L9

Ultra-high Energy Cosmic Rays: From GeV to ZeV

Gustavo Medina Tanco

Instituto de Ciencias Nucleares, UNAM, México
& Instituto Astronômico e Geofísico, USP
gmtanco@gmail.com

1 Introduction

Cosmic ray (CR) particles arrive at the top of the Earth's atmosphere at a rate of around 10^3 per square meter per second. They are mostly ionized nuclei - about 90% protons, 9% alpha particles traces of heavier nuclei and approximately 1% electrons. CRs are characterized by their high energies: most cosmic rays are relativistic, having kinetic energies comparable to or somewhat greater than their rest masses. A very few of them have ultra-relativistic energies extending beyond 10^{20} eV (tens of joules).

In this lecture we will give an overview of the main experimental characteristics of the cosmic ray flux and their astrophysical significance with a particular emphasis on the higher end of the spectrum. Unfortunately, due to space limitations, only a fraction of the content of the lectures is included in the present manuscript. In particular, the production mechanisms are not included and the fundamental topic of anisotropies is only dealt with in a very superficial way.

2 Energy Spectrum

Thus, the cosmic ray energy spectrum extends, amazingly, for more than eleven orders of magnitude. All along this vast energy span, the spectrum follows a power law of index ~ 2.7. Therefore, the CR flux decreases approximately 30 orders of magnitude from $\sim 10^3\,\mathrm{m^2sec^{-1}}$ at few GeV to $\sim 1\,\mathrm{km}^{-2}$ per century at 100 EeV.

The only spectral features are a slight bending at around few PeV, known as the first knee, another at approximately 0.5 EeV known as the second knee, and a dip extending from roughly the second knee up to beyond 10 EeV, known as the ankle (see figure 1). Note that in the right panel the spectrum is multiplied by E^3, a usual trick to highlight features that otherwise would be almost completely hidden by the rapidly falling flux.

A. Carramiñana et al. (eds.), Solar, Stellar and Galactic Connections between Particle Physics and Astrophysics, 165–196.

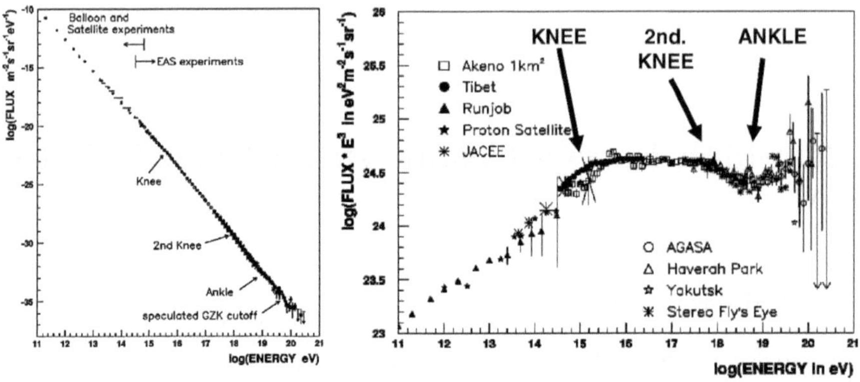

Fig. 1. The cosmic ray energy spectrum and its main features: [left] a remarkably uniform power law with [right] few bends knee (few PeV), second knee (0.5 EeV), ankle (EeV to few tens of EeV) and the still poorly known highest energy tail. Adapted from [1].

The second knee has been observed in the vicinity of 4×10^{17} eV by Akeno [2], Fly's Eye stereo [5, 6, 7], Yakutsk [3, 4] and HiRes [10]. The physical interpretation of this spectral feature is uncertain at present. It may be either the end of the Galactic cosmic ray component or the pile-up from pair creation processes due to proton interactions with the cosmic microwave background radiation during propagation in the intergalactic medium.

The ankle, on the other hand, is a broader feature that has been observed by Fly's Eye [5, 6, 7] around 3×10^{18} eV as well as by Haverah Park [8] at approximately the same energy. These results have been confirmed by Yakutsk [3, 4] and HiRes [10]. AGASA also observed the ankle, but they locate it at a higher energy, around 10^{19} eV [9]. As with the ankle, more than one physical interpretations are possible, which are intimately related with the nature of the second knee. The ankle may be the transition point between the Galactic and extragalactic components, the result of pair creation by protons in the cosmic microwave background, or the result of diffusive propagation of extragalactic nuclei through cosmic magnetic fields.

Composition

Certainly, one of the scientifically most relevant pieces of information inside the transition energy interval previously defined, is the precise chemical composition of the primary CR flux as a function of energy.

Several techniques have been used to determine the composition of cosmic rays along the spectrum and, in particular, in the highest energy region [12]: (i) depth of maximum of the longitudinal distribution, X_{max} [13, 14]; (ii) fluctuations of X_{max} [15, 16]; (iii) muon density [17]; (iv) steepness of the lateral distribution function [25, 26]; (v) time profile of the signal, in particular rise time of the signal [27]; (vi) curvature radius of the shower front; (vii) multi-parametric analysis, such as principal component analysis and neural networks [28], etc... Unfortunately, as is frequently the case in physics, whenever several techniques are applied to measure the same physical magnitude, correspondingly, several results are obtained and, not always agreeable among themselves. As will be shown below, this is critical to the understanding of the astrophysics of ultra high energy cosmic rays.

In order to analyze the astrophysical implications of the composition along the second knee/ankle energy region, it is more instructive to start from much lower energies. A significant point is the first knee. Following KASCADE [20], a gradual change in composition is observed through the knee, from a lighter to a heavier composition. The first knee is a broad feature which can be understood as a composition of power law energy spectra with breaks that are in agreement with a rigidity scaling of the knee position.

Therefore, at energies above few times 10^{16} eV, the flux is dominated by iron nuclei. These particles are of Galactic origin and what is being detected is, very likely, the end of the efficiency of supernova remnant shock waves as accelerators as the Larmor radii or characteristic diffusion scale lengths of the nuclei become comparable to the curvature radius of the remnants, breaking down the diffusive approximation. If there are not more powerful accelerators in the Galaxy, the Galactic cosmic ray flux continues dominated exclusively by iron above 10^{17} eV up to the highest energies produced inside the Milky Way. It must be noted that, even if the previous results are quantitatively dependent on the hadronic interaction model used in the data analysis, they are qualitatively solid and there is considerable degree of consensus on the existence of a progressive transition in composition through the knee. At higher energies, the composition has been measured by several experiments in the past, e.g., Haverah Park, Yakutsk, Fly's Eye, HiRes-MIA prototype and HiRes in stereo mode (see figure 2).

The X_{max} data suggest that, above $10^{16.6}$ eV, the composition changes once more progressively from heavy to light. At the lower limit of this energy interval, the composition is still heavy, i.e., iron dominated, in accordance to Kascade results. Nevertheless, at energies of 10^{19} eV, even if still showing signs of a contamination by heavy elements, it is more consistent with a flux dominated by lighter elements.

Despite the fact that there is a consensus among most of the experiments about the reality of this smooth transition, there is no consensus about the rate and extent to which the transition occurs. In fact, the combined data from the HiRes-MIA prototype and HiRes in stereo mode, signal to a much more rapid transition from heavy to light composition (see figure 3), starting 10^{17} eV but which would be over by 10^{18} eV [29]. Beyond that point, the composition would remain light and constant.

The later scenario, however, is not supported by the data of other experiments. Haverah Park, for example, shows a predominantly heavy composition

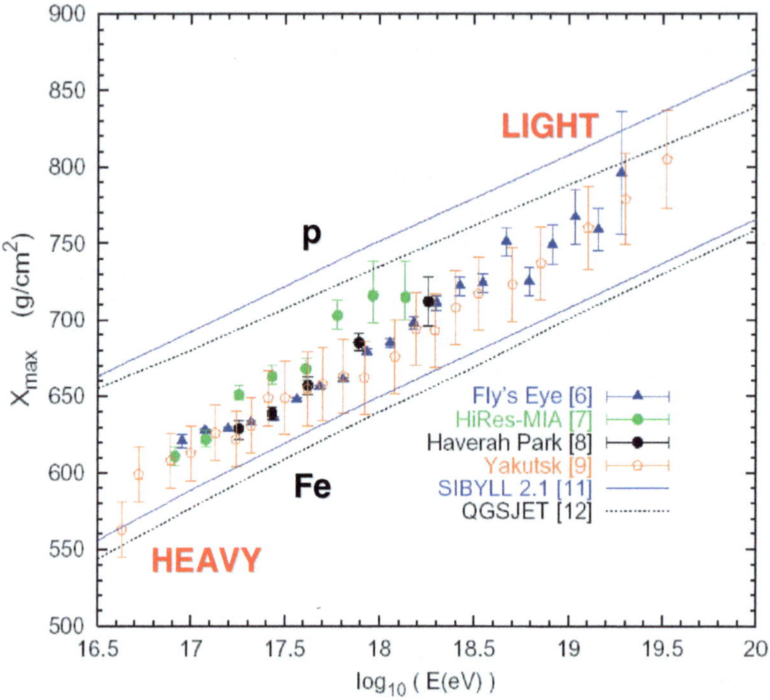

Fig. 2. Variation of X_{max} with energy (elongation rate) showing an apparent change in composition from heavy to light nuclei from $\sim$ 30 PeV to $\sim$ 30 EeV. This variation is at least possibly associated to the transition from Galactic to extragalactic cosmic rays. The lines indicate theoretical expectations corresponding to different hadronic interaction models.

up to 10^{18} eV, followed by an abrupt transition to lighter values compatible with HiRes stereo at around 10^{18} eV (see figure 4a). Volcano Ranch, even though there is a single experimental point, is compatible with a heavy composition still at 10^{18} eV, somewhat in accordance to Haverah Park data. Akeno (A1), on the other hand, is consistent with a continuation of the gradual transition from the second knee all across the ankle up to at least 10^{19} eV, only reaching there the same light composition that HiRes stereo claims from an order of magnitude below in energy. It must also be noted that above 10^{19} eV AGASA is only able to set upper limits for the fraction of iron, but these limits are high enough to leave room for much more complex astrophysical scenarios with a substantial added mixture of extragalactic ultra-high energy heavy elements [30].

Figure 4b shows a compendium of several of the available measurements of composition between $\sim 10^{17}$ and $\sim 10^{19}$ eV, with their corresponding error bars, under the simplistic assumption of a binary mixture of protons and

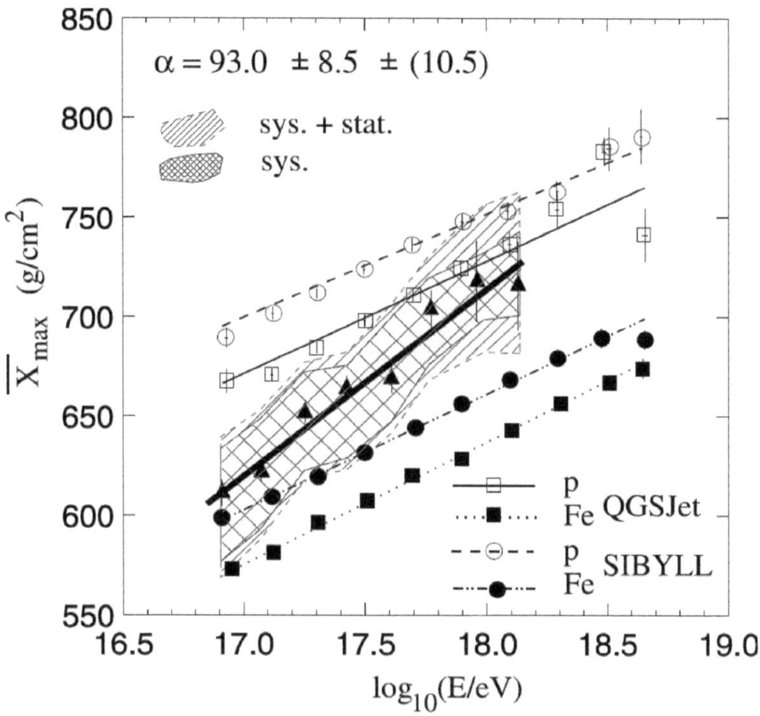

Fig. 3. Elongation rate measured by the HiRes-MIA prototype showing a rapid change to a light composition above 10^{18} eV.

iron nuclei. The emerging picture is one complete uncertainty, which has deep practical implications and imposes severe limitations to theoretical efforts.

At energies beyond $\sim$ 10 EeV the composition is essentially unknown. However, it seems compatible with hadrons even if some photon contribution cannot be discarded [31, 32, 33] (see figure 5). Although it is implicitly regarded as purely protonic in many theoretical works, only upper limit exists for the iron fraction (e.g. figure 4b) and therefore not much can be said until more accurate measurements are made available by new generation experiments like Auger and TA.

3 Galactic Propagation

The fundamental question of cosmic ray physics is, "Where do they come from?" and in particular, "How are they accelerated to such high energies?". These are difficult questions not fully answered after almost a century of history of the field. Some very general hints can be obtained, however, through

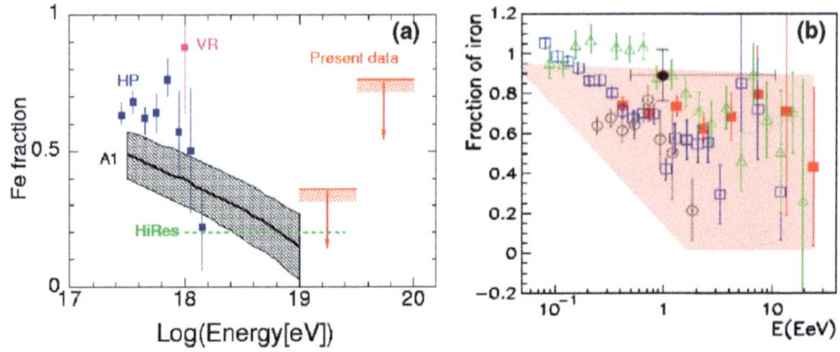

Fig. 4. (a) Variation of the iron fraction inside the transition energy interval for various experiments. (b) Idem, highlighting the uncertainties in composition inside the region encompassing the second knee and the ankle: literally almost any abundance of iron is allowed (adapted from [26]).

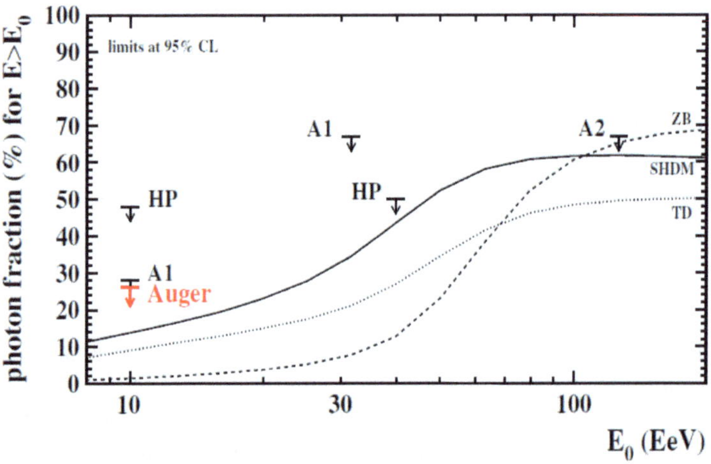

Fig. 5. Upper limits (95% CL) on cosmic-ray photon fraction for Auger [33], AGASA (A1) [31], (A2) [32] and Haverah Park (HP) [34, 35] data compared to some estimates based on TOP-DOWN models [36] (reproduced from [33]).

very simple arguments. The interstellar and intergalactic mediae are magnetized and, being charged, the CR are forced to interact with these fields.

From the point of view of propagation of charged particles, the Galaxy behaves as a magnetized volume, where the field is structured on scales of kpc, with typical intensities of the order of some few micro Gauss. The Larmor radius of a nucleus of charge Ze can be conveniently parameterized as:

$$r_{L,kpc} \approx \frac{1}{Z} \times \left(\frac{E_{EeV}}{B_{\mu G}} \right) \tag{1}$$

where E_{EeV} is the energy of the particle in units of 10^{18} eV and $r_{L,kpc}$ is expressed in kpc.

Equation (1) clearly shows that, given the typical intensity of the magnetic fields present in the interstellar medium (ISM), nuclei with energies below some few tens of EeV, regardless of their charge, have Larmor radii much smaller than the transversal dimensions of the magnetized Galactic disk. They must, therefore, propagate diffusively inside the ISM. This transforms the Galaxy in an efficient confinement region for charged particles with energies below the second knee. The confinement region is a flattened disk of approximately 20 kpc of radius and thickness of a few hundreds of pc.

Consequently, from the lowest energies and up to the second knee, the Galaxy is undoubtedly the source of the cosmic ray particles and of their kinetic energy. There is not a consensus about the actual source of the particles in itself, but the two main lines of thought propose either nuclei pre-accelerated at the chromospheres of normal F and G type stars or ambient electrically charged nuclei condensed in the dense winds of blue or red giant stars [37, 38]. On the other hand, several acceleration mechanism must be at play but it is widely expected that the dominant one is first order Fermi acceleration at the vicinity of supernova remnant shock waves. Nevertheless, theoretically, the Galactic accelerators should become inefficient between $\sim 10^{17}$ to $\sim 10^{18}$ eV. This upper limit could be extended to $\sim 10^{19}$ eV if additional mechanisms were operating in the Galaxy, e.g., spinning inductors associated with compact objects or cataclysmic events like acceleration of iron nuclei by young strongly magnetized neutron stars through relativistic MHD winds [39].

At energies above the second knee, particles start to be able to travel from the nearest extragalactic sources in less than a Hubble time. Consequently, at some point above $10^{17.5}$ eV a sizable cosmic ray extragalactic component should be detectable and become dominant above 10^{19} eV. Therefore, it is expected that the cosmic ray flux detected between the second knee and the ankle of the spectrum be a mixture of a Galactic and an extragalactic flux, highlighting the astrophysical richness and complexity of the region.

The type of propagation strongly depends on the charge of the corresponding nucleus. Protons with energies $\gtrsim 10^{17}$ eV have gyroradii comparable or larger than the transversal dimensions of the effective confinement region and, therefore, can easily escape from the Galaxy. On the other end of the mass spectrum, just the opposite occurs for iron nuclei that, even at energies of the order of 10^{19} eV, have gyroradii $< 10^2$ pc and must be effectively confined inside the magnetized ISM.

The previous results are based only on the consideration of the regular component of the Galactic magnetic field. However, there exist a superimposed turbulent component whose intensity is at least comparable to that of the regular field. Its spectrum seems to be of the Kolmogorov type, extending from the smallest scales probed, $\sim 10^0$ pc, to $L_c \sim 100$ pc, the correlation length of the turbulent field.

Wave-particle interactions between cosmic rays and MHD turbulence are resonant for wavelengths of the order of the Larmor radius, $\lambda \sim r_L$. This means that, for a nucleus of charge Z, a critical energy can be defined,

$$r_{L,kpc} \approx \frac{1}{Z}\left(\frac{E_{EeV}}{B_{\mu G}}\right) \approx L_c\,, \quad L_c \approx 10^2 pc \Rightarrow E_{c,EeV} \approx 0.5 \times Z\,, \qquad (2)$$

below which modes resonant with the particle gyroradius able to efficiently scatter the particle in pitch angle exist. Consequently, at energies below E_c the diffusion coefficient is small enough for the particle trajectory to be diffusive. Above E_c, on the other hand, the propagation is essentially ballistic.

Due to the interaction with the turbulent magnetic component, protons experience a propagation regime very different than iron inside the ankle energy region. Protons propagate ballistically in the ISM above $\sim 3 \times 10^{17}$ eV, while iron nuclei propagate diffusively even at energies $\gtrsim 10^{19}$ eV. Therefore, along the energy region extending from the second knee up to almost the end of the ankle, all nuclei from p to Fe, i.e. $1 < Z < 26$, experience a transition in their propagation regime inside the ISM changing gradually from diffusive to ballistic as the energy increases.

A pictorial example of how this transition takes place can be seen in figure 6a-d [40], which shown how protons with energies ranging from 0.5 to 6 EeV injected at the Galactic center propagate out of the Galaxy assuming a characteristic BSSS magnetic model. It must be noted that, despite the fact that the deflections induced by the Galactic magnetic field (GMF) diminish rapidly with energy for all nuclei, they can still be important even at the highest energies. This is more critical for heavier primaries and for all nuclei traversing the central regions of the Galaxy. Figure 7 illustrates this point by showing the intrinsic deflections experienced by proton and iron nuclei as a function of arrival Galactic coordinates [41].

Figure 8 shows deflections for 100 EeV protons for a more sophisticated GMF model inspired on Han's proposal [11]. The GMF is modelled by a disk BSSS component of 100 pc half thickness, embedded in an ASSA halo and a dipolar component originated in a Southward magnetic momentum anchored at the Galactic center.

4 Extragalactic Propagation

4.1 Superposition of the Extragalactic and Galactic fluxes

In the same way as the magnetic characteristic of the interstellar medium allow Galactic particles at these energies to escape into the extragalactic environment, extragalactic cosmic rays are also able to penetrate inside the Galactic confinement region. But, of course, extragalactic particles must first be able to reach us from the nearest Galaxies in less than a Hubble time.

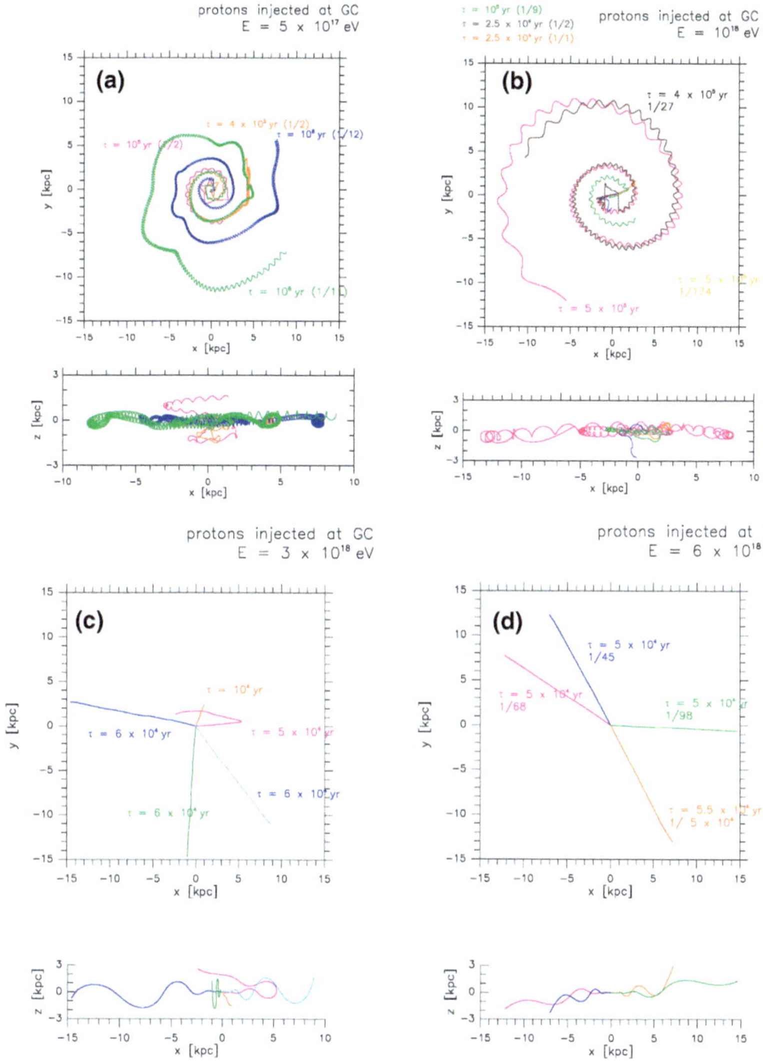

Fig. 6. Changes in propagation regime inside the Galaxy at energies of the second knee and ankle.

A crude approximation to this effect can be made in the following way. Faraday rotation measurements statistically impose to the extragalactic magnetic field the following restriction [42]:

$$B \times L_c^{1/2} \leqslant 1\,\mathrm{nG} \times \mathrm{Mpc}^{1/2} \tag{3}$$

where L_c is the correlation length of the magnetic field that we assume, somewhat arbitrarily, as being of the order of 1 Mpc. Assuming that the diffusion coefficient can be estimated by the Bohm approximation,

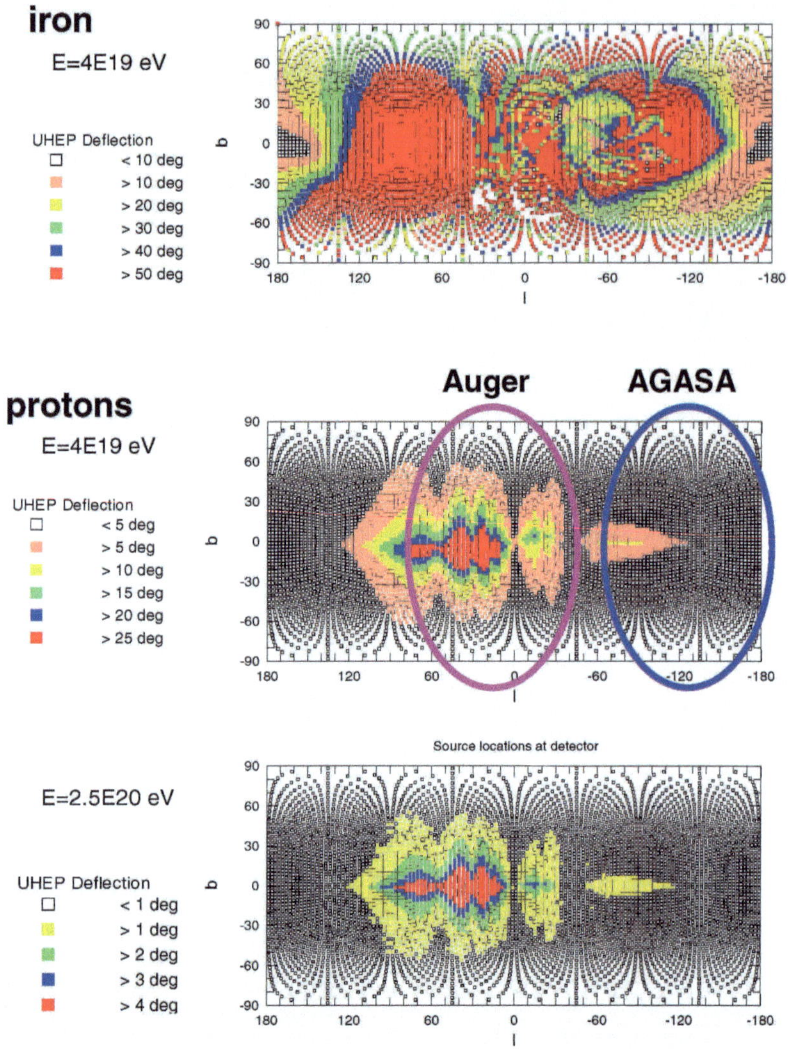

Fig. 7. Intrinsic deflections due to the GMF (BSSS model) suffered by protons and Fe nuclei at 4×10^{19} eV and protons at 2.5×10^{20} eV as a function of arrival direction. Galactic coordinates are used.

$$K \approx \frac{1}{3} r_L c \tag{4}$$

and using equation (1) the diffusion coefficient can be written:

$$K \approx \frac{0.1}{Z} \left(\frac{E_{EeV}}{B_{\mu G}} \right) \frac{\mathrm{Mpc}^2}{\mathrm{Myr}} \tag{5}$$

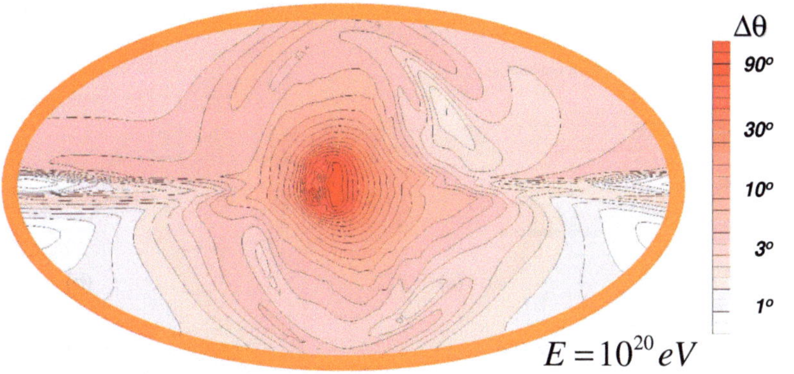

Fig. 8. Deflection suffered by 100 EeV protons due to a Han-type GMF (see text) as a function of arrival direction. Galactic coordinates are used in an Aitof projection.

The diffusive propagation time from an extragalactic source at a distance D can be estimated as:

$$\tau \approx D^2/K, \tag{6}$$

or, using equation (5):

$$\tau_{Myr} \approx 10 \times D_{Mpc}^2 \times Z \times \left(\frac{B_{nG}}{E_{EeV}}\right). \tag{7}$$

Equation (7) shows that there is a rather restrictive magnetic horizon. Basically, no nucleus with energy smaller than 10^{17} eV is able to arrive from regions external to the local group ($D \sim 3$ Mpc). Taking as a minimum characteristic distance $D = 10$ Mpc, which defines a very localized region completely internal to the supergalactic plane and even smaller than the distance to the nearby Virgo cluster, only protons with $E > 2 \times 10^{17}$ eV, or Fe nuclei with $E > 5 \times 10^{18}$ eV are able to reach the Galaxy in less than a Hubble time.

Therefore, it is at the energies of the second knee and the ankle that different nuclei start to arrive from the local universe. Concomitantly, at these same energies the magnetic shielding of the Galaxy becomes permeable to these nuclei, allowing them to enter the ISM and to eventually reach the solar system. Effectively, the energy interval from $\sim 2 \times 10^{17}$ to 10^{19} eV is the region of mixing between the Galactic and extragalactic components of cosmic rays.

Above few times 10^{17} eV, the dominant interactions experienced by cosmic rays are due to the cosmic microwave background radiation (CMBR) and, additionally in the case of nuclei, to the infrared background (CIBR) [43]. The diffuse radio background, despite its much lower density, must in turn become important at high enough energies.

At energies above $\sim 10^{19.2}$ eV, the dominant process is the photo production of pions in interactions with the CMBR (see figure 9a), which drastically

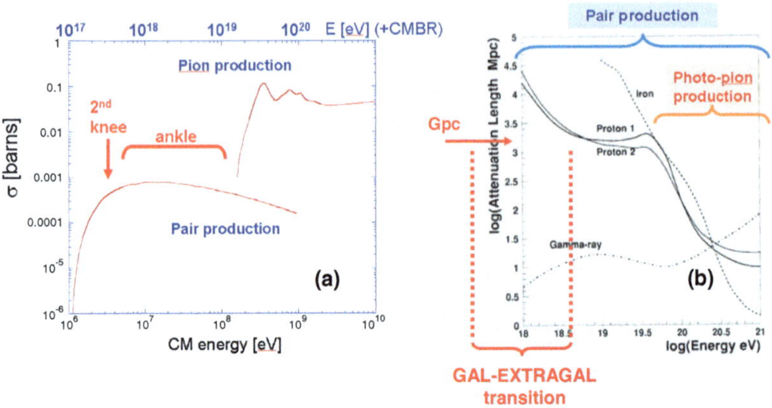

Fig. 9. (a) Cross section for pair production and pion production in interaction with the CMBR. The positions of the second knee and the ankle are also shown, demonstrating that electron positron pair production is the relevant interaction in the Galactic-extragalactic transition region. Note the similitude between the shape of the cross section for this interaction and the shape and location of the ankle. (b) Attenuation length in Mpc as a function of energy [44], showing how the universe, which is opaque for at energies above the photo-pion production threshold, becomes transparent to lower energy baryons.

reduces the mean free path of protons to a few Mpc, making the universe optically thick to ultra-high energy cosmic rays (figure 9b). This interaction, in the most conservative models, should produce a strong depression in the energy spectrum, with a major fall in the observed flux above 10^{20} eV, the so called GZK cut-off [45, 46].

At energies smaller than $\sim 10^{19.2}$ eV, the dominant process is the photo-production of electron-positron pairs in interactions with the CMBR. At these lower energies the attenuation length attain values of the order of Gpc and the universe is essentially optically thin to energetic baryons. CR observations at these energies sample the universe at cosmological distances, contrary to the highest energies, that only sample a sphere of a few tens of Mpc in diameter, a small portion of the local universe [49, 50, 51]. Therefore, strictly from the point of view of propagation in the extragalactic medium, in going down from the highest energies to the transition region, the observable horizon drastically increases from 10^1 to 10^3 Mpc, i.e., essentially the whole universe.

It can also be seen from figure 9a that the dependence of the cross section with energy is suggestive, since its shape resembles that of the ankle in the cosmic ray energy spectrum. In fact, the structure of the ankle can be explained exclusively as a result of pair photo-production by nucleons traveling cosmological distances between the source and the observer [47].

The energy region where the superposition of the Galactic and extragalactic spectra takes place is a theoretically challenging region, where the smooth

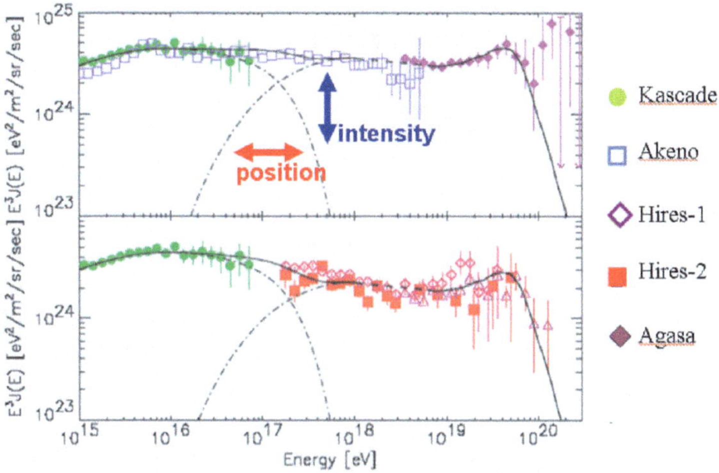

Fig. 10. Total cosmic ray spectrum from the combined data of several experiments. From the theoretical point of view, the transition region is highly complex and the Galactic and extragalactic models undergo the most critical test as fluxes must be simultaneously matched both in intensity and energy (adapted from [48]).

matching of the two rapidly varying spectra has yet to be explained. It must be noted that, even if the shape of the spectrum is important, it is by far insufficient to decipher the underlying astrophysical model. The Galactic magnetic fields are intense enough to dilute any directional information, which prevents the discrimination among the galactic and extragalactic components from the arrival direction of the incoming particles. The variation of the composition as a function of energy turns then into the key to discriminate both fluxes and to select among a variety of theoretical options.

As in the case of the ISM, it is expected that the intergalactic medium has a strong magnetic turbulent component which can severely affect propagation [52, 53, 54, 55, 56, 57]. The correlation length estimated from Faraday rotation measurements, L_c, is consistent with a maximum wavelength for the MHD turbulence determined by the largest kinetic energy injection scales in the intergalactic medium, $L_{max} \sim L_c \sim 1$ Mpc. In analogy to equation (2):

$$r_{L,kpc} \approx \frac{1}{Z} \times \left(\frac{E_{\rm EeV}}{B_{\rm nG}} \right) \approx L_{\rm max}\,, \quad L_{\rm max} \approx 1Mpc \Rightarrow E_{c,{\rm EeV}} \approx 1.0 \times Z\,. \quad (8)$$

For a given nucleus of charge Ze, the propagation is ballistic for $E > E_c$ being diffusive otherwise. Therefore, protons are ballistic above $\sim 10^{18}$ eV, but diffusive at the energies of the second knee. Iron nuclei, on the other hand, propagate diffusively along the ankle and even at energies as high as $\sim 5 \times 10^{19}$ eV. The boundaries for the transition between the ballistic and diffusive propagation regime for proton and Fe nuclei are shown in the figure 11. Furthermore, besides the total intensity and the minimum wavenumber, also the energy

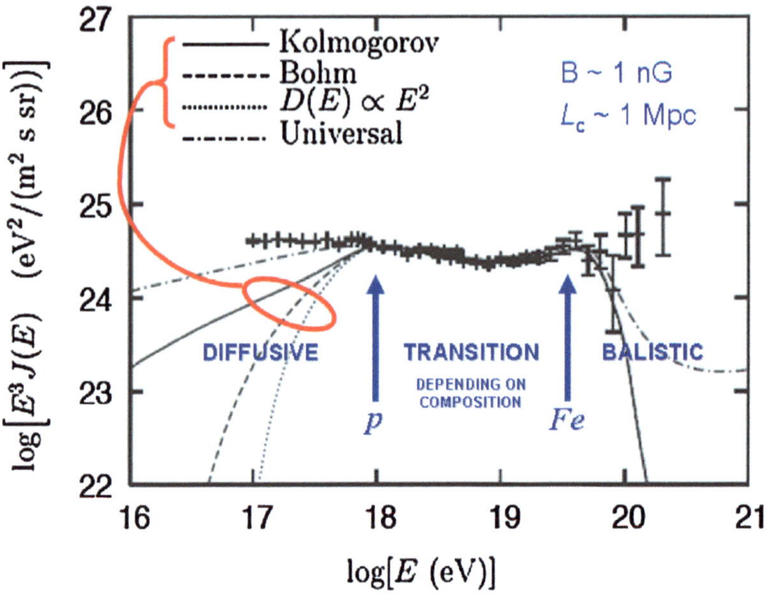

Fig. 11. Correlation between the detailed structure of the lower end of the extragalactic spectrum and the type of turbulence present in the ISM. This emphasizes the importance of the intergalactic turbulent component in the observed matching between the Galactic and extragalactic cosmic ray flux (adapted from [58]).

distribution among the different modes, that is the type of turbulence present in the intergalactic medium, has observational expression. In this case, the affected portion of the extragalactic spectrum is the lower energy region, where the flux is strongly suppressed by magnetic horizon effects. Figure 11 shows clearly this effect for three different assumptions for the diffusion coefficient.

Obviously, this has profound theoretical implications not only for the structure of the extragalactic magnetized medium, but also for cosmic ray acceleration conditions inside the Galaxy. This is exemplified in figure 12 where it is graphically illustrated that, by subtracting a given extragalactic spectrum from the observed total spectrum, conclusions can be drawn about relevant aspects of the Galactic component. For example, an extragalactic spectrum that has a small contribution at low energies, can imply the existence of additional acceleration mechanisms in the Galaxy other than the shock waves of supernova remnants.

4.2 The Highest Energies

Propagation of Protons

As was mentioned in the previous section 4.1 (see figure 9), above the threshold for photo-pion production by protons interacting with the CMBR, ~ 40 EeV

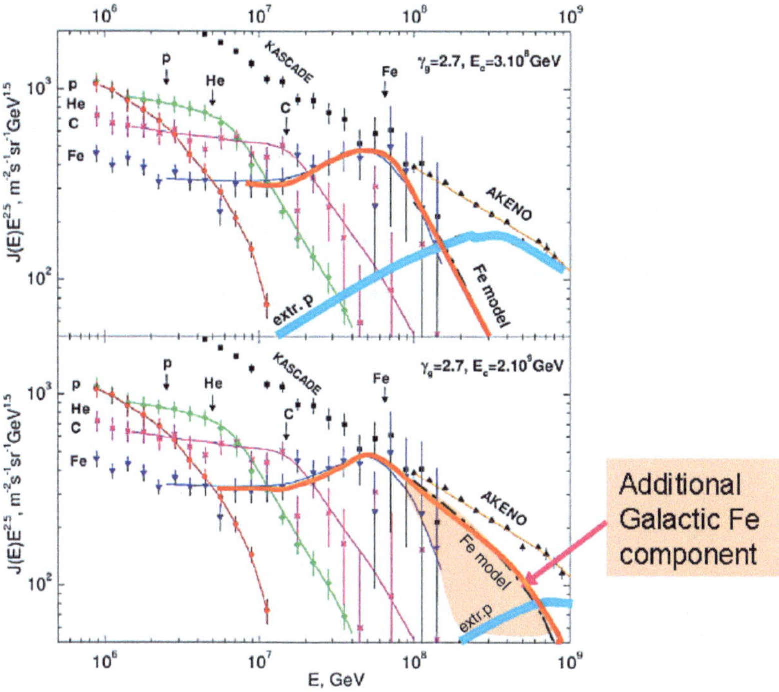

Fig. 12. Impact of the detailed characteristics of the extragalactic spectrum on our comprehension of the most powerful acceleration mechanisms in our Galaxy (adapted from [59]).

the universe becomes rapidly opaque for hadrons as the attenuation length goes down to values as low as $\sim$ 10 Mpc at few $\times 10^{20}$ eV. This determines a relatively small maximum distance scale, $R_{GZK} \simeq 50 - 100$ Mpc, to the sources that are able to contribute appreciably to the detected CR flux.

Under very general assumptions regarding the nature of the primaries and the cosmological distribution of the sources, photo-pion production should lead to the formation of a pile-up immediately followed by a severe reduction in CR flux, popularly known as the GZK cut-off. The existence of this spectral feature was proposed short time after the discovery of the CMBR [45, 46] but its actual existence is still a matter of considerable debate.

At present, there are conflicting measurements coming from two different experiments: AGASA and HiRes. The first one is a surface detector while the second one is a fluorescence detector, which further complicates the comparison of their results. As shown in figure 13, the differences are not only quantitative but, fundamentally, qualitative. While HiRes apparently shows the expected GZK flux suppression, AGASA seems incompatible with this result, showing an energy spectrum that extends undisturbed well beyond 100 EeV. There is also an apparently large difference in flux between both

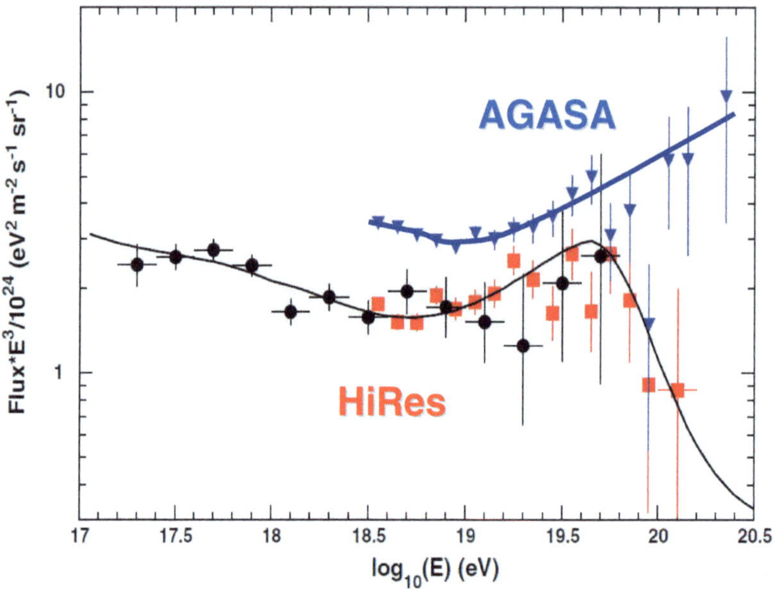

Fig. 13. Comparison between the AGASA and Hires monocular spectra (adapted from [60]).

experiments at energies below the cut-off. However, the fact that the energy spectrum has been multiplied by E^3 in figure 13 should be taken into account when assessing the significance of such difference.

Actually, both results may be reconciled at the $1.5\,\sigma$ level by re-scaling the energy of the experiments by 30% or both by 15% [61]. The Auger Observatory, being the largest detector ever built and having hybrid capacity i.e., simultaneous fluorescence and surface detection [62], has the potential -but not yet the statistics- to give a definite answer to this fundamental problem [63].

The absence of the GZK cut-off, if confirmed, could be compatible with a wide range of astrophysical scenarios. At least three possibilities can be considered, some rather conservative, some more exotic:

- The distance scale between sources could be large enough that, by chance, the few (or single) sources inside the GZK sphere dominate the flux, the rest of the population of ultra high energy cosmic ray (UHECR) accelerators being too distant to contribute appreciably to the observed flux at Earth.
- The primary CR might be particles that do not interact with photons or do so at much larger, and as yet unobserved, energies. These could be familiar standard model particles that present unexpected behavior at ultra-high energies, like neutrinos with hadronic cross section that can develop showers in the atmosphere resembling those expected from proton primaries [64, 65]. Another possibility could be a new stable hadron, heav-

ier than a nucleon, for which the threshold for photo-meson production would be at higher energy. An example of the latter would be uhecrons, e.g., a uds-gluino bound state [66].

- The primaries might be normal hadrons, but Lorentz invariance, never previously tested at $\gamma \sim 10^{11}$, could be violated at ultra-high energies, hampering photo-meson production [67, 68, 69]. The small violation of Lorentz invariance required might be result of Planck scale effects [70, 71].
- The observed spectrum could be the superposition of two components: (a) a hadronic component with a GZK cut-off and (b) a harder, top-down component that becomes dominant above ~ 100 EeV. These second component could be originated in the decay or annihilation of super heavy dark matter or topological defects [72, 73, 74, 76]. These scenarios have the general disadvantage of overproducing ultra-high energy neutrinos and photons rather than nucleons, which seems to be increasingly constrained by the observation [33, 81]. Nevertheless, there could still be models, like those involving necklaces, that could present an appreciable baryon content at energies $\gtrsim 100$ EeV [75]. In any case, these models suffer from a discomforting level of fine tuning with respect to the normalization of the intensities of the GZKed and the top-down spectra.

It must also be noted that the presence of the GZK flux suppression does not imply the non-existence of supra-GZK particles. These have certainly been detected by at least Volcano Ranch [77, 78], Fly's Eye [79], AGASA, HiRes and, more recently, Auger [80]. This means that, the detection of the GZK feature does not solve the puzzle about the generation of UHECR.

Propagation of UHE Photons

The propagation of photons is dominated by their interaction with the photon background. The main processes are photon absorption by pair-production on background photons ($\gamma\gamma_b \rightarrow e^+e^-$), and inverse Compton scattering of the resultant electrons on the background photons. These two processes acting in a chain are responsible for the rapid development of electromagnetic cascades in the intergalactic or interstellar media, draining energy to the sub-TeV region.

For a given UHE photon of energy E_γ, the minimum background photon energy, E_b, for electron-positron pair production is

$$E_b = \frac{m_e^2 c^4}{E_\gamma} \simeq \frac{2.6 \times 10^{11}\,\mathrm{eV}}{E_{\gamma,eV}} \tag{9}$$

and the corresponding cross section peaks near the threshold: $\sigma_{PP} \propto (m_e^2/s) * \ln(s/2m_e)$ (see Carramiñana, in this volume). Inverse Compton scattering, on the other hand, has no threshold but its cross section is also largest near the $\gamma\gamma_b$ pair production threshold. Therefore, the most efficient background for both processes is given by equation (9). For UHE this means that the cosmic radio background, whose magnitude is highly uncertain, is dominant followed by the CMBR below 10^{17} – 10^{18} eV. At progressively lower energies, the CIRB and optical background are important.

In the Klein-Nishina limit, $s \gg m_e^2$, one of the components of the $\gamma\gamma_b$-produced pair carries most of the energy of the energy of the UHE photon. This leading particle, afterwards, undergoes Compton scattering in the same limit, for which the inelasticity is very near 1. Therefore, the Compton up-scattered photon still has an appreciable fraction of the energy of the original UHE photon. The presence of magnetic fields in the medium may speed up the development of the cascade by draining the electron and positron energy due to synchrotron radiation. The larger the fields, the smaller the penetration. The electromagnetic cascades produced in this way can propagate an effective distance that is much larger than the interaction length yet, severely limit our UHE-γ horizon to the nearest regions of the supergalactic plane. Figure 14 [72] shows the effective penetration length of electromagnetic cascades for two different estimates of the cosmic radio background and two different average intergalactic magnetic field intensities.

Since the single pair cross section decreases as $\ln(s)/s$ for $s \gg m_e^2$, multiple pair production becomes important at extreme energies. Thus, double pair production ($\gamma\gamma_b \rightarrow e^+e^-e^+e^-$) begins to dominate above $\sim 10^{21}$ – 10^{23} eV. The relevant process for electrons is triple pair production ($e^\pm\gamma_b \rightarrow e^\pm e^- e^+$), whose attenuation becomes dominant at $\sim 10^{22}$ eV. Other processes (e.g., moun, tau or pion pair production, double Compton scattering, gamma scattering and pair production of single photons in magnetic fields) are in general negligible for electromagnetic cascade development. However, at energies in excess of 10^{24} eV, the pair production of single photons in the Galactic magnetic field should eliminate all the photons above that energy from specific lines of sight, generating an arrival direction anisotropy.

The penetration lengths shown in figure 14, combined with the threshold energies given by equation (9), imply that the injection of UHE-γ in the intergalactic medium results in the pile-up of photons at energies below 100 MeV, whose contribution to the diffuse cosmic γ background is already observationally constrained by EGRET. This overproduction of low energy diffuse photons is a strong restriction for top-down UHECR production models.

Besides the limitations imposed by the possible overproduction of low energy photons, the results in figure 14 have other profound implications for top-down scenarios. Models that claim decay or annihilation of dark matter, in general, tend to produce mainly photons and only a few percent baryons. Therefore, in such models, most of the detected CR should be photons from our own Galactic halo, with perhaps some localized contribution from Andromeda. The products of the decay of more distant dark matter would be cleared from photons, and only the small fraction of remaining baryons, suppressed by GZK effects, would give a positive contribution at Earth. In any case, for most top-down models, photons should be dominant above the GZK cut-off, but with perhaps a sizable baryonic component.

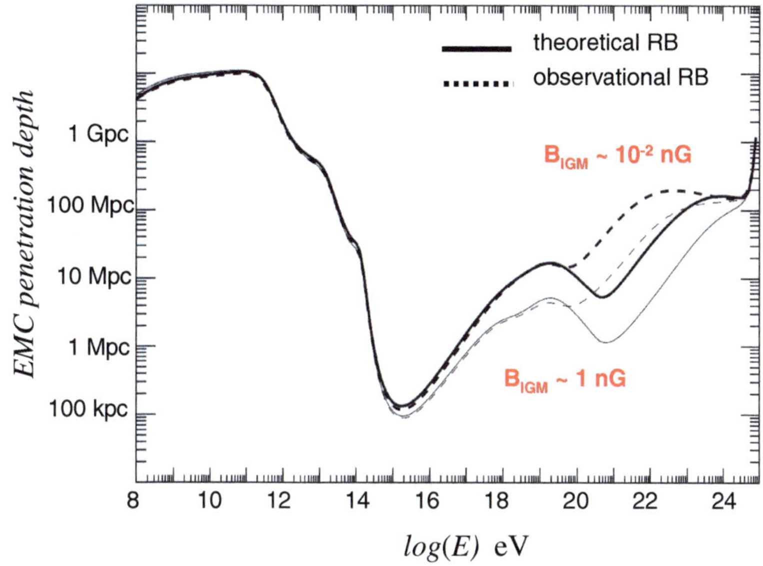

Fig. 14. Penetration of electromagnetic cascades in the intergalactic medium. Solid lines correspond to a fiduciary value of 1 nG for the intergalactic magnetic field (IGMF) and dotted lines to a very low IGMF, 10^{-2} nG. Thick and thin lines correspond to different estimates of the cosmic radio background (adapted from [72]).

Propagation of Nuclei

Heavy nuclei are attenuated by two basic processes: photodisintegration and electron-positron pair creation by interaction with background photons. For a given total energy, the threshold photo-pion production for a nucleus of mass A increases to $E_{th} \simeq 4 \times 10^{19} \times A$. Therefore, given the energies observed at present, pion production is not relevant for nuclei heavier than He. Figure 15 shows that below ~ 20 EeV, all nuclei are able to travel for virtually a Hubble time, while Fe can do the same up to ~ 100 EeV. Above that energy, nuclei start to disintegrate fast and the loss time is highly reduced.

It is clear from composition observations around the first knee of the cosmic rays spectrum that the acceleration mechanism by shock waves, either first order Fermi or drift are limited by the magnitude of the radius of curvature of the shock. This results in the preferential acceleration of large charge (Ze) nuclei, those with the smallest Larmor radius, to the highest energies.

Even if the mechanism responsible for the acceleration of the ultra-high energy extragalactic component is essentially unknown, the most conservative view points to bottom-up mechanisms. If the later is actually the case, then the most economic assumption is that shock wave acceleration. In this minimalistic, but still realistic, scenario the most likely high energy output would be heavy nuclei, very likely Fe as in the Galactic case. At lower energies,

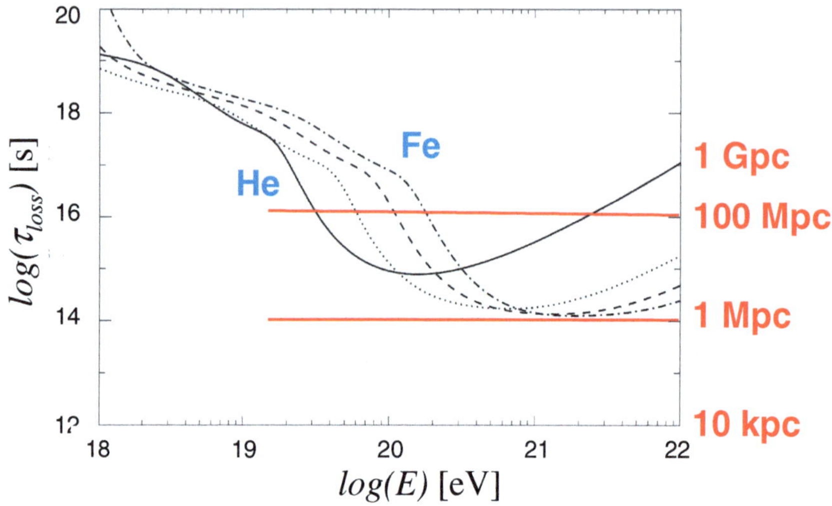

Fig. 15. Energy loss time (right axis: length) vs. energy for photodisintegration on background photons: radio, CMB and IR. Helium, Carbon, Silicon and Iron are shown. Single, double and multi-nucleon emissions are included (adapted from [82]).

progressively lighter nuclei should be observed due to two factors: (a) the acceleration process in itself and (b) the photo-disintegration on flight of the heavier nuclei due to their interaction with the CIBR. The latter process is very efficient, and can extract approximately one nucleon every few Mpc at the highest energies, depending on the CIBR level. Since photodisintegration occurs, to a good approximation, at constant energy per nucleon, disintegration of a nucleus A at energy E will produce light nuclei at energies nE/A and $(A-n)E/A$ with preferentially small n (e.g., $n = 1, 2, 3, 4$).

Figure 16 shows that power law spectra injected at cosmological sources with different compositions can produce experimentally very similar at the highest energies. Nevertheless, they can always be distinguished at smaller energies in the ankle region. In particular, figure 16a show a purely protonic flux can reproduce the ankle feature solely as an effect of photo production of electron-positron pairs in interactions with the CMBR. In this case, the transition between the Galactic and extragalactic fluxes must be located at the second knee or very near to it. Figure 16b shows that, on the other hand, for a heavier mixed composition the extragalactic spectrum falls down steadily with decreasing energy. In this scenario, the ankle must be the result of the composition between the Galactic and extragalactic spectra. Moreover, the composition will be a strong function of energy inside this interval, giving an additional tool to assess details of the astrophysical model.

The effects of photo-disintegration are clearly shown in figure 17, depicting the evolution of a pure iron injected spectrum as it propagates out from the source. Blue histograms correspond to p, white to the original surviving Fe

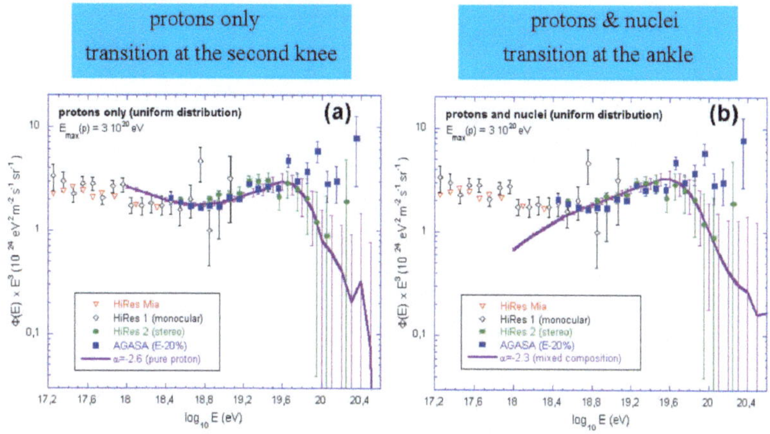

Fig. 16. Different primary compositions may produce extragalactic spectra essentially indistinguishable at the highest energies. However, as shown in (a) for protons and (b) for heavier mix, the spectra are considerably different at lower energies below the bottom of the ankle (adapted from [83]).

and red to intermediate mass nuclei. As the distance to the source increases, intermediate nuclei are produced at increasingly smaller masses and less total energy (fragmentation takes place at roughly constant mass per nucleon). At distances of the order of the GZK horizon, the region with larger mixing of nuclei, i.e., with a larger composition gradient, is the ankle. Consequently, this is the ideal region for the discrimination of the primary composition from local composition measurements.

5 Cosmic Magnetic Fields and Anisotropy

Luminous matter, as traced by galaxies, and dark matter, as traced by galaxies and clusters large scale velocity fields, are distributed inhomogeneously in the universe. Groups, clusters, superclusters, walls, filaments and voids are known to exist at all observed distances and are very well mapped in the local universe. Hence, the distribution of matter inside the GZK-sphere is highly inhomogeneous and so is, very likely, the distribution of UHECR sources.

Synchrotron emission and multi-wavelength radio polarization measurements show that magnetic fields are widespread in the Universe. But how do they encompass the structure seen in the distribution of matter we do not yet know [42]. The available limits on the intergalactic magnetic field (IGMF) come from rotation measure estimates in clusters of galaxies and suggest that $B_{IGM} \times L_c^{1/2} < 10^{-9}$ G $\times$ Mpc$^{1/2}$ [42], where L_c is the field reversal scale. Note, however, that this kind of measurement does not set an actual limit to the intensity of the magnetic field unless the reversal scale is known along a

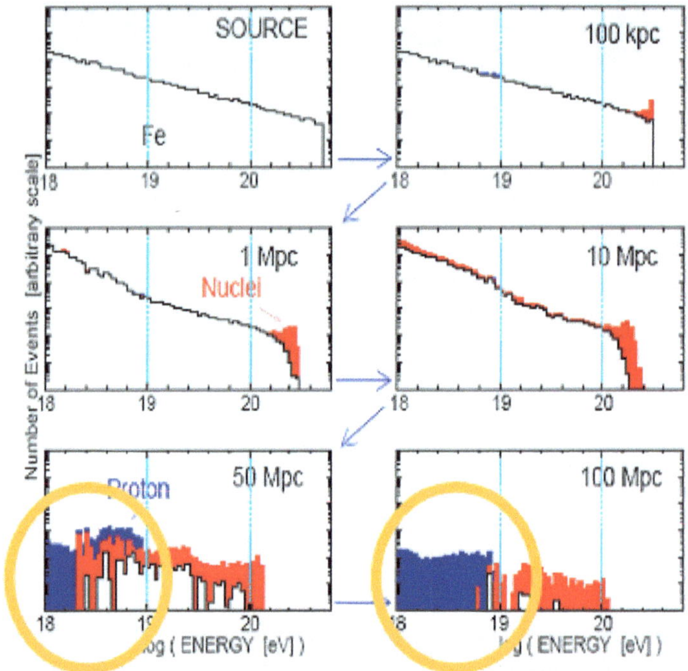

Fig. 17. Variation of the composition as a function of distance to the source for pure Fe power law injection. Blue histograms correspond to p, white to the original surviving Fe and red to intermediate mass nuclei (adapted from [84]).

particular line of sight. The latter means that, depending on the structure of the IGMF, substantially different scenarios can be envisioned that are able satisfy the rotation measure constraints.

Unfortunately, we do not know what is the actual large scale structure of the IGMF. But we can imagine two extreme scenarios that are likely to bound the true IGMF structure. In figure 18 calculations of large scale structure formation by Ryu and co-workers [99] have been modified by hand to exemplify these scenarios. The top frame displays Ryu's IGMF simulation results in the background showing how by $z = 0$ the magnetic field has been convected together with the accretion flows into walls, filaments and clusters, depleting the voids from field. According to these calculations, the magnetic field is confined in high density, small filling factor regions, bounded by a rather thin skin of rapidly decreasing intensity, surrounded by large volumes of negligible IGMF. As suggested by the free-hand lines on top of the figure, the IGMF inside these structures is highly correlated in scales of up to tens of Mpc. Furthermore, in order to comply with the rotation measurement constraints mentioned before, the intensity of the magnetic field inside the density structures must be

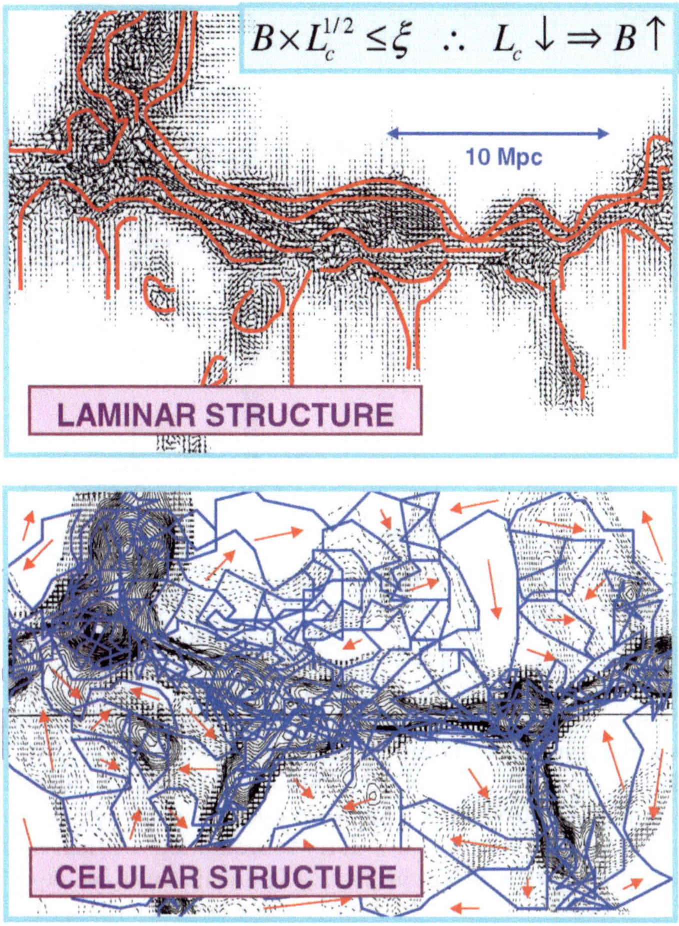

Fig. 18. Two possible extreme models for the IGMF structure: (top) laminar structure and (bottom) cellular structure. These schematic plots are adaptations by hand made on top of IGMF and density calculations by [99].

correspondingly high, 0.1–1 μG, which is comparable with values within the ISM. We will call the latter scenario *laminar-structure*.

The second model, that we will call *cellular-structure*, is depicted in the bottom panel of figure 18. We imagine the space divided into adjacent cells, each one with an uniform magnetic field randomly oriented. We identify the size of a cell with the magnitude of the local reversal scale. Furthermore, one can assume that the intensity of the magnetic field scales as some power of the local matter (electron) density and, consequently, the rotation measurement constraint $B_{IGM} \times L_c^{1/2} < 10^{-9}$ G $\times$ Mpc$^{1/2}$ tells us how the reversal scale, i.e., the size of the cells, should scale. A convenient reference, such as the IGMF in the Virgo [85] or Coma [89] cluster can be used for normalization. The

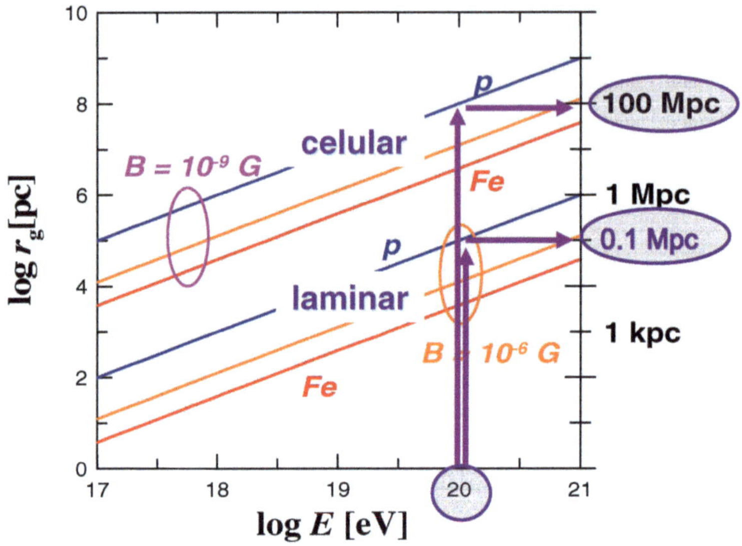

Fig. 19. Typical Larmor radii of nuclei in both IGMF scenarios, showing the very different scales involved in each model.

cellular-structure scenario leads to a more widespread IGMF, filling even the voids. The observational constraints imply then that the IGMF varies much more smoothly, from 10^{-10}G inside voids to a few times 10^{-9}–10^{-8}G inside walls and filaments, only reaching high values, $0.1 - 1$ μG, inside and around clusters of galaxies. Observations cannot presently distinguish between these two scenarios. Nevertheless, we can still try to asses what are their implications for UHECR propagation which, inspecting figure 19, must be important.

5.1 UHECR Propagation in a Laminar-IGMF

A laminar-IGMF is the most difficult scenario to dealt with because it does not accept a statistical treatment and results are very dependent on details about the exact magnetic field configuration inside the GZK-sphere, which is beyond our present knowledge.

A simple approach is to study the UHECR emissivity of a single wall surrounded by a void [94, 93]. Figure 20a shows the corresponding model for a wall immersed in a void. The magnetic field inside the wall has two components: an uniform field along the z-axis of intensity 0.1μG, plus a random field with a Kolmogoroff power law spectrum of amplitude equal to 30% of the regular component. One hundred UHECR sources are included inside the wall, and each one of them injects protons at the same rate and with the same power low energy spectrum, $dN_{inj}/dE \propto E^{-2}$. Pair-production and photo-pion production losses in interactions with the infrared and microwave backgrounds are also included. The wall has a radius of 20 Mpc, a thickness

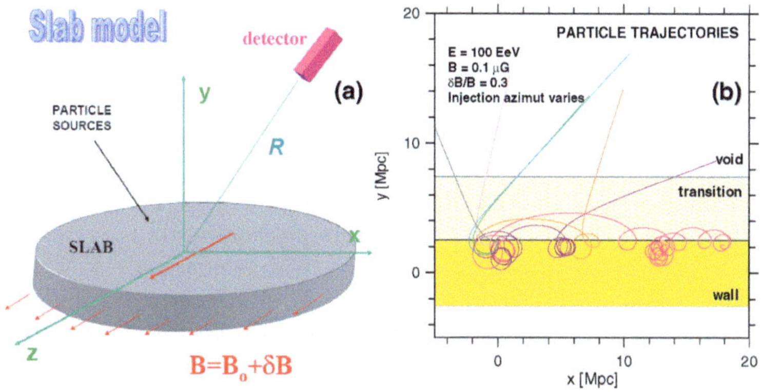

Fig. 20. (a) Simplified model of a wall, or slab, containing UHECR sources and surrounded by a void. The magnetic field configuration is representative of the laminar model. (b) Cross section of the wall in (a) at the plane $z = 0$. Several particles trajectories are shown for proton injection at $E = 100$ EeV. Adapted from [93].

of 5 Mpc and is sandwiched by a transition layer 5 Mpc in thickness where the magnetic field decreases exponentially up to negligible values inside the surrounding void. Once the system reaches steady state, a detector can be shifted around the wall to simulate observers at arbitrary positions with respect to the wall. In a real situation, this system could be representative, for example, of the supergalactic plane; in that case the Milky Way, i.e. we, the observers, should be located at some point on the $x - z$ plane (but we do not know at what angle with respect to the z-axis). The simulations show that the UHECR flux measured can vary by three orders of magnitude depending on the relative orientation between the wall, the field and the observer. At the same time, almost all directional information is lost, and the strength of the GZK-cut-off would vary considerably as a function of orientation [93].

The previous effects can be intuitively understood by looking at figure 20b, showing a cross section at $z = 0$ of the wall in figure 20a. Particle trajectories are shown for protons injected at $E = 100$ EeV, with different azimuthal angles and a slight elevation with respect to the $x - y$ plane. It can be seen that there is nothing like a random walk: particles tend to be trapped inside the wall and move in a systematic way. Most of the particles drift perpendicularly to the regular field while their guiding centers bounce along the field. It can also be seen how the gyroradii decrease as the particles lose energy in interactions with the radiation backgrounds. Even the few particles that escape from the wall, do so in a anisotropic manner (e.g., predominantly to the right for $y > 0$).

The laminar IGMF model is, actually, the worst scenario for doing some kind of astronomy with UHECR. It would be very difficult to interpret the UHECR angular data and to identify individual particle sources. Furthermore, the significance of any statistical analysis would be greatly impaired due to systematics. Further studies on this model can be found in [103, 104].

5.2 UHECR Propagation in a Cellular-IGMF

The cellular model is the easiest scenario to deal with numerically and, by far, the most promising from the point of view of the astrophysics of UHECR. This is also the IGMF model that has been used probably more frequently in the literature [86, 90, 53, 91, 92, 95, 97, 98].

The main assumption is that the intensity of the magnetic field scales with density. Indeed, for those spatial scales where measurements are available, the intensity of the magnetic fields seem to correlate remarkably well with the density of thermal gas in the medium. This is valid at least at galactic and smaller scales [100, 96]. It is apparent that B can be reasonably well fitted by a single power law over ~ 14 orders of magnitude in thermal gas density at sub-galactic scales. A power law correlation, though with a different power law index, is also suggested at very large scales (c.f. figure 5 in [96]), from galactic halos to the environments outside galaxy clusters, over ~ 4 orders of magnitude in thermal gas density. This view [100] is, however, still controversial [42]. In fact, magnetic fields in galaxy clusters are roughly $\sim 1\,\mu$G, of the same order as interstellar magnetic fields; furthermore, supracluster emission around the Coma cluster suggests μG fields in extended regions beyond cluster cores. The latter could indicate that the IGMF cares little about the density of the associated thermal gas density, having everywhere an intensity close to the microwave background-equivalent magnetic field, $B_{BGE} \simeq 3 \times 10^{-6}$ G.

Taking the view that a power law scaling exists, a model can be devised in which the IGMF correlates with the distribution of matter as traced, for example, by the distribution of galaxies. A high degree of non-homogeneity should then be expected, with relatively high values of B_{IGMF} over small regions ($\lesssim 1\,$Mpc) of high matter density. These systems should be immersed in vast low density/low B_{IGMF} regions, with $B_{IGMF} < 10^{-9}\,$G. Furthermore, in accordance with rotation measures, the topology of the field should be structured coherently on scales of the order of the correlation length L_c which, in turn, scales with IGMF intensity, $L_c \propto B^{-2}_{IGMF}(r)$. $\boldsymbol{B}_{IGMF}$ should be independently oriented at distances $> L_c$. Therefore, a 3D ensemble of cells can be constructed, with cell size given by the correlation length, L_c, and such that $L_c \propto B^{-2}_{IGMF}(r)$, while $B_{IGMF} \propto \rho^{0.35}_{gal}(r)$ [100] or $\propto \rho^{2/3}_{gal}(r)$ (for frozen-in field compression), where ρ_{gal} is the galaxy density, and the IGMF is uniform inside cells of size L_c and randomly oriented with respect to adjacent cells [91, 98]. The observed IGMF value at some given point, like the Virgo cluster, can be used as the normalization condition for the magnetic field intensity. The density of galaxies, ρ_{gal}, is estimated using either redshift catalogs (like the CfA Redshift Catalogue [92, 95] or the PSCz [88]), or large scale structure formation simulations [98]. This is a convenient way to cope with, or at least to assess, the importance of the several biases involved in the use of galaxy redshift surveys to sample the true spatial distribution of matter in 3D space. The spatial distribution of the UHECR sources is tightly linked to the nature of the main particle acceleration/production mechanism involved. However, in

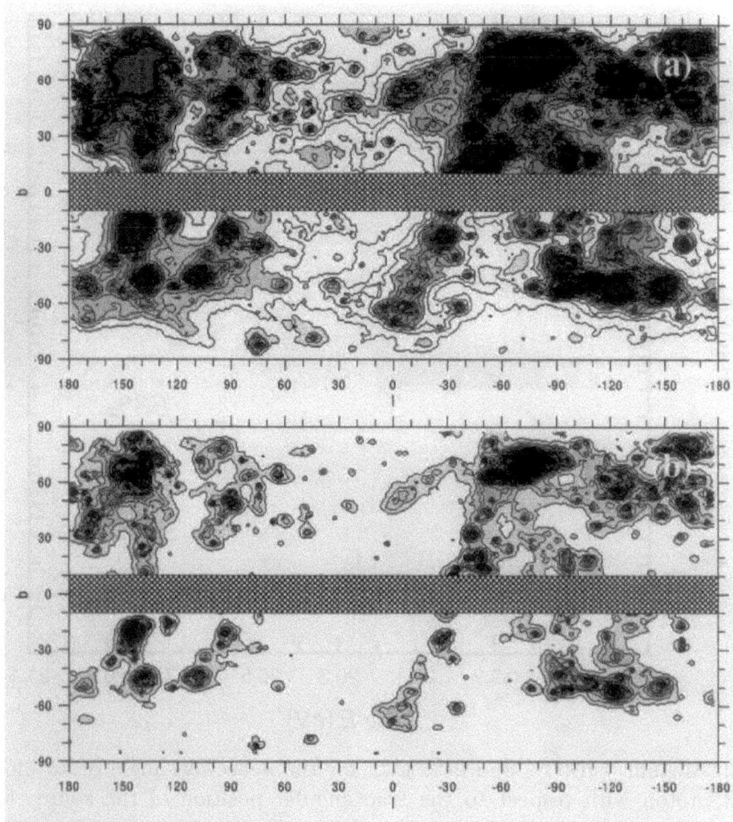

Fig. 21. Arrival probability distribution of protons (linear scale) as a function of Galactic coordinates for a distribution of sources inside 100 Mpc following that of luminous matter. $dN_{inj}/dE \propto E^{-2}$, with (a) $E_{inj} > 4 \times 10^{19}$ eV, (b) $E_{inj} > 10^{20}$ eV.

most models, particles will either be accelerated at astrophysical sites that are related to baryonic matter, or produced via decay of dark matter particles. In both cases the distribution of galaxies (luminous matter) should be an acceptable, if certainly not optimal, tracer of the sources.

The relevant energy losses for UHECR during propagation are $\gamma - \gamma$ pair production with CMB for photons, redshift, pair production and photopion production in interactions with the CMB for nuclei and, for heavy nuclei, also photo-disintegration in interactions with the infrared background. All of these can be appropriately included [87, 101, 102, 72].

Once the previously described scenario is built, test particles can be injected at the sources and propagated through the intergalactic medium and intervening IGMF to the detector at Earth. Figure 21a-b show the arrival probability distribution of UHECR protons in Galactic coordinates for a distribution of sources following the distribution of luminous matter in-

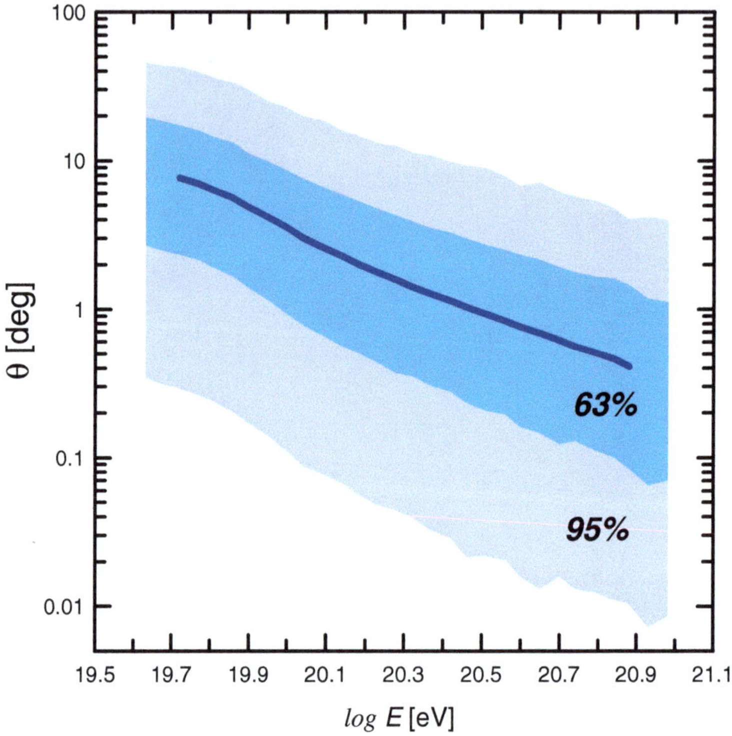

Fig. 22. Median and 63% and 95% C.L. for the deflection angle of an incoming UHECR proton with respect to the true angular position of the source for the example in figure 21. All sky average.

side 100 Mpc (CfA2 catalog). A power law injection energy spectrum at the sources is assumed, $dN_{inj}/dE \propto E^{-2}$, with (a) $E_{inj} > 4 \times 10^{19}$ eV and (b) $E_{inj} > 10^{20}$ eV respectively. It can be seen that, in contrast to the laminar IGMF case, in this scenario information regarding the large scale distribution of the sources inside the GZK-sphere is easily recoverable. The supergalactic plane and the Virgo cluster, in particular, are clearly visible between $\ell \simeq 0 - 100$. The increase in resolution as the energy reaches the 100 EeV range and the gyroradii of UHECR protons become comparable to the size of the GZK-sphere can also be appreciated. It is in the cellular model that the deflection angle of the incoming particle with respect to the true angular position of the source is small enough for UHECR astronomy to develop at the largest energies (figure 22).

5.3 Anisotropy Observations and Magnetic Fields

CR anisotropies can be divided, in principle, according to the angular scales they affect at a given energy region: (a) large scales, extending across several tens of degrees in the sky, (b) medium scales, $\lesssim 10°$ and (c) small scales, of the order of the angular resolution of the experiment, i.e, $\sim 1-3^{\circ}$. At all energies CR are remarkably isotropic. At energies below $\sim$ 100 TeV, a first harmonic analysis shows an anisotropy amounting to less than 0.07%. At large energies, on the other hand, measurements are increasingly difficult due to lowering statistics and ambiguities in the interpretation of the data due to the non-uniformity of the detectors' acceptance [105, 106]. Nevertheless, all the data available are consistent with large scale isotropy [107, 108, 109, 80]. In fact, besides the AGASA experiment, neither HiRes or Auger have been able so far of detecting anisotropy at any energy or angular scale [109, 80, 110].

At energies around 1 EeV, i.e., the beginning of the ankle where there should still be a sizable Galactic contribution to the flux, Fly's Eye has encountered a statistically significant correlation with the Galactic plane in the energy range between 0.2 and 3.2 EeV [111]. They assessed the probability of this result being a statistical fluctuation of an isotropic distribution at $< 0.06\%$. The most significant enhancement is in the interval 0.4–1.0 EeV. AGASA, on the other hand has reported a $> 4\sigma$ excess towards the direction of the Galactic center [112]. The excess was confirmed in two independent data sets with 18274 and 10933 events in the 1–2 EeV region, with chance probabilities of 0.3 and 0.5% respectively. The 4.5σ effect observed corresponds to 506 events detected in a region of the sky where only 413.6 were expected. Associated with this enhancement was a probably dipolar signal towards the inner regions of the Galaxy of amplitude 0.04 [113]. An independent confirmation of an anisotropy possibly related to this one comes from the SUGAR experiment [114] which, different from AGASA, was able to look directly at the Galactic center. Unfortunately, these findings have not been confirmed so far by either HiRes or Auger [115].

At small scales, $< 2.5^{\circ}$, comparable with the resolution of the experiment, AGASA has reported the existence of pairs and triplets of events, which have grown in number over the years to a total of 7 pairs and 1 triplet (or 9 pairs if the triplets is counted as 2 pairs) above 40 EeV [119, 120, 121, 122]. Since, following AGASA estimations, a total of 1.7 pairs was expected at this separation, the results has a chance probability of less than 0.1%. This result has not been confirmed by HiRes in the combined AGASA-HiRes data set [123], but still remains a topic of hot debate due to its enormous astrophysical significance. It must also be noted that it is difficult to understand simultaneously the existence of at least the three original pairs when the actual distribution of matter inside the GZK sphere is taken into account [92].

Finally, another very promising anisotropic signal coming from the AGASA experiment is found as an alignment in the relative orientation of pairs of incoming events above 10 EeV in the $\Delta\ell - \Delta b$ plane (Galactic coordinates) on scales of $\lesssim 10^{\circ}$ [116, 117]. It must be noted that this anisotropy is fundamentally different from a simple clustering of events in a given angular scale, since it is limited to an aligned structure in the two point correlation function. This signal can only be produced as the result of charged CR bending their trajectories in the Galactic magnetic field. The astrophysical implications of this observation have been analyzed in detail in [118]. CR polarization, if confirmed, could turn into a powerful tool for the determination of the number of nearby CR point sources and for imposing constraints on the intensity and topology of the Galactic and extragalactic magnetic fields.

Summing up the results on anisotropy at the highest energies so far, and remembering our discussion about the effects of different spacial structures and intensities of the IGMF (sections 5.1 and 5.2), the high degree of isotropy observed by most experiments, seems to favor a laminar IGMF structure. However, it must be remembered that local coherent magnetized structures, as our own halo may be, could de-focus particles coming from point sources into an apparently isotropic flux further complicating the analysis [124, 125].

References

1. Nagano M. and Watson A. A., Rev. Mod. Phys. 72 (2000) 689.
2. Nagano M. et al., J. Phys. G 10, (1984) 1295.
3. Pravdin M. I. et al., Proc. 26th ICRC (Salt Lake City) 3 (1999) 292.
4. Pravdin M. I. et al., Proc. 28th IRCR (Tuskuba) (2003) 389.
5. Bird D. J. et al., Phys. Rev. Lett. 71 (1993) 3401.
6. Bird D. J. et al., Astrophys. J. 441 (1995) 144.
7. Abu-Zayyad T. et al., Astrophys. J. 557 (2001) 686.
8. Ave M. et al., Proc. 27th Int. Cosmic Ray Conf. (Hamburg) 1 (2001) 381, See also astro-ph/0112253.
9. Takeda M. et al., Astropart. Phys. 19 (2003) 447.
10. HiRes Collaboration, Phys. Rev. Letters 92, 151101 (2004).
11. Han J. L. (2001) astro/ph0110319.
12. Watson A. A. Nuclear Physics B (Proc. Suppl.) 136 (2004) 290.
13. Linsley J. Proc 15th Int Cos Ray Conf (Plovdiv) 12 (1977) 89
14. Zha M., Knapp J., Ostapchenko S., Proc 28th ICRC (Tsukuba) 2 (2003) 515.
15. Gaisser T. K. et al., Phys Rev D 47 (1993) 1919.
16. Archbold G. et al., Proc. 28th Int Cos Ray Conf (Tsukuba) 1 (2003) 405.
17. Shinosaki K. for the AGASA group, Nucl. Phys. B (Proc. Suppl.) (2004) 136.
18. Antoni et al., Nucl. Instr. and Methods in Phys. Res. A 513 (2003) 490-51.
19. Badea A. F. et al. Nuclear Physics B (Proc. Suppl.) 136 (2004) 384-389.
20. Kampert K.-H. et al. Nuclear Physics B (Proc. Suppl.) 136 (2004) 273-281.
21. Antoni T. et al., Nucl. Inst. Meth. A 513 (2003) 490.
22. Navarra G. et al., Nucl. Inst. Meth. A 518 (2004) 207.
23. Linsley J. Proc 18th ICRC, Bangalore, 12 (1983) 144.
24. Song C. et al., Astropart Phys 14 (2000) 7.
25. Ave M., et al., Astroparticle Physics 1961 (2003), and astro-ph/0203150
26. Dova M. T. et al., Astropart. Phys. 21, 597 (2004).
27. Ave M. et al., Proc 28th Int Cos Ray Conf (Tsukuba) 1 (2003) 349.
28. Tiba A., Medina Tanco G. A., and Sciutto S. J. astro-ph/0502255.
29. Abu-Zayyad T. et al., ApJ 557 (2001) 686.
30. Shinozaki K. & Teshima M., Nuclear Physics B (Proc. Suppl.) 136 (2004) 18.
31. Shinozaki K. et al., ApJ Lett., 571 (2002) 117.
32. Risse M., 29th ICRC, Pune, 7 (2005) 143.
33. Pierre Auger Collaboration, 29th ICRC, Pune, 7 (2005) 147.
34. Ave M. et al., Phys. Rev. Lett. 85, (2000) 2244.

35. Ave M. et al., Phys. Rev. D65, (2002) 063007.
36. Gelmini G., Kalashev O. E., and Semikoz D.V. astro-ph/0506128 (2005), and references therein.
37. Ptuskin V.S. Rapporteur Talk, 29th ICRC, Pune, 10 (2005) 317.
38. Meyer J. P. et al., ApJ 487 (1997) 182.
39. Blasi P., Epstein R. I. and Olinto A. V. astro-ph/9912240.
40. Medina Tanco G. A. and Watson A. A. 27th ICRC, Hamburg (2001) 531.
41. Medina Tanco G. A. et al., ApJ, 492 (1998) 200.
42. Kronberg P. P. Rep. Prog. Phys. 57 (1994) 325.
43. Berezinsky V. S., Bulanov S., Dogiel V., Ginzburg V., and Ptuskin V. "Astrophysics of Cosmic Rays" 1990, North-Holland Publishing Company, Amsterdam.
44. Berezinsky V. S. and Grigoréva S. I. A & A 199 (1988) 1.
45. Greisen K. Phys. Rev. Lett., 16 (1966) 748
46. Zatsepin G. T. and Kuzmin V. A. PisÏma Zh. Eksp. Teor. Fiz., 4 (1966) 114
47. Berezinsky V., Gazizov A. Z., and Grigorieva S. I. astro-ph/0502550.
48. Lemoine M. Phys. Rev. D71 (2005) 083007
49. Stanev T. arXiv:astro-ph/0303123;
50. Stanev T., Seckel D., and Engel R. Phys. Rev. D 68 (2003) 103004.
51. Stanev T. et al., Phys. Rev. D 62 (2000) 093005.
52. Medina Tanco G. in Physics and Astrophysics of UHECRs, eds M. Lemoine & G. Sigl,Lect. Notes in Phys., 576 (2001) 155.
53. Medina Tanco G. A. et al., Astroparticle Physics, 6 (1997) 337.
54. Medina Tanco G. A. and Ensslin T. A. Astroparticle Phys. 16 (2001) 47.
55. Medina Tanco G. A. Ap. J. 549 (2001) 711.
56. Medina Tanco G. A. Ap. J. Letters, 505 (1988) L79.
57. Medina Tanco G. A. proc. 25th ICRC, Durban, South Africa, 4 (1997) 477.
58. Aloisio R. and Berezinsky V. S. Astrophys. J., 625 (2005) 249.
59. Berezinsky V. S., Grigorieva S. I., and Hnatyk B.I., Astropart. Phys., 21 (2004) 617.
60. Abbasi R. U. et al., Astroparticle Phys., 23 (2005) 157.
61. De Marco D. et al., Astroparticle Phys. 20 (2003) 53.
62. Auger Collaboration, Nuc. Inst. Methods A, 523 (2004) 50.
63. Auger Collaboration, 29th ICRC Pune, 7 (2005) 387.
64. Bordes J. et al., Astropart. Phys. 8 (1998) 135.
65. Jain P., McKay D.W., Panda S., and Ralston J.P., Phys. Lett. B484 (2000) 267.
66. Chung D. et al., Phys. Rev. D57 (1998) 4606.
67. Sato H., Tati T., Prog. Theor. Phys., 47 (1972) 1788.
68. Kirzhnits D. A., Chechin V. A., Yad. Fiz. 15 (1972) 1051.
69. Coleman S. and Glashow S. L., Phys. Rev. D 59 (1999) 116008.
70. Aloisio R. et al., Phys. Rev. D 62 (2000) 053010.
71. Alfaro J. and Palma G., Phys. Rev. D, 67 (2003) 083003.
72. Bhattacharjee P. and Sigl G., Phys. Rept. 327, 109 (2000) [arXiv:astro-ph/9811011].
73. Berezinsky V. S. Nucl. Phys. (Proc. Suppl) B87, 387 (2000).
74. Kuzmin V.A. and Tkachev I. I. Phys. Rep. 320, 199 (1999).
75. Aloisio R., Berezinsky V., and Kachelriess M. Nucl. Phys. B, 136 (2004) 319.
76. Sarkar S. and Toldra R. Nucl. Phys. B 621 (2002) 495.
77. Linsley J. Phys. Rev. Lett. 10 (1963)146.

78. Linsley J. AIP Conf. Proc. 433 (1998) 1.
79. Bird D. et al., 1995, Astrophys. J. 441, 144.
80. Auger Collaboration, 29th ICRC Pune, 10 (2005) 115.
81. Rubtsov G. I. et al., PRD 73 (2006) 063009.
82. Bertone G., Isola C., Lemoine M., and Sigl G. Phys. Rev. D66 (2002) 103003.
83. Allard D. et al. astro-ph/0505566.
84. Yamamoto T. et al. Astropart.Phys. 20 (2004) 405.
85. Arp H. Phys. Lett. A 129 (1988) 135.
86. Auger Design Report, 1996.
87. Berezinsky V. S. and Grigoreva S. I. A & A, 199 (1988) 1.
88. Blanton M., P., B., and V., O. A., 2000, astro-ph/0009466
89. Kim K.-T., Kronberg, P. P., G., G., and T., V., 1989, Nature 341, 720.
90. Medina Tanco G. A., 1997b, in 25th ICRC, Vol. 4, p. 481.
91. Medina Tanco G. A., 1997c, in 25th ICRC, Vol. 4, p. 477.
92. Medina Tanco G. A., 1998a, Astrophys. J. Lett. 495, L71.
93. Medina Tanco G. A., 1998b, Astrophys. J. Lett. 505, L79.
94. Medina Tanco G. A., 1999, in Proc. 16th ECRC, Alcalá de Henares, Espanha (J. Medina ed.), pp 295–298.
95. Medina-Tanco G. A., 1999, Astrophys. J. Lett. 510, L91.
96. Medina Tanco G. A., 2000b, astro-ph/9809219,in Topics in Cosmic Ray Astrophysics, M. Duvernois (ed.),Nova Science Pub. Inc.,New York. p. 299.
97. Medina Tanco G. A., 2001, ApJ 549 (2001) 711, astro-ph/0011454.
98. Medina Tanco G.A., Enßlin , T.A., 2001, Astropart. Phys. (in Press), astro-ph/0011454.
99. Ryu D., Kang, H., Biermann, P. L., 1998, Astronomy and Astrophys. 335, 19.
100. Vallee J. P., 1997, Fundamentals of Cosmic Physics 19, 1.
101. Yoshida, S. and Teshima, M., Prog. Theor. Phys. 89 (1993) 833.
102. Stecker F. W. and H., S. M., 1999, ApJ 512 (1999) 521.
103. Sigl G., Lemoine M. and Biermann P. L., Astropart. Phys. 10 (1999) 141.
104. Lemoine M., Sigl G., and Biermann P. L., astro-ph/9903124.
105. Watson A. A., Adv. Sp. Res. 4 (1984) 35.
106. Watson A. A., Nucl. Phys. B Suppl, 22B (1991) 116.
107. Takeda M. et al., ApJ 522 (1999) 225.
108. Shinozaki K., Teshima M., Nuclear Physics B (Proc. Suppl.), 136 (2004) 18.
109. HiRes Collaboration, Nuclear Physics B (Proc. Suppl.) 138 (2005) 307.
110. Auger Collaboration, 29th ICRC Pune, 7 (2005) 75.
111. Bird D. J. et al., ApJ, 511 (1999) 739.
112. AGASA coll. (1998), Astropart.Phys. 10 (1999) 303-311, astro-ph/9807045.
113. Teshima M. et al., 27th ICRC Hamburg (2001).
114. Bellido J. A. et al., Astropart.Phys.,15 (2001)167
115. Auger Collaboration, 29th ICRC Pune, 7 (2005) 67.
116. Takeda M. et al., 27th ICRC. Hamburg, 1 (2001) 345.
117. Teshima M. et al., 28th ICRC. Tsukuba, 1 (2003) 401.
118. Medina Tanco G. A., Teshima M. Takeda M., 28th ICRC Tsukuba (2003)
119. Takeda M. et al., Phys. Rev. Lett. 81 (1998) 1183.
120. Takeda M. et al., Astrophys.J. 522 (1999) 225.
121. Takeda M. et al., Proc. 27th ICRC Hamburg 1 (2001) 345.
122. Teshima M. et al., Proc. 28th ICRC Tsukuba, 1 (2003) 401.
123. Abbasi R. U., ApJ 623 (2005)164.
124. Ahn E. J., Medina-Tanco G., Biermann P. L., Stanev T., astro-ph/9911123.
125. Biermann P. L., Ahn E. J., Medina Tanco G., Stanev T., Nucl. Phys. Proc. Suppl. 87 (2000) 417 astro-ph/0008063].

Part II

Astronomical Technical Reviews

Radio Astronomy: The Achievements and the Challenges

Luis F. Rodríguez[1]

Centro de Radioastronomía y Astrofísica, UNAM, Apdo. Postal 3-72, Morelia, Michoacán 58089 México l.rodrigueze@astrosmo.unam.mx

Radio astronomy was the second electromagnetic window used to study the Universe and its development opened the door to the multiwavelength astronomy that characterizes much of the present day research. I briefly review the early story and achievements of radio astronomy. I also present recent astronomical discoveries obtained at radio wavelengths, as well as the characteristics and key science drivers of the next generation of radio telescopes, now under construction or design in various countries. I will also discuss the situation of this area of astronomical research in Mexico and the possibilities for the future.

1 Introduction

Radio astronomy is the science that studies the Universe in electromagnetic radiation with wavelengths between approximately 20 meters and 0.3 millimeters. Radio astronomy then covers a range of about 10^4 in the size of the waves studied (as opposed to a factor of only 2 in the visible waves). This has an immediate consequence: since it is impossible in practice to build an instrument that will detect efficiently waves that are so different in size, it is necessary to have different radio telescopes to cover the whole of the radio window. In what respects to the atmospheric transparency, at the long wavelengths the limit is set by the ionosphere, that reflects radiation of longer wavelengths coming into the earth's surface (or going out). At the short wavelengths the limit is set by absorption in the the troposphere by molecules like H_2O (water vapor) and O_2 (molecular hydrogen).

The first detection of extraterrestrial radio waves was made by Karl Jansky in 1931. Jansky was a physicist working for the famous Bell Labs, with the assignment of identifying the sources of static that might interfere with radio voice transmissions. He built an antenna, shown in Figure 1, designed to receive radio waves at a frequency of 20.5 MHz (a wavelength of about 14.5 meters). It was mounted on a turntable that allowed it to rotate in any

A. Carramiñana et al. (eds.), Solar, Stellar and Galactic Connections between Particle Physics and Astrophysics, 199–213.

direction, so that one could find the direction of origin in azimuth of the radio signals. Jansky soon found that most of the static was coming from thunderstorms, both nearby and distant. But in addition he found a faint steady hiss of unknown origin. Over the years it became clear that Jansky had detected emission from the plane of our Galaxy, that emits synchrotron radiation (produced by relativistic electrons spiraling in the magnetic fields that exist in interstellar space). The emission was strongest in the direction of the center of our Galaxy.

Jansky wanted to investigate the nature of the radio waves from space and proposed to Bell Labs the building of a 30 meter diameter dish antenna. However, Jansky was assigned to other projects and did no more research in radio astronomy, dying in 1950 at the early age of 44 years.

Fig. 1. Karl Jansky with his radio telescope. Image courtesy of NRAO/AUI.

Professional astronomers in the epoch of Jansky did not take seriously the fact that the Universe could be studied at wavelengths other than those of the visible light and did not make any serious follow up of his work. Grote Reber, a ham radio operator, learned about Karl Jansky's discovery and constructed in his back yard in Wheaton, Illinois a radio telescope (see Figure 2) to learn more about cosmic radio waves. He was able to confirm the radiation detected by Jansky and to show that it was stronger at the longer wavelengths, a key

Fig. 2. The parabolic dish built by Grote Reber (who appears in the inset in a 1937 photograph, the same year when he built the radio telescope) in his backyard. Images courtesy of NRAO/AUI.

observational fact to interpret its nature as synchrotron radiation. Reber lived to be 90, dying in Tasmania in 2002.

Radio astronomy became institutional only after the Second World War, when the developments in radar technology opened the doors to the construction of large radio telescopes in many countries in the world. Since radio astronomy started working first in the long wavelengths (meter and centimeter), where atmospheric weather affects very little the radio observations, this new field was particularly attractive for countries like England or Holland, where secular poor weather and lack of high elevation sites had difficulted the development of modern optical observatories. Radio astronomy became an important alternative for these countries. With time, radio astronomical research included also the short wavelengths (millimeter and submillimeter), where the atmosphere affects seriously the observations and high and dry sites, similar to those for optical observatories, are required.

In the beginnings, angular resolution was a serious limitation of radio astronomy. The angular resolution of a telescope is given by

$$\theta \simeq \frac{\lambda}{D}, \tag{1}$$

where θ is the angular resolution in radians, λ is the observed wavelength, and D is the diameter of the telescope, in the same units that the λ. For a radio telescope with $D = 24$ m observing at $\lambda = 21$ cm, the angular resolution is 9×10^{-3} radians or about half a degree, the angular diameter of the Moon. This is a very poor angular resolution and optical astronomy had been achieving for decades angular resolutions thousands of times better (about one arc second). This lack of angular resolution in the early years of radio astronomy also helped explain the modest interest that the new science produced in the consolidated community of optical astronomers. Paradoxically, with time radio astronomy strongly developed interferometry, where at least two and preferably many radio telescopes observing simultaneously the same region in the sly are spread over a large region. The angular resolution of an interferometer is given by

$$\theta \simeq \frac{\lambda}{B}, \tag{2}$$

where now B is the maximum separation (or baseline, as it is usually called) between the radio telescopes that form the interferometer. B can be kilometers or even thousands of kilometers long and over the years radio astronomy eventually surpassed all other astronomies in angular resolution. Now intercontinental interferometers routinely reach angular resolutions of milliarc seconds.

2 The Achievements of Radio Astronomy

Beyond the scientific contributions that we will review later, the major achievement of radio astronomy is that it persuaded the astronomical community (then mostly optical astronomers) that it was worthwhile to look at the Universe at other wavelengths. Nowadays, we observe the Universe in all the windows of the electromagnetic spectrum: radio, infrared, optical, ultraviolet, X rays, and γ rays. Of these waves, only the visible light, part of the infrared, and the radio waves can be observed from the surface of the earth (see Fig. 3). The other electromagnetic windows had to wait for the development of space technology to blossom. Certainly, it was radio astronomy the one that opened the possibility of multiwavelength astronomy, that is now more the rule than the exception in attacking an astronomical problem. At present many studies

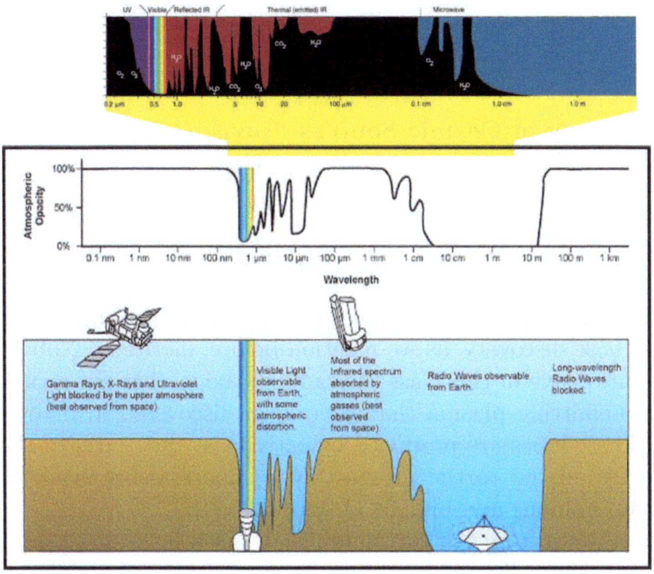

Fig. 3. Atmospheric opacity as a function of wavelength. Only the visible light, part of the infrared, and the radio waves can be observed from the surface of the earth. Other wavelengths are observed from space. Image courtesy of IPAC/NASA.

are being made combining results from several windows of the spectrum (plus a theoretical interpretation).

2.1 Nobel Prizes of Physics to Radio Astronomers

Perhaps one indicator that we could use to quantify the importance of radio astronomy is to look for results that deserved the Nobel Prize in Physics. On my count, there have been seven Nobel Prizes in Physics given to astronomers; of these three have been to radio astronomers:

• 1974: Martin Ryle and Antony Hewish ("for their pioneering research in radio astrophysics: Ryle for his observations and inventions, in particular of the aperture synthesis technique, and Hewish for his decisive role in the discovery of pulsars")

• 1978: Robert W. Wilson and Arno Penzias ("for their discovery of cosmic microwave background radiation")

• 1993: Russell A. Hulse and Joseph H. Taylor Jr. ("for the discovery of a new type of binary pulsar, a discovery that has opened up new possibilities for the study of gravitation")

But it would be a mistake to judge radio astronomy (or other branch of astronomy) by the number of Nobel Prizes in Physics received. Astronomy has its own goals and priorities that are key to astronomy and not necessarily so to physics. In what follows, we will try to present a wider panorama of the achievements of radio astronomy.

2.2 The Discovery of Cosmic Sources "Invisible" at Other Wavelengths

One on the most important contributions of radio astronomy was the discovery of new types of sources that since they do not emit at detectable levels at visible wavelengths, had remained "invisible" even for the most powerful optical telescopes. In this class we have the cosmic background radiation and the pulsars (whose discovery, as we mention before, merited a Nobel Prize in Physics). We also have in this class the radio galaxies, that are gigantic lobes of synchrotron-emitting plasma that have been discovered in many galaxies (see Fig. 4). These lobes are produced by activity at the center of the galaxy almost certainly related to the presence of a supermassive black hole with masses between millions and billions of solar masses.

Relativistic ejecta from the surroundings of black holes are usually traced via radio observations. The ejected plasma carries relativistic electrons and magnetic fields and this produces synchrotron radiation that is typically brightest in the radio regime. In addition to the radio galaxies and quasars, now it is known that also galactic black holes with masses of a few times that of the Sun can also produce relativistic ejecta that can be detected and studied in the radio. These black holes are part of a binary system and are called microquasars to emphasize that they reproduce in small scale the phenomena seen in quasars (see Fig. 5)

Another type of sources that are "invisible" at other wavelengths and where radio astronomy provides unique information are those sources embedded in large amounts of cosmic dust and gas, such as forming stars and forming galaxies. In these sources dust opacity can be so large that even infrared radiation is severely attenuated on its way out and only radio can penetrate the obscuration. These are sources that emit at wavelengths other than radio, but where dust obscuration renders them 'invisible". The origin and formation of sources can take place in conditions of enormous dust obscuration and its study benefits greatly from the radio contributions.

3 Some Recent Contributions

As in all fields of astronomy, at radio wavelengths the scientific results have continued to flow with great continuity. In the present section I have listed some recent results that sample the great variety of research being undertaken at present. By far, the list is not complete and only tries to convey the vitality and versatility of the field.

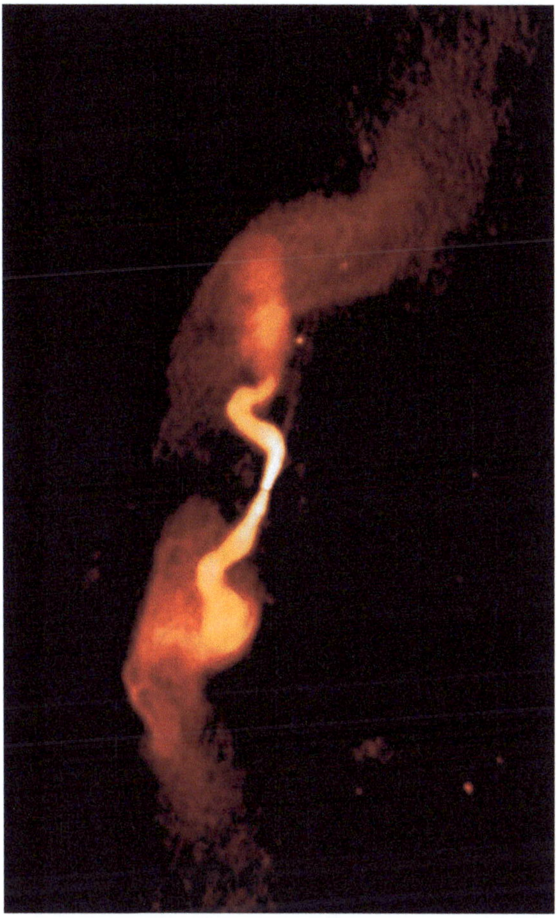

Fig. 4. This image shows the radio morphology of the radio galaxy 3C31. The bright twin jets emanate from the center of the galaxy (not visible in this radio image) and develop into distorted lobes at a distance of 300 kpc from the center. The radio structure is very large; our Galaxy has a diameter of about 25 kpc, about a tenth of the size of this radio galaxy. Image courtesy of NRAO/AUI.

3.1 Our Galaxy

The Sun and Stars

For a summary of the type of radio observations of the Sun been carried out in recent years, we refer the reader to section 2.5.3 of [1].

[2] were able to follow the radio emission in the wind-collision region of the archetype W-R+O star colliding-wind binary system WR 140. They find that the region is a bow-shaped arc that rotates as the orbit progresses and are able to model it.

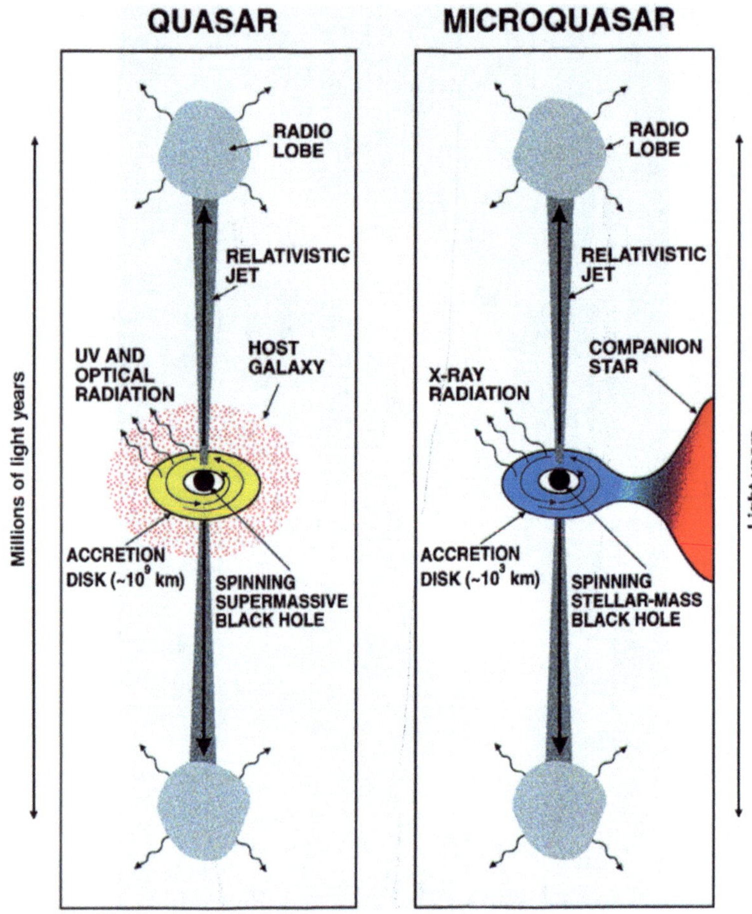

Fig. 5. The quasar-microquasar analogy. Diagram illustrating current ideas about quasars and microquasars (not to scale). As in quasars, the following three basic 'ingredients' are found in microquasars: (1) a spinning black hole, (2) an accretion disk heated by viscous dissipation, and (3) collimated jets of relativistic particles. But in microquasars the black hole is only a few solar masses instead of several million solar masses; the accretion disk has mean thermal temperatures of several million degrees instead of several thousand degrees; and the particles ejected at relativistic speeds can travel up to distances of a few light years only, instead of the several million light years as in some giant radio galaxies. In quasars matter can be drawn into the accretion disk from disrupted stars or from the interstellar medium of the host galaxy, whereas in microquasars the material is being drawn from the companion star in the binary system. In quasars the accretion disk has a size of 10^9 km and radiates mostly in the ultraviolet and optical wavelengths, whereas in microquasars the accretion disk has a size of 10^3 km and the bulk of the radiation leaves as X-rays. Image from [3].

Using VLBI observations, [4] have made very precise (2%) determinations of the trigonometric parallax of T Tau South. This technique could be extended to other gyrosynchrotron stars in regions of star formation.

Molecular Clouds and Star Formation

A recent review on ultra-compact H II regions and massive star formation has been made by [5]. An update of more recent results is presented in [6].

Observing at millimeter wavelengths, [7] and [8] have studied the density structure of molecular cores to test the "inside-out" collapse models. A detailed chemical study of the molecular core in 25 transitions of 9 molecules was presented by [9].

[10] presented images and kinematical data of a disk of dust and molecular gas around a high-mass protostar. This result is important because it has been proposed that high-mass stars form through accretion of material from a circumstellar disk, in essentially the same way as low-mass stars form. However, the alternative possibility that high-mass stars form through the merging of several low-mass stars should be further explored. Along these lines, the observations of proper motions in the BN object and the I source in the Orion KL region suggest the possible formation of a close binary or even a merger after a three-body encounter between young massive stars ([11]).

Glycine, the simplest of aminoacids, may have been detected by [12] using an improved search strategy for intrinsically weak molecular lines. However, [13] have questioned the detection arguing that several key lines necessary for a rigorous interstellar glycine identification have not yet been found.

Pulsars and Supernovae

[14] reported the detection of the 2.8-second pulsar J0737–3039B as the companion to the 23-millisecond pulsar J0737–3039A in a highly relativistic double neutron star system. It is expected that this true binary pulsar system will allow unprecedented tests of fundamental gravitational physics.

[15] report the discovery of isolated, highly polarized, two-nanosecond subpulses within the giant radio pulses from the Crab pulsar. The plasma structures responsible for these emissions must be smaller than one meter in size, making them by far the smallest objects ever detected and resolved outside the Solar System, and the brightest transient radio sources in the sky.

[16] identified 21 new millisecond pulsars in the globular cluster Terzan 5, bringing the total of known such objects in Terzan 5 to 24. These discoveries confirm fundamental predictions of globular cluster and binary system evolution.

[17] detected a radio counterpart to the 27 December 2004 giant flare from SGR 1806−20 and were able to obtain a high-resolution 21-cm radio spectrum that traces the intervening interstellar neutral hydrogen clouds. Analysis of this spectrum yields the first direct distance measurement of SGR 1806−20:

the source is located at a distance greater than 6.4 kpc and the authors argue that it is nearer than 9.8 kpc. If correct, this distance estimate lowers the total energy of the explosion and relaxes the demands on theoretical models.

[18] measured the proper motion and parallax for the pulsar B1508 + 55, leading to model-independent estimates of its distance (2.37 $\pm$ 0.22 kpc) and transverse velocity (1083 $\pm$ 100 km s^{-1}). This is the highest velocity directly measured for a neutron star.

Microquasars

The microquasar GRS 1915 + 105 has been studied exhaustively and the combination of radio and X ray data has advanced our understanding or the coupling between inflow (in an accretion disk) and outflow (in the relativistic jets) in this source ([19]). [20] find in the galaxy 3C120 that, as has been observed in microquasars, the dips in the X-ray emission are followed by ejections of bright superluminal knots in the radio jet. This result points to an accretion-disk origin for the radio jets in active galaxies.

The Galactic Center

The studies of the linear and circular polarizations associated with Sgr A* are expected to reveal crucial information with regard to the radio source [21, 22]. A radio image of Sgr A* at a wavelength of 3.5 mm, reported by [23], demonstrated that its size is ~1 AU. When combined with the lower limit on its mass, the lower limit on the mass density is 6.5×10^{21} $M_{\odot}$ pc^{-3}, which provides strong evidence favoring Sgr A* as a supermassive black hole.

[24] report a transient radio source, GCRT J1745-3009, which was detected during a moderately wide-field monitoring program of the Galactic Centre region at 0.33 GHz. The characteristics of its bursts are unlike those known for any other class of radio transient.

3.2 Extragalactic Sources

Galaxies

[25] report the detection in HI of what appears to be a dark halo that does not contain the expected bright galaxy. A galaxy with the observed velocity width would be expected to be 12 mag or brighter; however, deep CCD imaging has failed to turn up a counterpart down to a surface brightness level of 27.5 B mag $arcsec^{-2}$. However, [26] argue that this object is not really a dark halo but most likely tidal debris from the nearby galaxy NGC 4254.

[27] have reviewed the sample of 36 detected galaxies that have molecular masses in the range of 4×10^9 to 1×10^{11} $M_{\odot}$ and star formation rates derived from their FIR luminosities in the range of 300 to 5000 $M_{\odot}$ yr^{-1}. These objects

are generally starbursts in centrally concentrated disks, sometimes, but not always, associated with active galactic nuclei.

[28] measured the angular rotation and proper motion of the Triangulum Galaxy (M33) with VLBI observations of two H_2O masers on opposite sides of the galaxy. By comparing the angular rotation rate with the inclination and rotation speed, they obtain a distance of 730 ± 168 kiloparsecs. This distance is consistent with the most recent Cepheid distance measurement.

Gamma Ray Bursts

Radio astronomy played a key role in the elucidation of the nature of cosmic gamma ray bursts [29, 30].

Active Galaxies and Quasars

A review on mega-masers in external galaxies was completed by [31]. [32, 33] have questioned the so-called radio loud/quiet dichotomy by finding many "intermediate" radio galaxies in their samples. [34] show evidence that binary supermassive black holes may be produced by galactic mergers as the black holes from the two galaxies fall to the center of the merged system and form a bound pair. They propose that the winged or X-type radio sources are galaxy pairs in which this merging has occurred.

3.3 Cosmology

A major development was the accurate measurement and interpretation of the anisotropies in the cosmic microwave background [35]. The spectrum of amplitudes of temperature (or brightness) fluctuations expanded in multipole moments reported by the Wilkinson Microwave Anisotropy Probe (WMAP) satellite is a remarkable achievement that allowed accurate determination of the main cosmological constants [36, 37]. The detection of polarization in the cosmic microwave background [38] confirmed the predictions of the standard theory.

There also significant advances in the measurement of the Sunyaev-Zeldovich effect in clusters of galaxies [39, 40]. The realization that the first generation of stars may be detectable by means of radio observations of red-shifted atomic hydrogen [41] triggered the development of ad hoc radio telescopes for this purpose and results may be obtained in a few years.

4 The Challenges of the Future

4.1 International Level

At the international level the radio community faces two major challenges: 1) the construction and operation of the Atacama Large Millimeter Array

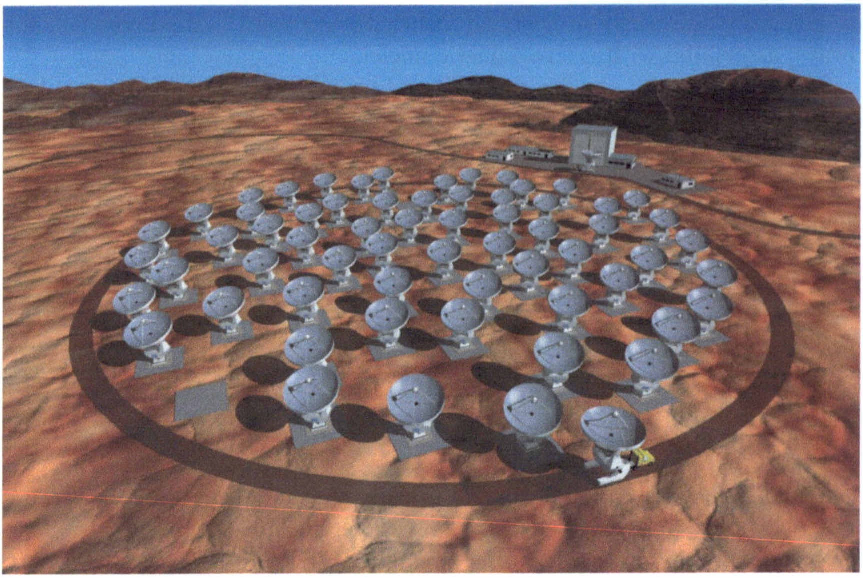

Fig. 6. Artistic depiction of the Atacama Large Millimeter Array. Image courtesy of NRAO/AUI and ESO.

(ALMA, see Fig. 6) and 2) the design and financing of the Square Kilometer Array (SKA).

The Atacama Large Millimeter Array (ALMA) project is a millimeter and sub-millimeter interferometer originally planned to be constituted by 64 radio antennas, each 12 meters in diameter. The project will be located in the Chajnantor plateau in northern Chile, one of the highest (5,000 meters), driest places on Earth. It is well underway, with major contributions from the USA, Europe and Japan, as well as other countries. This interferometer will revolutionize all we know from studies in that region of the electromagnetic spectrum. Perhaps the major challenge is, with the rising costs of petroleum and steel, to avoid a downsizing of the design (from example, reducing the number of antennas from 64 to 50) that could compromise the expected high performance (see [42] for an update on this issue).

The Square Kilometer Array (SKA) project has as its major goal to be the next generation interferometer for meter and centimeter wavelengths. As you may remember at the beginning of this article we mentioned that since radio wavelengths vary by a factor of 10^4, it is impossible to do all radio astronomy with a single instrument. ALMA will then cover the short wavelengths, from a few millimeters and shorter, and SKA will cover the longer wavelengths. At present, the Very Large Array (VLA) is the major centimeter interferometer in the world and is being upgraded, with collaborations from the USA, Canada, and México to become the Expanded Very Large Array (EVLA). New receivers and wider bandwidths will give one or two more decades of premier science with the EVLA. However, the total area of the EVLA is at present some 13,000 square meters. The SKA project proposes to build a new interferometer with a total area of about one square kilometer, about two orders of magnitude larger than the EVLA. The challenge here is that some new solution for the building of large surfaces in a relatively inexpensive way is needed. Several countries are interested in hosting the SKA.

4.2 Mexican Level

In a developing country like México, radio astronomy and all other scientific activities are also just developing. At present, we count with about one dozen radio astronomers distributed in a handful of institutions. The Center of Radio Astronomy and Astrophysics (CRyA) of the National University has led a successful effort to participate in the EVLA project, getting in exchange competitive access to both the EVLA and ALMA, in the same conditions of a

Fig. 7. The Large Millimeter Telescope in early 2006. Image courtesy of INAOE.

USA university. Another institution in México, the National Institute for Astrophysics, Optics, and Electronics has undertaken, in collaboration with the University of Massachusetts, the construction of the Large Millimeter Telescope (LMT, see Fig. 7). This telescope is a 50-m diameter single-dish telescope optimized for astronomical observations at millimeter wavelengths (0.85 mm$\leq \lambda \leq$4 mm). A principal scientific goal of the LMT is to understand the physical process of structure formation and its evolutionary history throughout the Universe. The telescope site is Volcán Sierra Negra (lat. of $+19°$), situated about 100 km east of INAOE, in the Mexican state of Puebla, at an altitude of 4,600 m. This project is well underway and it is expected that it will see first light in the year 2008. The LMT, in combination with the competitive access to the EVLA and ALMA as well as with the share of time that mexican astronomers have in the Gran Telescopio Canarias (Large Canary Islands Telescope) in Spain, open many observational possibilities that the young generation of mexican astronomers will seize.

References

1. V. Trimble, M. J. Aschwanden: Publications of the Astronomical Society of the Pacific **116**, 187 (2004)
2. S. M. Dougherty et al.: The Astrophysical Journal **623**, 447 (2005)
3. I. F. Mirabel, L. F. Rodríguez: Nature **392**, 673 (1998)
4. L. Loinard et al.: The Astrophysical Journal **619**, L179 (2005)
5. E. Churchwell: Annual Reviews of Astronomy and Astrophysics **40**, 27 (2002)
6. L. F. Rodríguez et al.: The Astrophysical Journal **627**, L65 (2005)
7. D. W. A. Harvey et al.: The Astronomical Journal **123**, 3325 (2002)
8. Y. L. Shirley et al.: The Astrophysical Journal **575**, 337 (2002)
9. N. J. Evans et al.: The Astrophysical Journal **626**, 919 (2005)
10. N. Patel et al.: Nature **437**, 109 (2005)
11. L. F. Rodríguez: Radio observations of ultracompact HII regions. In *Massive star birth: A crossroads of Astrophysics, IAU Symposium Proceedings of the international Astronomical Union 227*, ed by Cesaroni, R.; Felli, M.; Churchwell, E.; Walmsley, M. Cambridge (Cambridge University Press, 2005) pp. 120-127
12. Y.-J. Kuan et al.: The Astrophysical Journal **593**, 848 (2003)
13. L. E. Snyder et al.: The Astrophysical Journal **619**, 914 (2005)
14. A. G. Lyne et al.: Science **303**, 1153 (2004)
15. T. H. Hankins et al.: Nature **422**, 141 (2003)
16. S. M. Ransom et al.: Science **307**, 892 (2005)
17. P. B. Cameron et al.: Nature **434**, 1112 (2005)
18. S. Chatterjee et al.: The Astrophysical Journal **630**, L61 (2005)
19. R. Fender, T. Belloni: Annual Reviews of Astronomy and Astrophysics **42**, 317 (2004)
20. A. P. Marscher et al.: Nature **417**, 625 (2002)
21. G. C. Bower et al.: The Astrophysical Journal **571**, 843 (2002)
22. G. C. Bower et al.: Science **304**, 704 (2004)
23. Z.-Q. Shen et al.: Nature **438**, 62 (2005)

24. S. D. Hyman et al.: Nature **434**, 50 (2005)
25. R. Minchin et al.: The Astrophysical Journal **622**, L21 (2005)
26. B. Bekki et al.: Monthly Notices of the ROyal Astronomical Society **363**, 21 (2005)
27. P. M. Solomon, P. A. vanden Bout: Annual Reviews of Astronomy and Astrophysics **43**, 677 (2005)
28. A. Brunthaler et al.: Science **307**, 1440 (2005)
29. P. Mészáros: Annual Reviews of Astronomy and Astrophysics **40**, 137 (2002)
30. K. W. Weiler, N. Panagia, M. J. Montes, R. A. Sramek: Annual Reviews of Astronomy and Astrophysics **40**, 387 (2002)
31. K. Y. Lo: Annual Reviews of Astronomy and Astrophysics **43**, 625 (2005)
32. C. L. Drake et al.: The Astronomical Journal **126**, 2237 (2003)
33. L. C. Bassett et al.: The Astronomical Journal **128**, 523 (2004)
34. D. Merritt, R. D. Ekers: Science **297**, 1310 (2002)
35. W. Hu, S. Dodelson: Annual Reviews of Astronomy and Astrophysics **40**, 171 (2002)
36. C. L. Bennett et al.: The Astrophysical Journal (Supplement Series) **148**, 1 (2003)
37. D. N. Spergel et al.: The Astrophysical Journal (Supplement Series) **148**, 175 (2003)
38. J. M. Kovac et al.: Nature **420**, 772 (2002)
39. J. E. Carlstrom, G. P. Holder, E. D. Reese: Annual Reviews of Astronomy and Astrophysics **40**, 643 (2002)
40. K. Lancaster et al.: Monthly Notices of the Royal Astronomical Society **359**, 16 (2005)
41. V. Bromm, R. B. Larson: Annual Reviews of Astronomy and Astrophysics **42**, 79 (2004)
42. J. Kanipe: Nature **439**, 526 (2006)

Gamma-ray Astrophysics - Before GLAST

Alberto Carramiñana

Instituto Nacional de Astrofísica, Óptica y Electrónica, Luis Enrique Erro 1, Tonantzintla, Puebla 72840, México
alberto@inaoep.mx

Introduction: γ-rays and γ-ray Astronomy

Gamma-rays were first found as photons with energies in the range of tens of keV to several MeV produced by unstable nuclei. High energy photons can be produced by other types of processes, like inverse Compton scattering, expanding the γ-ray spectrum to arbitrarily high energies. A photon with energy $E \geq m_e c^2 \simeq 0.511\,\mathrm{MeV}$ can be defined as a γ-ray, making an exception at lower energies for (line) photons produced by radioactive decay. A finer division of the spectrum is often made on instrumental basis: low energy γ-rays are detectable via the photoelectric ($E \lesssim 1$ MeV) or Compton effects (1–30 MeV); high energy γ-rays (30 MeV–30 GeV) with pair production telescopes in orbit; very high γ-rays (100 GeV–10 TeV) are observed through the Čerenkov emission of secondary particles in the atmosphere; finally, ultra high energy γ-rays ($E \gtrsim 10$ TeV) can be studied by the direct detection of secondary particles in extended air shower arrays or the fluorescence emission they cause in atmospheric nitrogen nuclei. This review addresses the pair production and Čerenkov regimes, with lower emphasis on the low energy band.

Gamma-ray astronomy had an early -but slow- start with the first detection of Čerenkov light from atmospheric cascades, reported in 1953 [1]. In the 1960s the technique was already in use for the search of celestial γ-ray sources [2], but today we know that those instruments were too crude to permit real detections, as an effective method for separating cosmic-rays from γ-rays is mandatory. Solid evidence for celestial sources came from the *OSO-III* satellite, a spark chamber which detected the Galactic plane in the early 1970s [3]. Shortly after, *SAS-II* confirmed the *OSO-III* results, with evidence for a handful of point sources [4], prior to the second *COS-B* catalog of high energy γ-ray sources which contains 25 entries [5]. At present more than 250 sources of photons with $E > 100$ MeV are known, as listed in the third catalog compiled with EGRET on-board of the *Compton Gamma-Ray Observatory (CGRO)* [6]. An attempt to join results from different experiments lead to the general γ-ray source catalog [7]. In the low energy γ-ray band the

A. Carramiñana et al. (eds.), Solar, Stellar and Galactic Connections between Particle Physics and Astrophysics, 215–229.

first COMPTEL catalog reports about 30 steady sources in the 0.75–30 MeV interval [8]. In the hardest X-ray bands, the IBIS-*Integral* team compiled a catalog of over 200 sources at 20–100 keV from their Galactic plane survey [9].

In the absence of MeV and GeV telescopes in orbit, γ-ray astronomy is presently relying on the soft γ-ray data of *Integral*, in the 100s keV ,and on the new generation of Čerenkov telescopes, like the productive HESS array working around 300 GeV [10]. HESS is been joined by the MAGIC large aperture telescopes and the Veritas array [11, 12]. These pointed air Čerenkov telescopes are complementary to water Čerenkov experiments, like MILAGRO [13], that continuously cover a sizable fraction of the sky, although at lower point source sensitivity. HAWC, a proposed 90,000 m^2 high altitude water Čerenkov tank to be located above 4000 meter sea-level, has the potential of a wide area survey with a point source sensitivity equivalent to the Whipple telescope and prompt coverage of TeV transients [14]. New pair production telescopes, based on solid state technology, are also coming along. *AGILE* is an Italian compact telescope of sensitive area comparable to EGRET but much improved off-axis performance, almost ready for launch [15]. It is similar in concept, but smaller than *GLAST*, the *Gamma-ray Large Aperture Space Telescope* that will be in orbit by the end of 2007 [16]. *GLAST* will have a large collecting area, field of view and improved tracker reconstruction, resulting in over an order of magnitude increase in sensitivity compared to EGRET. This will result in an energy coverage overlap with ground based telescopes.

I will describe physical absorption processes behind γ-ray telescopes, often relevant in astrophysical scenarios (§1); then proceed to the most common production mechanisms of γ-rays, like neutral pion production and inverse Compton scattering (§2); this will lead to an outline of some relevant results in γ-ray astronomy and related astrophysical scenarios (§3).

1 Gamma-ray Absorption and Detection

The three basic interaction processes for high energy photons are the photoelectric effect, Compton scattering and pair production, each dominant in a different energy band (Figure 1). Photoelectric absorption is the basic process behind hard X-ray and low energy γ-ray telescopes; Compton scattering, the preferred absorption process at 1–10 MeV, is the basis for Compton telescopes; and pair production is the underlying process behind high energy space telescopes and very high energy ground-based telescopes and experiments.

1.1 Photoelectric Effect

In the photoelectric effect a photon of energy E is absorbed by an atom of ionization potential $\phi \leq E$. Most hard X-ray and low-energy γ-ray telescopes are based on photoelectric absorption in crystals like NaI. These have intrinsic poor angular resolution, requiring collimators or other devices to improve their

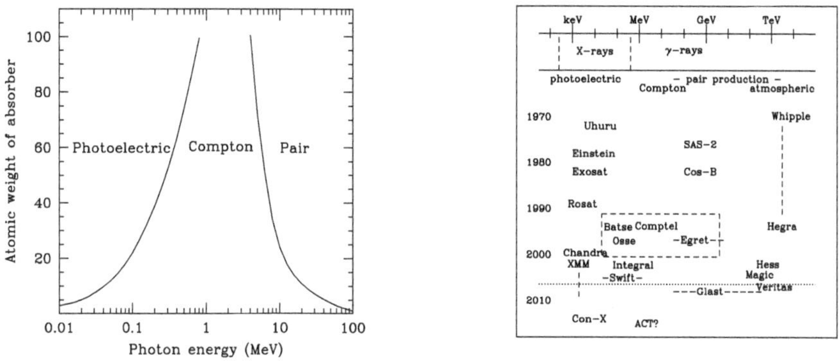

Fig. 1. *Left:* regimes of the dominant high energy absorption mechanisms. *Right:* some of the most relevant high energy astrophysics missions classified by date and energy range. The box denotes the CGRO satellites, with its four instruments inside.

response. The development of codded mask apertures has given a noticeable improvement in angular resolution, $\gtrsim 1$ arc-min for telescopes like *Swift* and *Integral*, wide field of view monitors to search for γ-ray bursts (GRBs) [17, 18]. OSSE and *Integral* are different low energy telescopes conceived with sufficient energy resolution and mapping capabilities to study the annihilation line at 0.511 MeV and radioactive decay lines in the interstellar medium [19, 20].

1.2 Compton Scattering

Compton scattering is the energy exchange between a photon and a charged particle, usually an electron, $e\gamma \to e\gamma$. Seen in the rest frame of the electron prior to the interaction, a photon of initial energy $\hbar\omega$ acquires an energy

$$\hbar\omega_1 = \frac{\hbar\omega}{1 + (\hbar\omega/mc^2)\,(1 - \hat{k}\cdot\hat{k}_1)}\,, \tag{1}$$

with $\hat{k}$ and $\hat{k}_1$ the initial and final direction of the photon momentum. Expression (1) describes the loss of energy of a photon scattered by an electron at rest. The differential cross section, $d\sigma/d\Omega$, is given by the Klein-Nishima expression, with dipolar shape at low energies and a well-defined preference for a photon recoil at high energies [21]. The total cross section decreases slowly from $\sigma = 8\pi r_e^2/3$ at low energies to $\sigma = r_e^2(1/2 + \ln 2x)/x$ at high energies, with $r_e = e^2/mc^2$ the classical electron radius and $x = \hbar\omega/mc^2$.

The Compton cross section dominates for 0.5 MeV $\lesssim \hbar\omega \lesssim$ 30 MeV (Fig. 1), with a dependence on the composition of the target [22]. This is a very thought energy regime technically with the added problem of a very high background induced by cosmic-ray collisions in the spacecrafts. COMPTEL, the only telescope to perform a sky survey in the low MeV range, followed a two level configuration design, with Compton scattering occuring in the upper D_1 layer of NaI detectors followed by photoelectric absorption in the lower D_2 layer. The locations and energy deposits of the interactions, E_1 and E_2, were measured and the scattering angle was recovered from eq. 1, with the total photon energy given by $E_\gamma = E_1 + E_2$. However, the scattering angle by itself is insufficient to point to the celestial source and telescopes like COMPTEL suffered from a severe azimuthal indetermination, resulting in a ring shaped point spread function of tens of degrees radius and over a degree wide. Another drawback was that $E_\gamma = E_1 + E_2$ assumes the scattered photon to be absorbed in D_2, rather than re-scattered outside the instrument. This was known not to be the general case and was considered in the COMPTEL analysis software. Even through the use of the Compton effect has proven to be much harder than the photoelectric effect or pair production, the increased cross section at MeV energies demands pursuing the design of new types of Compton telescopes. To obtain a significant improvement in sensitivity, Compton telescopes must include features like measuring the recoil electron to suppress the azimuthal uncertainty, as in the design of MEGA [23].

1.3 Pair Production

Two photon pair production, like $\gamma\gamma \to e^+e^-$, is an important process in regions with high photon density, like the environment GRBs or the vicinity of accretion disks. It also has been found to have an important effect over cosmological distances. In the center of momentum reference frame (CM), two photons of identical energy ω produce a particle pair of same energy[1], $\gamma = \omega$. Invariance under Lorentz transformations gives $\omega = \sqrt{\omega_1\omega_2(1-\cos\theta)/2}$, where ω_1 and ω_2 are the photon energies in the observer frame and θ is the angle between their momenta. The total cross section, of the form $\sigma = r_e^2 f(\omega)$, has a relatively narrow profile, decaying rapid from $\omega = 1$ to $\omega \sim 10$. The differential angular cross section is roughly isotropic for pair production just above threshold, $\omega \gtrsim 1$, becoming biased to pairs moving in the same directions as the original photons (in the CM!) at high energies.

Direct one photon pair production $\gamma \to e^-e^+$, forbidden in vacuum, can occur in the presence of a field or medium able to warrant energy-momentum conservation. In matter, pair production dominates over Compton scattering for energies above 30 or 50 MeV (Fig. 1). At high energies the cross section is given by the Bethe-Heitler cross section [21],

$$\sigma = \frac{28}{9}\alpha\, Z^2 r_e^2 \left\{ \ln(2\omega) - \frac{109}{42} \right\}, \quad \text{for} \quad \omega \gg 1\,. \tag{2}$$

Pair production telescopes have been the most successful γ-ray instruments so far. Their design is based on combining a converter system, where electron-

[1] in units where $\hbar/mc^2 = 1$, used hereafter, except when energy units are quoted.

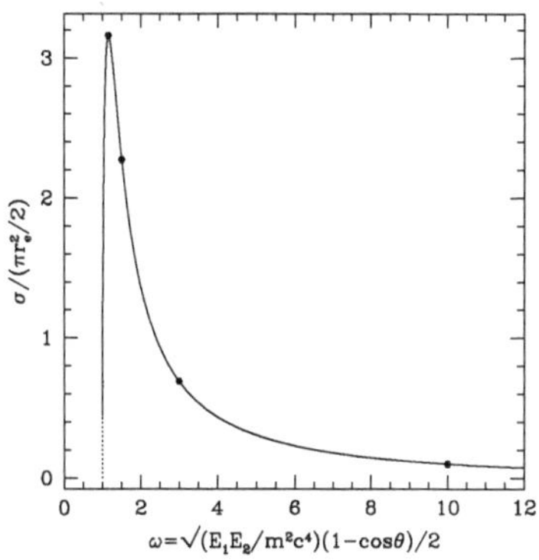

Fig. 2. The total $\gamma\gamma \to e^+e^-$ cross section as a function of $\omega = \sqrt{\omega_1\omega_2(1-\cos\theta)/2}$, the energy of each photon -and lepton- in the center of momentum reference frame.

positron pairs are materialized, with a tracker system, to reconstruct the trajectories of each $e^\pm$, and a calorimeter, to measure the total energy of the pair. *GLAST* incorporates sensitive solid state elements in the tracker, avoiding the use of expendable gas needed in previous spark chamber based telescopes.

When applied to air, expression (2) gives an absorption path of $\xi_p = \mu m_H/\sigma \approx 37$ g cm^{-2} at $\hbar\omega \simeq 1$ TeV. With a total depth of 1032 g cm^{-2}, the Earth atmosphere is about 28 absorption lengths thick, ensuring the absorption of incoming γ-ray photons. The bremsstrahlung (§ 2.3) $e \to \gamma e$ and pair production cross sections are intimately related at high energies, resulting in very similar paths for secondary high energy electrons to produce new photons and photons to produce new pairs. The succession of pair production and bremsstrahlung processes results in the development of electromagnetic cascades, which grows until Compton scattering takes over pair production ($\hbar\omega \sim 10$ MeV) and photo-ionization over bremsstrahlung ($\gamma mc^2 \sim 84$ MeV). Secondary particles of a cascade can reach sea level for primaries with 10^{15}eV, allowing the use of arrays of particle detectors as γ-ray detectors. In practise the atmosphere acts as a γ-ray converter and ground-based telescopes serve as calorimeters, detecting secondaries either directly or through the Čerenkov radiation produced when moving faster than light in the medium. The index of refraction of air is $n \simeq 1 + 3 \times 10^{-4}$, and the Čerenkov condition $v > c/n$ becomes $\gamma > 1/\sqrt{2(n-1)}$ for $(n-1) \ll 1$, or $\gamma mc^2 \gtrsim 20$ MeV for air.

One of the main problems for ground-based γ-ray telescopes resides in the high flux of cosmic-rays, which also produce atmospheric cascades. The first interaction of a cosmic-ray nucleon in the atmosphere differs in producing charged and neutral pions, leading to a more complex cascade with muonic, nucleonic and electromagnetic components. Cosmic-rays can be rejected through numerically modelling nucleon initiated and photon initiated cascades and comparing the predicted and measured distributions of Čerenkov light or particles on the ground. Imaging the Čerenkov light provides an effective method of cosmic-ray rejection, as first demonstrated by the 1989 Whipple high significance detection of the Crab nebula [24].

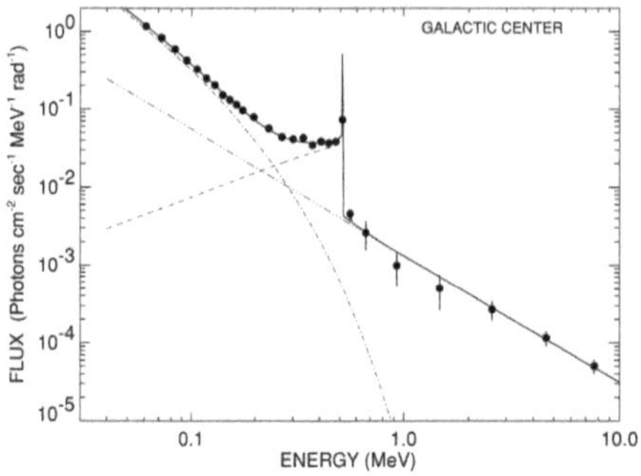

Fig. 3. The narrow annihilation 0.511 MeV line overimpossed on the positronium continuum, observed by OSSE in the Galactic plane.

A different kind of very high energy γ-ray detector is made taking into account that the index of refraction of water is 1.33, implying a threshold $\gamma > 1.52$. Practically any cascade $e^{\pm}$ reaching water radiates therein. Water Čerenkov detectors are better suited for high altitude sites in order to maximize the number of cascade $e^{\pm}$. MILAGRO is among the first of these TeV monitors that observe a large portion of the sky and, through precise timing of signals in its phototubes, they reconstruct the direction of arrival of events [13]. Larger area detectors, like the 150×150m MiniHawc proposal, may be able to provide the sensitivity required to a proper monitoring of flaring celestial sources like active galactic nuclei (AGNs) and GRBs.

2 Gamma-ray Production

2.1 γ-rays from Nuclear Decay

The production of nuclear γ-rays in the interstellar medium can be due to the excitation of atomic nuclei by cosmic-rays or through direct production of unstable species in violent events, like novae or supernovae. A particular case of interest in astrophysics is the neutron capture reaction, $n + \mathrm{H} \to \mathrm{D} + \gamma$, which gives 2.23 MeV photons, as observed in solar flares [25]. Gamma-ray line astronomy has succeeded in detecting the decay of radioactive species of particular astrophysical interest [20]. For example, COMPTEL and *Integral* have mapped the distribution of the 1.8 MeV line produced by the decay of ^{26}Al in the Galactic plane. The short lived ^{44}Ti, with a decay time of just 80 years, was detected in the young supernova remnant Cas A [26].

2.2 Electron - Positron Annihilation

Electron-positron annihilation, usually $e^+e^- \to \gamma\gamma$, can proceed through two photon or three photon production, $e^+e^- \to \gamma\gamma\gamma$, the later via the intermediate creation of a positronium e^+e^- bound system. In the CM the cross section is inversely proportional to the speed of each of the particles, $\sigma \propto 1/\beta$, favouring annihilation of low speed pairs and the emission of a thin line spectrum.

The decay of the positronium into three photons gives an underlying continuum of photons with energies $\hbar\omega \lesssim mc^2$. This Galactic plane emission was first detected with a balloon experiment, then measured with OSSE on board *CGRO* (Fig. 3) and more recently better mapped with *Integral* [27, 19, 28].

2.3 Bremsstrahlung

Bremsstrahlung radiation is produced when an electron of energy $E = \gamma mc^2$ is deflected during encounters with atomic nuclei of charge Ze. A relativistic electron travelling through a medium of nuclei number density n suffers an exponential energy loss, $dE/dt = -(c/x_0)E$, of characteristic length,

$$x_0 = \left[\frac{4Z(Z+1.3)e^6}{\hbar c\, m^2 c^4}\, n \left(\ln\left(\frac{183}{Z^{1/3}}\right) + \frac{1}{8}\right)\right]^{-1}, \tag{3}$$

with $x_0^{-1} = \sum_i x_i^{-1}$ for a mixture of media. For air this gives $\xi_b \simeq 36$ g/cm^2, close to the pair production value, both processes working at the same scales in atmospheric cascades. The relativistic bremsstrahlung energy spectrum is flat, equivalent to a photon number spectrum $N(\hbar\omega) \propto 1/\hbar\omega$. When integrating over a power-law distribution of electrons with $E \geq E_0$, spectral index p and energy density u_e one obtains a power-law spectrum

$$I_\gamma(\hbar\omega) = c/x_0 \left(u_e/E_0^2\right) \left(\hbar\omega/E_0\right)^{-p} . \tag{4}$$

2.4 Inverse Compton Scattering

Compton scattering can be a mechanism of photon energy loss or gain, depending on the electron involved. Equation (1) applies in the original rest frame of the electron and does not describe photons interacting with relativistic electrons, i.e. inverse Compton scattering. This is better described with transformations to and from the CM frame, where the energies of the photons (ω_{cm}) and electrons (γ_{cm}), related by $\gamma_{cm}^2 = 1 + \omega_{cm}^2$, remain unchanged through the interaction, which consists in a re-orientation of the momenta weighted with the CM differential cross section. The scattering of a photon from ω to ω_1 seen by an arbitrary observer, can be translated to the CM using,

$$\omega_0' = \gamma^* \omega (1 - \boldsymbol{\beta}^* \cdot \hat{k}), \quad \omega_1 = \gamma^* \omega_1' (1 + \boldsymbol{\beta}^* \cdot \hat{k}_1'), \tag{5}$$

with $(\omega_0', \hat{k}_0')$ and $(\omega_1', \hat{k}_1')$ describing to the photon before and after the interaction as seen in the CM, and $\boldsymbol{\beta}^* = (\gamma\boldsymbol{\beta} + \omega\hat{k})/(\gamma + \omega)$ defining the transformation from the observer frame to the CM.

In the CM $\omega_0' = \omega_1' = \omega_{cm}$, and $\hat{k}_1'$ relates to $\hat{k}_0'$ probabilistically via the cross section. If $\omega_0' \ll 1$ the Thompson cross section, $\propto [1 + (\hat{k}_0' \cdot \hat{k}_1')^2]$, dominates favoring $\hat{k}_2' \cdot \boldsymbol{\beta}^* \approx 0$ and $\omega_1 \approx \gamma^2 \omega (1 - \boldsymbol{\beta} \cdot \hat{k})$ for a relativistic electron. In the opposite case ($\omega_0' \gg 1$) a perfect recoil ($\hat{k}_1' = -\hat{k}_0'$) is the preferred interaction, leading to

$$\omega_1 = \frac{\gamma^2\omega\ [\hat{k} - \boldsymbol{\beta}]^2}{1 + 2\gamma\omega(1 - \boldsymbol{\beta}\cdot\hat{k})} \longrightarrow \frac{2\gamma^2\omega\,(1-\cos\theta)}{1+2\gamma\omega\,(1-\cos\theta)} \longrightarrow \gamma\,, \tag{6}$$

with θ the interaction angle (observer) and $\rightarrow$ indicating the $\beta \rightarrow 1$ limit.

The γ-ray spectrum arising from the interaction of isotropic populations of monoenergetic electrons and photons is obtained considering the $\hat{k}_0' \rightarrow \hat{k}_1'$ distribution consistent with the angular cross section, averaged and integrated over incidence angle θ. This mono-mono spectrum can be weighted with input photon and electron populations, like a black body and power-law combination, to give the Compton component of a γ-ray spectrum model.

The Synchrotron Connection and Curvature Radiation

The distribution of cosmic-ray electrons is generally much harder to know than that of nucleons. As their energy losses are much more rapid, cosmic-ray electrons are short lived and hardly propagate after acceleration, as shown by the low flux of cosmic-ray electrons above Earth's atmosphere, $\sim 1\%$ of the total cosmic-ray flux [29]. Considering the energy dependence of electron synchrotron lifetime, $\tau = 0.93 \times 10^6$ yrs $(B/3\mu\mathrm{G})^{-2}(\gamma mc^2/\mathrm{TeV})^{-1}$ and their highly diffusive propagation, their density is bound to be a complicated function of the distance to cosmic-ray sources, magnetic field and energy. On the other hand, the synchrotron radiation of cosmic-ray electrons in galactic magnetic fields is the origin of the radio emission of the galaxy. The spectral index of the optically thin region of the radio spectrum, s, is directly related to the spectral index of the electron distribution, p, through $s = (p-1)/2$.

Synchrotron emission works mostly below the γ-ray range. However, the same basic process is behind curvature radiation, a relevant emission mechanism in highly magnetized neutron stars where energetic charged particles are constrained to move along magnetic field lines. Relativistic $e^{\pm}$ moving along a field line have significant radial acceleration due to the large curvature radius, R_c, defined by the assumed dipolar geometry ($R_* \lesssim R_c \lesssim c/\Omega$). This gentle radial acceleration translates into radiation in the γ-ray range. Energy losses, $d(\gamma mc^2)/dt = \gamma^4\left(2e^2c/3R_c^2\right)$, occur in timescales larger than the $e^{\pm}$ travel time along the magnetic field lines for $\gamma mc^2 \lesssim 1$ TeV -assuming $R_c = R_* = 10$ km- with the production of GeV photons. Given an acceleration mechanism along the magnetic field lines, the curved motion ensures MeV to GeV emission through the standard synchrotron formulas. Elaborate pulsar models have been constructed following this principle [30, 31].

2.5 Hadronic Production of γ-rays

Energetic hadrons produce γ-rays during nuclear collisions, the most relevant process for high energy astrophysics been the intermediate production and decay of neutral pions, $\pi^0 \rightarrow \gamma\gamma$. In normal galaxies most of the γ-rays come from cosmic-ray nucleons colliding with interstellar medium particles. More violent scenarios involve hadronic collisions in mildly relativistic shocks apply to supernovae, while highly relativistic shocks occur in GRBs or blazars.

Neutral pion decay into two photons of identical energy, $\omega = m_\pi/2$ and opposite momenta in the pion rest frame. In the observer frame the photon energies are are $\omega_\pm = \gamma_\pi m_\pi(1 \pm \beta_\pi \cos\theta)/2$, with $\boldsymbol{\beta}_\pi$ the pion velocity and θ the angle between the photon momenta in the pion frame and the pion velocity in the observer frame. Averaging over solid angle, photons are produced with a flat spectrum in the range $\frac{1}{2}\gamma_\pi m_\pi(1-\beta_\pi) \leq \omega \leq \frac{1}{2}\gamma_\pi m_\pi(1+\beta_\pi)$. Taking one step back, neutral pion production is mediated by the production of the intermediate $\Delta(1232)$ particle, conserving energy and momentum at head step. This has an impact on the π^0 and γ-ray spectra produced by protons or nucleons of given kinetic energy, as calculated for the cosmic-ray population of the Milky Way [32].

2.6 Particle Acceleration

High energy electrons and nucleons are the basic ingredient for γ-ray production and the fundamental connection between cosmic-ray and γ-ray astrophysics. Except for the low energy range, γ-ray production requires a particle acceleration mechanism. In 1949 Fermi proved that a series of collisions between a macroscopic system and a microscopic charged particle can lead to a power-law spectrum of high energy particles, similar to the observed cosmic-ray spectrum [33]. Fermi first consideration of particle acceleration due to interstellar turbulence evolved to the present paradigm of diffusive shock acceleration in supernova fronts as the source of Galactic cosmic-rays [29]. With the important exceptions of electrodynamical particle acceleration models in pulsars, either in polar caps or outer gaps [34, 35], or in supermassive black holes [36], most γ-ray production scenarios involve shock acceleration. Acceleration models can be divided as leptonic or hadronic [37, 38]. Electrons can be accelerated up to radiative loss limit, while acceleration of proton and nuclei is limited to their Larmor radii reaching the size of the accelerating region, ℓ. Hadrons can reach larger energies, but -even assuming no losses- they are ultimately limited by the accelerating region, the velocity of the front shock, βc, and the magnetic field, B, to $E_{max} \lesssim \beta \ell B$, as illustrated by Hillas [39].

3 Celestial Sources of γ-rays

The third EGRET catalog contains 271 entries of which close to two thirds are unidentified γ-ray sources of $E > 100$ MeV photons [6]. Non catalogued known EGRET sources are the Galactic plane, GRBs, the Moon and the extragalactic γ-ray background [40, 41, 42, 43]. The signification non detection of the quiet Sun has a bearing on normal stars as a class [42]. Identified sources

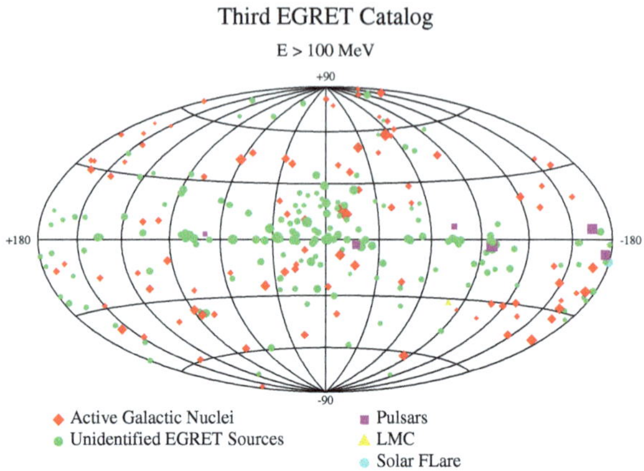

Fig. 4. The Third EGRET Catalog of γ-ray sources [6]. Note the two groups of unidentified sources, in (green) circles: those of the first type are preferentially distributed along the Galactic plane; the second group appears clustered around the Galactic center.

include solar flares, pulsars, normal galaxies and radio loud flat spectrum blazars [44, 45]. We believe unidentified sources include Galactic sources like radio quiet pulsars, supernova remnants and, maybe, black hole related objects like miniquasars [46]. Recently HESS has found several unidentified γ-ray sources along the Galactic plane which are under investigation [47]. Another active topic of research is pair production absorption of TeV photons, an unexpected mechanism to explore the far infrared extragalactic background [48].

3.1 The Galactic Plane and Star Forming Galaxies

The most prominent feature in the 100 MeV sky is the diffuse emission of the Galactic plane found by the *OSO-III* satellite [3]. This emission is fairly well understood and modelled as the interaction of cosmic-rays with interstellar gas [40]. Supernova are believed to provide the cosmic-rays which create the Galactic γ-ray emission through bremsstrahlung, π^0 production and inverse Compton scattering [49]. This process must also be in action in other normal galaxies. If cosmic-rays are produced locally within the galaxies, their γ-ray emission must scale with supernova rate and interstellar medium density. The non-detection of M 31 and the SMC suggests their cosmic-ray density is lower than that of the Milky Way and the overall paradigm of the local origin of cosmic-rays. The detection of some nearby normal galaxies seems to be certain during the *GLAST* era, at least for the LMC, the SMC, M 31 and maybe even M 33 [50]. Within the same scenario, starburst galaxies must possess a high density of cosmic-rays and be γ-ray sources. EGRET was not able to detect a sample of starbursts galaxies but the upper limits do not constrain significantly the properties of these, like the transfer of SNe energy into cosmic-rays [51]. *GLAST* will have good prospects for detecting starburst galaxies, from NGC 253 to Arp 220, and luminous infrared galaxies also [52].

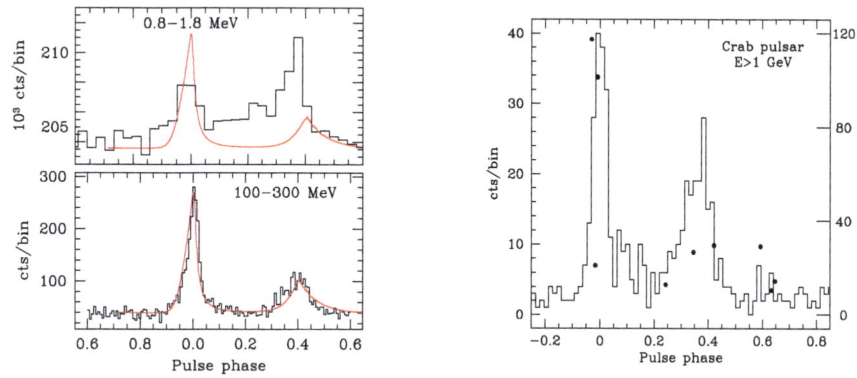

Fig. 5. *Left:* the optical light curve of the Crab pulsar -solid line- compared to the COMPTEL and EGRET histograms. *Right:* the GeV light curve of the Crab pulsar. Dots indicate the energies -right-hand side axis- and phases of $E > 10$ GeV photons.

3.2 Pulsars

Pulsars were the first type of γ-ray source identified, their pulsed signal providing a perfect identification signature. The Crab pulsar was found as a γ-ray source in balloon experiments [53], while the Vela pulsar required the confirmation of marginal balloon detections by *SAS-II* [54, 55]. Both pulsars were clearly seen by *COS-B* [56] and remained the only pulsars firmly established as high energy γ-ray emitters until the launch of *CGRO*. EGRET detected over half a dozen pulsars above 100 MeV [57], with COMPTEL supplying significant detections in the 0.75–30 MeV range for the Crab and Vela and significant *non-detections* of Geminga and PSR 1706–44 [58]. Of particular relevance during the *CGRO* era was the discovery that Geminga is a radio-quiet γ-ray loud pulsar [59]. This had led to the idea that of most of the unidentified EGRET sources might be radio-quiet γ-ray pulsars [60].

Crab, Vela and Geminga are the brightest sources above 100 MeV. As pulsars they show a double peaked light curve with $\Delta\phi \simeq 0.4$ peak-to-peak separation for Crab and Vela, and $\Delta\phi \simeq 0.5$ for Geminga. But they do show their own peculiarities. For example, the secondary peak of the Crab light curve becomes dominant in the MeV region to return to a secondary status above 100 MeV [61]. The Crab light curve keeps a similar shape over more than nine orders of magnitude, from the near infrared to 1 GeV, although the pulse separation has a slight but real increase with energy (Fig. 5). PSR B1706–44 is a relatively bright pulsar coincident with a *COS-B* source which shows a single broad pulse [62]. The detection of the old pulsar PSR B1055–52 implies a high degree of beaming together with a high efficiency in converting rotational energy into γ-rays, prompting for an age-efficiency relation. Interesting physics apply to PSR B1509–58, a pulsar with a large magnetic field, $B \sim B_{cr} \simeq 4.4 \times 10^{13}$ G, detected by COMPTEL up to $\gtrsim 10$ MeV, where photon splitting appears to play a major role in its magnetosphere [64, 65].

Čerenkov telescopes have detected a few plerions, most notably the Crab, but have found no pulsar to date. Interestingly, the Crab is a pulsed source in the highest end of the EGRET range, certainly up to $E \gtrsim 4$ GeV, with marginal evidence above 10 GeV. In fact the highest energy photon detected by EGRET, at 120 GeV, is in phase with the Crab main pulse (Fig. 5). This is below but close to the 0.25–4 TeV range where the Whipple telescope find no pulsations [66] and *inside* the energy interval of the Celeste experiment, which set a limit of 12% to the pulsed fraction of the Crab signal above 60 GeV [67]. *GLAST* will be able to resolve this near-conflict situation, measuring the pulsed and unpulsed components of the Crab spectrum between 10 and 100 GeV. Although the common consensus is that the Crab emission seen by Čerenkov telescopes originates in the nebula, it is theoretically feasible that pulsars might produce unpulsed high energy photons [68].

3.3 Supernova Remnants and Unidentified EGRET Sources

Supernova remnants (SNR) have been expected to be γ-ray sources since the 1949 paper of Fermi -who associated particle acceleration to interstellar turbulence in 1954 [33, 69]. Moving shocks are able to accelerate charged particles with a power-law spectrum close to E^{-2}. Indirect evidence that SNRs are the sources of cosmic-rays resides in the energetics of supernova explosions, which release energy at a rate large enough to sustain the cosmic-ray energy density. SNR have the power and the means to produce Galactic cosmic-rays, with energies up to $E \lesssim 10^{15}$ eV [70]. There is indeed a positional correlation between EGRET sources and SNRs [71]. However, the physical association is not fully confirmed by the EGRET data, which cannot rule out a pulsar origin for the GeV emission. A powerful diagnostic would be for *GLAST* to resolve the extended emission structure of a SNR. In that respect the most clear evidence for diffusive shock acceleration of cosmic-rays comes from Čerenkov telescopes, like HEGRA and Hess, which have been able to map the extended TeV emission in some SNR [72].

Most of the 3EG catalog entries are unidentified γ-ray sources within our Galaxy, with at least two populations of objects: bright sources in the Galactic plane, $|b| \lesssim 5°$, form a first group of objects; a second group is made of fainter sources forming an halo around the Galactic center direction [73]. The first group matches the distribution of known γ-ray pulsars and is believed to correspond to young objects: the second group might be formed by older recycled pulsars, like millisecond pulsars, as suggested by Grenier [74]. Other Galactic sources could be black holes, miniquasars and molecular clouds. Although most of the unidentified sources are Galactic, a smaller component of extragalactic objects is not ruled out.

Motivated by the serendipitous discovery of an unidentified TeV sources, the HESS collaboration has been performing a Galactic Plane survey which has uncover more unidentified γ-ray sources. *GLAST* might accurately locate some of these to encourage their multiwavelength identification [75, 47].

3.4 Blazars and the Extragalactic Background

Prior to the 1991 launch of *CGRO* only one extragalactic γ-ray source was known, 3C273 as observed by *COS-B* [76]. Aside from a single radio galaxy detection, Cen A, EGRET detected fifty to eighty radio loud flat spectrum sources [45]. These radio sources correspond with quasars or Bl Lac objects, a large fraction of which show strong variability and/or superluminical motions. The data support scenarios based on the supermassive black hole paradigm of AGNs. Particles accelerated in relativistic shocks inside jets pointed towards the observer produce γ-rays either through π^0 decay of hadrons or inverse Compton scattering of X-ray or optical photons from an accretion disk, or photons produced by synchroton from the same relativistic electrons, forming the Synchrotron Self Compton picture. The present data cannot distinguish between hadronic and leptonic models.

During the 1990s, Čerenkov telescopes failed to detect most of EGRET blazars, except for the nearest of them, Mk 421. The TeV detections of Mk 501 (undetected by EGRET) prompted the suggestion that TeV photons could be absorbed by the IR extragalactic background [77]. Further observations provided blazar spectra fitted with a power law spectrum attenuated with a reasonable infrared background model [78]. In fact the background responsible of attenuating TeV photons has not been directly measured and the γ-ray data are providing lower limits, still in debate [48]. *GLAST* will be able to detect over a thousand blazars, providing numerous lines of sight to test extragalactic pair absorption, in coordination with HESS, MAGIC and VERITAS.

The extragalactic γ background was found and measured by *SAS-II* [79]. EGRET determined the 30 MeV–100 GeV spectrum, with COMPTEL covering 0.8–30 MeV [43, 80]. Although the flux and spectral index are consistent with non resolved blazars, much interest in its study with γ-ray instruments remains due to the possibility of dark matter signatures [81].

3.5 Gamma-ray Bursts in the GeV Regime

GRBs are extremely brief and intense burst of high energy radiation. They are isotropically distributed and have been associated to high redshift galaxies which, together with their non-repeatability, led to models which consider catastrophic events, either highly beamed supernova explosions or mergers of compact objects, like neutron stars and/or black holes. They often show afterglow emission at lower energies, which allowed the identification of long duration (> 2 s) events with high redshift galaxies. They are extensively reviewed in the literature and will not be described in detail here (Ramirez Ruiz in this volume). As they are more easily monitored in the tens to hundreds of keV, their behaviour in the actual γ-ray regime is deduced from a relatively small sample of events detected by COMPTEL and EGRET. GRBs show power-law, or broken power-law spectra in the MeV region [82]. Of particular interest was GRB 940217, where GeV photons were detected over an hour after the BATSE trigger, probably as a high energy afterglow [83]. These γ-ray data established the need of anisotropic emission to overcome pair production absorption in the highly photon dense environment [84]. The detection of a 18 GeV photon in this event suggests that Čerenkov detections might be feasible. Another burst of interest was GRB 941017, modelled with a distinct high energy component extending beyond 200 MeV, not seen before in previous GRBs [85]. *GLAST* is expected to observe dozens of GRBs, probing their high energy emission, providing excellent targets for ground-based telescopes and allowing the study of extragalactic pair absorption to these objects.

References

1. Galbraith, W. Jelley, 1953, Nature 171, 349.
2. Fazio, G.G. et al. 1968, ApJ 154, L83.
3. Kraushaar, W.L., et al. 1972, ApJ 177, 341.
4. Fichtel, C.E., et al. 1975, ApJ 198, 163.
5. Swanenburg, B.N, et al. 1981, ApJ 243, L69.
6. Hartmann, R.C. et al. 1999, ApJS, 123, 79.
7. Macomb, D.J., Gehrels, N. 1999, ApJS 120, 335.
8. Schönfelder, V. et al. 2000, A&A Suppl. 143, 145.
9. Bird, A.J., et al. 2006, ApJ 636, 765.
10. http://www.mpi-hd.mpg.de/hfm/HESS/HESS.html
11. http://wwwmagic.mppmu.mpg.de/
12. http://veritas.sao.arizona.edu/
13. http://www.lanl.gov/milagro/
14. Sinnis, G., Smith, A., McEnery J.E 2004, astro-ph/0403096.
15. http://agile.rm.iasf.cnr.it/
16. http://glast.gsfc.nasa.gov/
17. http://swift.gsfc.nasa.gov/
18. http://integral.esa.int/
19. Kiner, R.L., et al. 2001, ApJ 559, 282.
20. Diehl, R. 2005, Nuclear Physics A. 758, 225.
21. Berestetskii, V.B., Lifshitz, E.M., Pitaevskii, L.P., *Quantum electrodynamics*, ed. Pergamon Press, Oxford 1980.
22. Fichtel, C.F., Trombka, J.I., 1997, *Gamma-Ray Astrophysics: New Insight into the Universe*, NASA Technical Report 1386.
23. Ryan J.M. et al. 2004, Proc. SPIE 5488, 977.
24. Weekes T.C., et al. 1989, ApJ 342, 379.
25. Rank, G. et al. 2001, A&A 378, 1046.
26. Iyudin, A.F. et al. 1994, A&A 284, L1.
27. Johnson, W.N., Harden, F.R., Haymes, R.C. 1972, ApJ 172, L1.
28. Churazov, E., Sunyaev, R., Sazonov, S. et al., 2005, MNRAS 357, 1377.
29. Gaisser, T. 1991, *Cosmic-rays and particle physics*, Cambridge University Press.
30. Cheng, A.F., Ruderman, M.A. 1977, ApJ 212, 800.
31. Harding, A.K. Tademaru, E., Esposito, L.W. 1978, ApJ 225, 226.
32. Dermer, C.D., 1986, A&A 157, 223.
33. Fermi, E., 1949, Phys. Rev. 75, 1169.
34. Ruderman, M.A., Sutherland, P.G., 1975, ApJ 196, 51.
35. Cheng, K.S., Ho, C., Ruderman, M., 1986, ApJ 300, 500.

36. Blandford, R.D., Znajek, R.L., 1977 MNRAS 179, 433.
37. Pohl, M., Esposito, J.A., 1998, ApJ 507, 327.
38. Kazanas, D., Ellison, D.C., 1986, ApJ 304, 178.
39. Hillas, A.M., 1984 ARA&A 22, 425.
40. Hunter, S.D., et al. 1997, ApJ 481, 205.
41. Jones, B.B., et al. 1996 ApJ 463, 565.
42. Thompson, D.J., et al. 1997, J. Geophys. Res. 102, 14735.
43. Sreekumar, P., et al. 1998, ApJ 494, 523.
44. McLaughlin, M.A., Cordes, J.M. 2000, ApJ 538, 818.
45. Mattox, J.R., Hartmann, D.C., Reimer O. 2001, ApJS 135, 155.
46. *The Nature of Unidentified Galactic High-Energy Gamma-Ray Sources*, ed. A. Carramiñana, O. Reimer, Thompson D.J., Kluwer Academic Press 2001.
47. Aharonian, F., et al. 2006, ApJ 636, 777.
48. Aharonian, F., et al. 2006, Nat. 440, 1018.
49. Bertsch, D.L., et al. 1993, ApJ 416, 587.
50. Pavlidou, V., Fields, B. 2001, ApJ 558, 63.
51. Blom, J.J., Paglione, T.A.D., Carramiñana, A. 1999, ApJ 516, 744.
52. Torres, D.F. 2004, ApJ 617, 966.
53. Browning, R. Ramsden, D., Wright, P.J. 1971, Nat. Phys. Sci. 232, 99.
54. Albats, P. et al. 1974, Nat. 251, 400.
55. Thompson, D.J. et al. 1975, ApJ 200, L79.
56. Bennett, K., et al. 1977, A&A 61, 279.
57. Nolan, P.L. et al. 1996 A&A Suppl. 120, 61.
58. Carramiñana, A., et al. 1995, A&A 304, 258.
59. Halpern, J.P., Holt, S.S. 1992, Nat. 357, 222.
60. Dermer, C.D., Sturner, S.J. 1994, ApJ 420, L75.
61. Kuiper, L., et al. 2001, A&A 378, 918.
62. Thompson, D.J., et al. 1992, Nat. 359, 615.
63. Fierro, J.M., et al. 1993, ApJ 413, L27.
64. Kuiper, L., et al. 1999, A&A 351, 119.
65. Baring, M.G., Harding, A.K. 2001, ApJ 547, 929.
66. Lessard, R.W., et al. 2000, ApJ 531, 942.
67. de Naurois, M., et al. 2002, ApJ 566, 343.
68. Cheung, W.M., Cheng, K.S. 1994, ApJS 90, 827.
69. Fermi, E. 1954, ApJ 119, 1.
70. Hayakawa, S. 1956, Prog. Theo. Phys. 15, 111.
71. Sturner, S.J., Dermer, C.D., 1995, A&A 293, L17.
72. Aharonian, F. et al. 2006, A&A 449, 223.
73. Gehrels, N., et al. 2000, Nature, 404, 363.
74. Grenier, I.A. 2000, A&A 364, L93.
75. Aharonian, F., et al. 2005, A&A 439, 1013.
76. Swanenburg, B.N., et al. 1978, Nat. 275, 298.
77. Stecker, F.W., de Jager, O.C., Salamon, M.H. 1992, ApJ 390, L49.
78. Kneiske, T.M., Mannheim, K., Hartmann, D.H. 2002, A&A 386, 1.
79. Fichtel, C.E., et al. 1977, ApJ 217, L9.
80. Kappadath, S.C. et al 1996, A&AS 120, 619.
81. Ullio, P., et al. 2002, Phys. Rev. D. 66, 123502.
82. Hanlon, L.O., et al. 1994, A&A 285, 161.
83. Hurley, K. et al. 1994, Nat. 372, 652.
84. Baring, M.G., Harding, A.K., 1997, ApJ 491, 663.
85. González, M.M., et al. 2004, AIP Conf. Proc. 727, 203.

Gravitational Wave Detectors: A New Window to the Universe

Gabriela González, for the LIGO Scientific Collaboration

Department of Physics and Astronomy, Louisiana State University
202 Nicholson Hall, Tower Drive, Baton Rouge, LA 70803
gonzalez@lsu.edu

Summary. The LIGO gravitational wave detectors have achieved their designed sensitivity, and are currently in operation. We describe the technology of the detectors, as well as results from the analysis of some of the data collected so far.

1 Introduction

The existence of gravitational waves is a beautiful, if straightforward, prediction of Einstein's theory of relativity, arising from the deep relationship between space and time: dynamic changes in matter distribution will distort the space time, and the space-time "ripples" will travel outwards from the source, carrying energy and precious information about the astrophysics of the source. Many black hole scenarios would not emit electromagnetic waves and thus be invisible to instruments detecting different wavelengths of light; they would, however, produce gravitational waves traveling at the speed of light. Many other astrophysical sources (supernovae, collisions of neutron stars,...) would produce both electromagnetic and gravitational waves, but they would carry very different information: gravitational waves would tell us about the macroscopic structure of the mass of the source and the effects on the space time produced by the large relativistic fields. Gravitational waves interact very weakly with matter: most of the universe is essentially transparent to the traveling waves; the information encoded in them is pristine.

Gravitational waves distort space-time, changing distances between freely falling objects (acting as coordinate markers) by an amount proportional to the gravitational wave strength, and the distance between the objects: $\Delta L = hL$. Gravitational waves have a transverse and quadrupolar nature: a plane wave would change distances in the plane perpendicular to the direction of propagation, and it would make distances shorter in one direction, and longer in the perpendicular direction, as shown in Figure 1. There are two polarizations of a plane wave, called "$\times$" and "$+$" corresponding to maximum distortions along two directions 45° apart.

A. Carramiñana et al. (eds.), Solar, Stellar and Galactic Connections between Particle Physics and Astrophysics, 231–239.

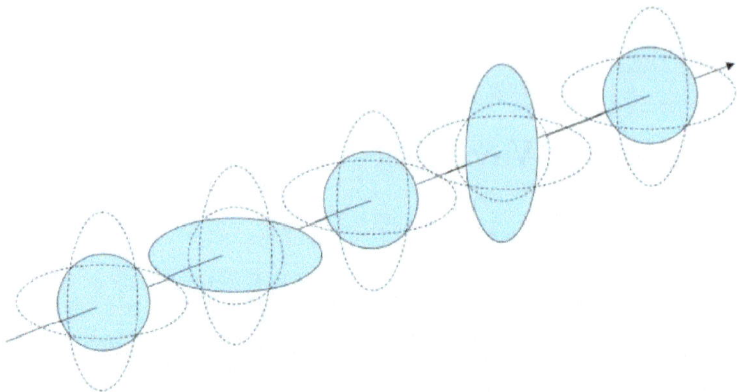

Fig. 1. Spatial distances changed by a propagating gravitational wave with a "+" polarization.

Gravitational waves are produced by accelerated mass quadrupoles Q_{ij}, and the strength of the wave is proportional to the second derivative of the mass quadrupole, and inversely proportional to the distance r to the source: $h \approx 2G\ddot{Q}/c^4r$. The rate of energy radiated away from the source is $P = (G/5c^5)(d^3Q/dt^3)^2$: the source will be changed due to the loss of radiated energy. An orbiting system of the most compact of stars, neutron stars, is known to exist from pulsar observations, and forms a radiating quadrupole, thus emitting gravitational waves. The energy of the system decreases due to the emission of gravitational waves: the orbit shrinks, and the orbiting frequency of the system increases (from Newton's law $\omega^2 = GM/r^3$). The agreement of the predicted change in orbit parameters has been beautifully demonstrated with observations of the first pulsar binary system PSR 1913+16, discovered by Hulse and Taylor in 1974 [1].

Since gravitational waves do exist, as proven by the Hulse Taylor system, it is the nature of human curiosity to try to directly detect them. However, the effect of gravitational waves is very small: a binary system of neutron stars, about 10 diameters or 200 km away from each other, at a distance r from Earth, would emit gravitational waves of about 300 Hz with an amplitude of $h \sim 10^{-22}(20\text{MPc}/r)$. The changes in distance produced by a source in the Virgo Cluster is an atom diameter for a distance of several million kilometers! This shows the incredible challenge to measuring such small effects, even from large astrophysical systems. However, we will show that present detectors can achieve measurements of sub-nuclear distances, over distance of kilometer scale, making the direct observation of gravitational waves a plausible, and very exciting, enterprise.

2 Interferometric Gravitational Wave Detectors

The quadrupolar nature of gravitational waves seems naturally appropriate to be measured by some of the oldest precision measurement instruments, Michelson interferometers. Such a detector measures differences in length between perpendicular arms, so it can be naturally adapted to measure the effect of a passing gravitational wave that changes the length of its arms. However, even for a 4km long interferometer (as they now exist!), a gravitational wave with strength $h \sim 10^{-22}$ produces a difference in arm length of $\Delta L = hL \leq 10^{-18}$m, or a thousandth of a nucleon diameter. The measurement of such a small quantity, with an instrument with km scale, seems to defy quantum mechanics, not just common sense. Are gravitational waves detectable? The answer is yes, if the question is well defined, and the instrument sensitive enough. First, even though we talk about sub-nuclear length scales, the question does not enter the realm of quantum uncertainty, because we are not measuring the position of any one nucleon, or atom, but instead we are measuring changes in distances defined by macroscopic objects, whose position is well defined, well beyond nuclear distances: in other words, we are measuring the average position of many atoms, which is better defined than the position of any one of the atoms forming the system. The quantum nature of the world does limit the sensitivity of the measuring instruments, but the limitations depend on the instrumental set up.

Technology has also been available to measure such small distances, in more than one way. Resonant bar detectors, pioneered by Joseph Weber in the '70s and still in use in the US and Italy, have achieved sub-nuclear displacement sensitivities, even if not reaching their quantum limit (most are limited by the noise in their transducers). These detectors consist of a large resonant masses of meter scale in length, and 1-2 tons in mass, placed in vacuum, at low temperatures, with very sensitive transducers to measure the differential displacement of the ends of the mass. The measurements are most sensitive near the resonance frequency bars, about a kHz.

The LIGO interferometric detectors, through very different measurement techniques than resonant bars, achieve similar precision for displacement sensitivity, but over longer length scales (kilometers!), which then makes for more sensitive detectors to *strain*, the natural measure of gravitational wave strength. Interferometers are also most sensitive at lower frequencies ($\sim$100 Hz), and have a broader response, which makes for a better chance of measuring signals from several other astrophysical sources, other than collisions or explosions of stars.

2.1 The LIGO Detectors

The LIGO detectors [2] use interferometric techniques: they are essentially Michelson interferometers that use coherent light and an optical readout to deduce, from the interference of the beams returning from each arm, the difference in arm length. In the famous Michelson-Morley experiment, such interference would be caused by the different light speed in each arm, presumably affected by ether. In a gravitational wave detector, the difference in arm length would ideally be caused by the distance between the beamsplitter and the mirrors at the ends of the arms being affected differentially by a gravitational wave. In the LIGO detectors, a coherent laser source is used (a NdYAF laser with $\lambda = 1064$nm wavelength), and the signal detected at the antisymmetric port is the power on a photodiode, measuring the phase difference between beams that travel in the different arms of the detector. The antisymmetric port is kept "dark" with feedback controls, which push on the mirrors to make the interference between the returning beams to be destructive. In this case, the beams returning to the laser source have constructive interference, making the whole detector behave like a mirror. In order to enhance the signal, the light in each arm is stored in a Fabry-Perot optical resonant cavity, using partially transmissive input mirrors; two more feedback loops are needed to keep these cavities resonant. The circulating power in the detector is increased by making another optical resonant cavity between the light reflected by the detector, and a partially transmissive mirror at the input, or a "recycling" mirror. A schematic drawing of the optical topology used in the LIGO detectors is shown in Fig. 2.

In order to allow an approximation of free masses for the mirrors (and to improve seismic isolation), the mirrors are suspended as pendulums by single looping wires. The mirrors are cylindrical, made of fused silica, 25 cm in diameter, 10cm thick, and 10 kg heavy. To avoid spurious phase differences due to varying index of refraction, the light travels in vacuum beam tubes: this is the largest volume high vacuum system in the world!

Of course, even in the absence of a gravitational wave, the signal at the output is not identically zero: there is a certain amount of "noise" that will then limit the magnitude of detected gravitational waves. The noise detected will be the sum of several different noise sources: some sources of noise make the actual distance between mirrors change (like seismic noise and brownian motion of the mirrors), and some sources of noise affect the readout (like shot noise of the laser light). The different noise sources have each their own spectral features, with different power at different frequencies: seismic noise is largest at low frequencies, brownian motion is largest at the resonances of the pendulum systems and the mirror masses, shot noise is largest at frequencies above the optical cavity pole frequency ($\sim$100 Hz). The resulting sum of all the noise sources makes the detectors most sensitive at frequencies near 100-200 Hz, but have a broad sensitive band between $\sim$50 Hz and a few kHz, as shown in Fig. 3.

There are two 4km long LIGO detectors in the United States, one in the LIGO Livingston Observatory, in the state of Louisiana, and another in the LIGO Hanford Observatory, in the state of Washington; they are about 3000 km away. This will allow increased confidence in an eventual detection, since the false alarm rate is greatly reduced by requiring coincidence between the

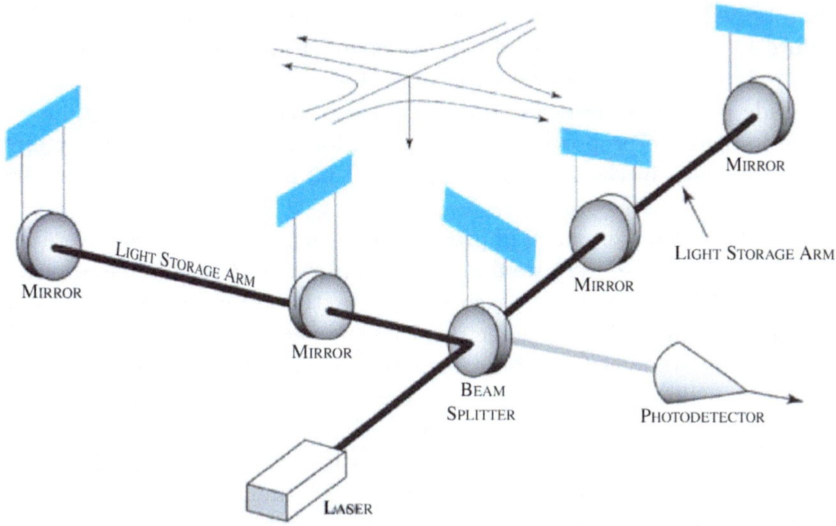

Fig. 2. Optical topology used by the current LIGO detectors

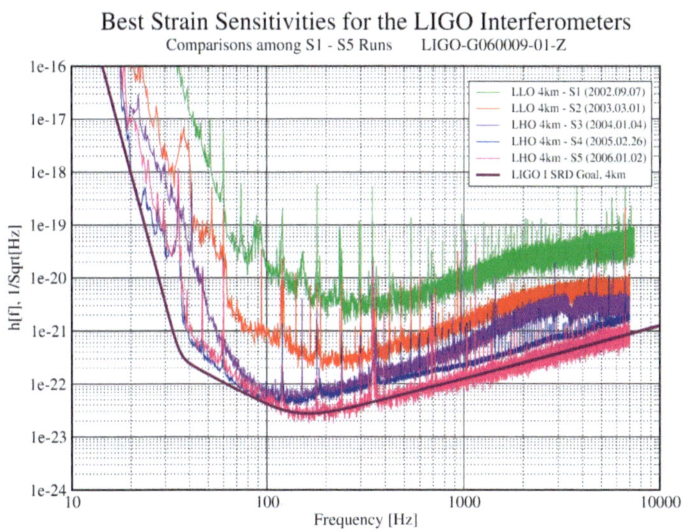

Fig. 3. Improvement of sensitivity of the LIGO detectors in the different Science Runs S1-5. The solid line represents the goal for the detectors' sensitivity, taking into account fundamental noise sources like seismic noise, brownian motion of the suspended mirrors, and shot noise in the detected light.

detectors, within the maximum 10ms of light travel distance. In the LIGO Hanford Observatory there is also an independent 2km long detector, which again reduces the false alarm rate, and allows for a consistency check on amplitude of a possible detection: since the gravitational wave produces a change in distance proportional to distance, for a true signal, the measured change in length in the 2km detector should be half as large as in the 4km detector.

The LIGO detectors have improved their sensitivity since they were first turned on, as noise sources were identified and reduced or eliminated one by one. Since 2002, the detectors reached significantly better sensitivity than any previous gravitational wave detector in its frequency band, and work in the detectors was stopped four times to allow for data taking. These "Science Runs" were called S1, S2, S3 and S4, and happened for 17, 61, 70, and 30 days respectively, starting in Aug 23 2002, Feb 14 2003, Oct 31 2003 and Feb 22 2005, also respectively. Not only the sensitivity, but also the duty cycle improved in S4 with respect to previous runs, since an improved, active seismic isolation system was installed in the LIGO Livingston Observatory to allow daytime operations (the LIGO Hanford Observatory is more isolated from human noise sources). In fall of 2005, the detectors achieved their designed sensitivity, and starting taking data in continuous mode since November 2005 for an extended period of time, which will end when a year of coincident data is obtained.

3 Astrophysical Sources of Gravitational Waves

There are several different astrophysical sources of gravitational waves that may produce signals in the LIGO detectors' sensitive frequency band. According to their spectral content, we classify them in four groups: continuous signals from rotating stars; signals from binary systems; stochastic signals from a cosmological background; and burst signals from collisions and explosions of stars.

Rotating stars will produce gravitational waves if they are not perfectly spherical, and have a mass quadrupole. The signal produced at the source is monochromatic, with the frequency of the gravitational wave being twice the rotation frequency of the star. There are many neutron stars in our Galaxy that are pulsars, emitting radio waves that can be detected on Earth by radio telescopes. From the detected radio signals, we know their position in the sky and their rotational frequency. Some of these sources are also known to be slowing down: they are "spinning" down. If we attribute the loss of energy to the emission of gravitational waves, we obtain from the known spin derivative an upper limit on the magnitude of the gravitational waves. With the LIGO detectors, we can obtain a direct observational limit for the known pulsars, since the predicted gravitational waves are in the detectors' band. The data is searched for a periodic signal with the appropriate Doppler shift for the star's position in the sky and Earth's rotation; in the absence of a signal, an upper limit can be deduced on the strength of the gravitational waves emitted by the source. This upper limit can also be translated into an upper limit for the ellipticity of the star. With S2 LIGO data, 28 isolated radio pulsars were studied, and limits were set in strain as low as few times 10^{-24} and in ellipticity of 10^{-5} [3]. The search for rotating stars at all positions in the whole sky is computationally more challenging, and must be tackled by different techniques [4], or using shared resources. The American Physical Society sponsored an exciting project, "Einstein@Home", which uses people's idle computers to search for gravitational waves from rotating stars in LIGO data.

The emission of gravitational waves from binary star systems is well understood as long as the objects are far enough away from each other for Post-Newtionina approximations to apply: the signal emitted will have a frequency equal to twice the orbital frequency, and will increase in frequency and amplitude as the system loses energy. Binary neutron star and small black holes systems ($< 20M_{\odot}$) will emit waves in the LIGO detectors' band. We can in fact translate a sensitivity curve into a distance at which we would detect a binary neutron star system with average position in the sky and average orientation, with a signal to noise larger than 8. The curves shown in Figure 3 correspond to a range of 80 kpc (S1), 1Mpc (S2) , 6.5 Mpc (S3), 8.4 Mpc (S4) and 12 Mpc (S5). With optimal orientation and position in the sky, systems from distances up to 2.2 times farther away could be detected: in S5, we are observing a fraction of the systems in the Virgo Cluster of galaxies. The search in S2 data for neutron stars and black holes smaller than $1M_{\odot}$ resulted in no detections, and the first direct upper limits on galactic and extra-galactic systems [5, 6].

Looking for signals from violent events such as collisions of stars (the final stage of a binary system) or supernova explosions does not have models to use in the search, so they rely on techniques looking for excess power in the data, as measured by Fourier transforms, wavelet transforms, or other appropriate methods. Searches for "bursts" in S2 data with frequency content in the 100-1100 Hz data yielded no candidates. The sensitivity of this search, measured in *root-sum-square* of the strain of possible waveforms, lies in the range of $h \sim 10^{-20} - 10^{-19}/\sqrt{\text{Hz}}$.

Sources of Gamma Ray Bursts are known to be supernova explosions, at least for a large fraction of the "'long" bursts (more than two seconds long): depending on the asymmetry of the mass distribution of the star and the explosion, these sources can also originate gravitational waves in the LIGO detectors' frequency band. During one of the brightest Gamma Ray bursts, GRB030329, the LIGO Hanford detectors were in operation, taking data for the Second Science Run, and a dedicated search of the data at the time of the Gamma Ray Burst yielded no detection, and an upper limit on the emitted strain by an optimally polarized source of $h_{rss} \sim 10^{-20}$ was obtained [8].

The superposition of many unresolved burst signals results in a continuous random signal, or a "stochastic background". These signals can be generated by astrophysical sources such as the ones considered earlier, or to cosmological processes, similar to the cosmic microwave background. Although the signal in a single detector would be undistinguishable from other random noise sources, the signal in a *network* of independent detectors can be detected by finding correlated noise. The correlation will get weaker and eventually vanish for signals with wavelengths shorter than the distance between the detectors. A stochastic background can be characterized by a dimensionless function of frequency $\Omega_{gw}(f)$, the gravitational wave energy density per unit logarithmic frequency, divided by the critical energy density to close the universe; if the spectrum is flat, the quantity Ω_{gw} is a constant Ω_0 independent of frequency. The analysis of LIGO S3 data resulted in an upper limit $\Omega_0 < 8.4 \times 10^{-4}$ in the frequency band between 60 Hz and 156 Hz [9].

4 Present and Future of Gravitational Wave Astrophysics

Although there has not been any direct observation of gravitational waves yet, the data being taken now with the LIGO detectors in the Fifth Science Run that started in November 2005 shows enormous promise: even if no signal is found, the observational upper limits on the strength of different sources of gravitational waves will be orders of magnitude better than previous published results. The prediction for the rate of observation of signals from binary neutron systems, extrapolated from the few known pulsar binary systems known in the galaxy, is low enough so that no signal is expected in a year of operation [10] (barring serendipity, never out of the question). However, extrapolations from recent observations of short gamma ray burst implying an association with the coalescence of compact binary systems [11], suggest that the rates, especially for black holes, may be high enough to either expect direct observations in a year of data, or, in the absence of signals, to begin ruling out some possible evolutionary astrophysical scenarios. A detector in Europe built by the VIRGO French-Italian collaboration [12], with topology and sensitivity similar to the LIGO detectors, may also begin operations in the near future; the existence of a network with four detectors will not only lower the frequency of possible false alarms, but also, in the case of detections, help identify physical parameters of the source, such as polarization and location in the sky.

The most exciting prospect, however, is that now that we know that the basic technologies work in detectors of kilometer scale (a non trivial task!), new and better technologies can be used to improve the sensitivity of the LIGO detectors by about an order of magnitude. Since the reach in distance is proportional to the sensitivity, the volume surveyed increases with the cube of the sensitivity, and the rate of sources could be as much as 1,000 times higher than in the present LIGO detectors. The predicted rate for such Advanced LIGO detectors from binary neutron star systems extrapolated from galactic systems [10] is a detection every few days! The Advanced LIGO detectors could be operating at the beginning of the next decade. Even a few months of observations will result in a significant advance in our knowledge of the Universe: a new window will be opened, and we cannot expect less than a few surprises.

References

1. R. A. Hulse and J. H. Taylor, 1974, ApJ, 191, L59.
2. B. Barish, R. Weiss, Phys. Today, 52, 44-50 (1999)
3. B. Abbott et al. (LIGO Scientific Collaboration), Phys. Rev. Lett. 94, 181103 (2005)
4. B. Abbott et al. (LSC), Phys. Rev. D 72, 102004 (2005)
5. B. Abbott et al. (LSC), Phys. Rev. D. 72, 082001 (2005)
6. B. Abbott et al. (LSC), Phys. Rev. D. 72, 082002 (2005)
7. B. Abbott et al. (LSC), Phys. Rev. D 72, 062001 (2005)
8. B. Abbott et al. (LSC), Phys. Rev. D 72, 042002 (2005)
9. B. Abbott et al. (LSC), Phys. Rev. Lett. 95, 221101 (2005)
10. V. Kalogera et al., Astrophys.J. 601 (2004) L179-L182; Erratum-ibid. 614 (2004) L137
11. E. Nakar, A. Gal-Yam, D. Fox, preprint astro-ph/0511254, (2005)
12. F. Acernese et al., Class. Quantum Grav. 22 No 18 (21 September 2005) S869-S880

Part III

Research Short Contributions

Hybrid Extensive Air Shower Detector Array at the University of Puebla to Study Cosmic Rays

O. Martínez[1], E. Pérez[1], H. Salazar[1] and L. Villaseñor[2]

[1] Facultad de Ciencias Físico-Matemáticas, BUAP, Puebla Pue., 72570, México, `hsalazar@fcfm.buap.mx`
[2] Instituto de Física y Matemáticas, Universidad Michoacana, Edificio C3 Ciudad Universitaria, Morelia, Mich., 58060, México, `villasen@ifm.umich.mx`

Summary. We describe the design of an extensive air shower detector array built in the Campus of the University of Puebla (located at 19°N, 90°W, 800 gcm^{-2}) to measure the energy and arrival direction of primary cosmic rays with energies around 10^{15} eV. The array consists of 18 liquid scintillator detectors (12 in the first stage) and 6 water Cherenkov detectors (one of 10 m^2 cross section and five smaller ones of 1.86 m^2 cross section), distributed in a square grid with a detector spacing of 20 m over an area of 4000 m^2. In this paper we discuss the calibration and stability of the array, and discuss the capability of hybrid arrays, such as this one consisting of water Cherenkov and liquid scintillator detectors, to allow a separation of the electromagnetic and muon components of extensive air showers. This separation plays an important role in the determination of the mass and identity of the primary cosmic ray. This facility is also used to train students interested in the field of cosmic rays.

1 Introduction

The collisions of primary cosmic rays with nitrogen and oxygen nuclei high in the Earth atmosphere give rise to extensive air showers (EAS).The latter are composed of a large number of of secondary particles which penetrate the atmosphere. EASs can be studied by measuring their particle densities as they arrive at the ground by means of ground detectors or their particle densities as they traverse the atmosphere by means of fluorescence or Cherenkov light telescopes on the ground.

EASs consist of four components depending on the type of secondary particles: hadronic, electromagnetic, muonic and neutrino component; out of these components only the electromagnetic and the muonic are detected with ground detectors, because the hadronic components die away converting their energy into the other components soon after the primary collisions take place.

A. Carramiñana et al. (eds.), Solar, Stellar and Galactic Connections between Particle Physics and Astrophysics, 243–251.

In turn, the neutrino component is undetected because neutrinos interact only weakly.

The energy spectrum of cosmic rays has been studied extensively by direct measurements with detectors on balloons and satellites for the energy range 10^9 - 10^{14} eV, where their flux is large enough to allow direct detections with light-weight small-area detectors. For higher energies of the primaries, only indirect measurements with detectors on the ground are possible due to the small fluxes involved, i.e., arrays of particle detectors on the ground or arrays of telescopes that measure the flux of fluorescence or Cherenkov light produced by the charged particles in the EAS as they interact with the atmosphere.

It has been found that the energy spectrum of primary cosmic rays is well described by a power law, i.e., $dE/dx \sim E^{-\gamma}$, over many decades of energy with the spectral index γ approximately equal to 2.7, and steepening to $\gamma = 3$ at $E = 3 \times 10^{15}$ eV [1]. This structural feature is known as the "knee" of the cosmic ray spectrum.

The nature of the knee is still a puzzle despite the fact that it was discovered more than 46 years ago [2]. Most theories consider its origin as astrophysical and relate it to the breakdown of the acceleration mechanisms of possible sources within our galaxy or to a leakage during propagation of cosmic rays in the magnetic fields within our galaxy; in particular, these theories lead to the prediction of a primary composition richer in heavy elements around the knee due to the decrease of galactic confinement of cosmic rays with increasing energy of the primary cosmic rays. Alternatively, there are scenarios where a change in the hadronic interaction at the knee energy gives rise to new heavy particles [3] which produce, upon decay, muons of higher energies than those produced by normal hadrons.

The best handle to study the composition of primary cosmic rays by using ground detector arrays is the measurement of the ratio of the muonic to the electromagnetic component of EAS; in fact, Monte Carlo simulations show that heavier primaries give rise to a bigger muon/EM ratio compared to lighter primaries of the same energy [4]. In fact, evidence for such variations has been reported recently [5].

The extensive air shower detector array at University of Puebla (EAS-UAP) was designed to measure the lateral distribution and arrival direction of secondary particles for EAS in the energy region of $10^{14} - 10^{16}$ eV. The special location of the EAS-UAP array; 2200 m above sea level; and all the facilities coming from the Campus of the University of Puebla make it a valuable apparatus for the long term study of cosmic rays and at the same time an important training center for new physics students interested in getting a first class education in the field of cosmic rays in Mexico. In this paper we describe the experimental setup of the EAS-UAP array and discuss the ability of arrays like this one to measure the independent contributions of the muonic and EM components of EASs.

2 Experimental Setup

The EAS-UAP array is located in the campus of the University of Puebla in Mexico (UAP) at 19° N, 89° W and 800 gcm^{-2}; it consists of 18 liquid scintillator detectors distributed uniformly on a square grid with spacing of 20 m, and six water Cherenkov detectors (one of 10 m^2 cross section and five smaller ones of 1.86 m^2), as shown in Fig. 1, where liquid scintillator detectors are represented by black cylinders and water Cherenkov detectors by stars (the bigger one is close to tl 6).

Each of the liquid scintillator detectors consists of a cylindrical container of 1 m^2 cross section made of polyethylene and filled with 130 l of liquid scintillator up to a height of 13 cm. As sensor we use a 5" photomultiplier (PMT) located inside each tank along the axis of the cylinder and facing down with the photo-cathode 70 cm above the surface of the liquid scintillator. We used commercial liquid scintillator manufactured by Bicron.

Out of a total of six water Cherenkov detectors, the array has one detector bigger than the other five; it consists of a cylindrical tank made of roto-molded polyethylene with a cross section of 10 m^2 and a height of 1.5 m. This tank is filled with purified water up to a height of 1.2 m and has three 8" PMTs looking downwards at the tank volume from the water surface. The five smaller water Cherenkov detectors consist of cylindrical tanks made of polyethylene with an inner diameter of 1.54 m and a height of 1.30 m filled with with 2300 l of purified water up to a height of 1.2 m.

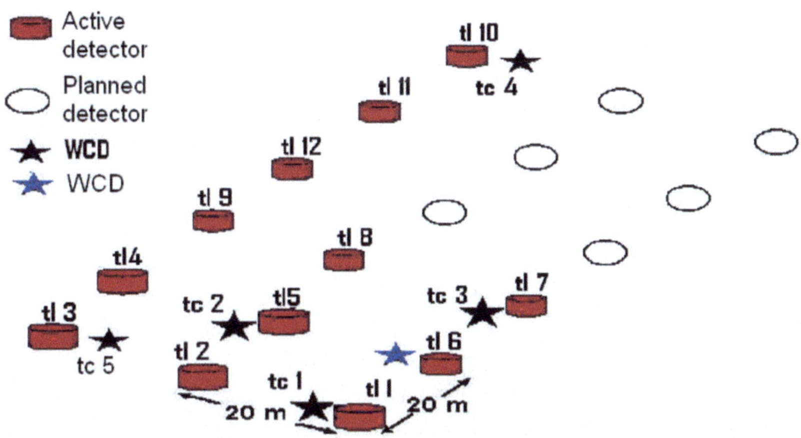

Fig. 1. EAS-UAP array located on the Campus of the University of Puebla. Stars represent Cherenkov detectors filled with 2230 l of water and cylinders represent detectors filled with 130 l of liquid scintillator. The star by tl 6 represents a bigger water Cherenkov detector filled with 12 000 l of water.

2.1 Data Acquisition System

The trigger we use is flexible enough, one of its options requires the coincidence of signals from the four central liquid scintillator detectors (tl 1, tl 3, tl 7 and tl 7) which form a rectangular sub-array with an area of 40 × 40 m^2. This trigger sub-array enhances the events in which the shower core falls inside this sub-array. The measured trigger rate in this case is 80 events per hour. The data acquisition system consists of a set of digital oscilloscopes that digitize the signals from the PMTs of the liquid scintillator detectors and the water Cherenkov detectors. All the digital oscilloscopes are connected to the GPIB port of a PC in a daisy chain configuration.

The system is controlled by the PC running a custom-made acquisition program written in a graphical language called LabView [6]. We used commercial NIM modules to discriminate the PMT signals at a threshold of -30 mV and to generate the coincidence trigger signal. The DAQ system acquires all the PMT traces for each triggered event. The acquired traces are used by the PC to perform on-line measurements of the integrated charges, arrival times, amplitudes and widths of all signals the PMTs, these data are saved into a hard-disk file for further off-line analysis.

2.2 Monitoring and Calibration

Single particle triggers are used simultaneously with EAS triggers for monitoring the stability of the array and for obtaining the calibration constants for each detector. We make use of the natural flux of background muons and electrons to monitor and calibrate our detectors.

Calibration of the detectors is essential as it allows the conversion of the electronics signals measured in each detector into the number of particles in the EAS that reach the detectors and finally into the energy of the primary cosmic ray. For the location of the EAS-UAP, muons are the dominant contribution to the flux of secondary cosmic rays for energies above 100 MeV with about 300 muons per second per m^2 and a mean energy of 2 GeV; at lower energies, up to 100 MeV, electrons dominate with a flux 1000 times bigger and a mean energy around 10 MeV [7].

It is important to keep in mind that a 2 GeV muon can cross the detector ($dE/dx \sim 2MeV/cm$) whereas a 10 MeV electron cannot; therefore muons produce more Cherenkov light than lower energy electrons (the range of 10 MeV electrons in a water-like liquid is about 5 cm). A vertical equivalent muon (VEM)is the integrated charge on a PMT pulse as a consequence of a vertical muon traversing the detector. Techniques to measure the values of one VEM are reported elsewhere for water Cherenkov detectors [8] and liquid scintillator detectors [9]. Thanks to these measurements we have a reliable way of converting the charge deposited in each detector into a number of equivalent particles (electrons for liquid scintillator and muons for water Cherenkov detectors) [10].

3 Performance of the EAS-UAP Detector Array

We have reported on the performance of the EAS-UAP array elsewhere [11, 12, 13]. The direction of the primary cosmic ray is inferred directly from the relative arrival times of the shower front at the different detectors. The core position, lateral distribution function and total number of shower particles on the ground N_e are reconstructed from a fit of the measured electron-positron densities to the NKG [14] expression

$$\rho(S,r) = K(S)(\frac{r}{R_0})^{S-2}(1+\frac{r}{R_0})^{S-4.5} \quad (1)$$

where S is the shower age, r the distance of the detector to the shower core, $K(S)$ is a normalization constant and R_0 is the Moliere radius (90 m for an altitude of 2200 m a.s.l.) [15]. This fit is done on an event-by-event basis. The shower energy is obtained by using the following relation [16]

$$N_e(E_0) = 117.8E_0^{1.1} \quad (2)$$

where N_e is the total number of particles on the ground obtained by integrating Eq. (1) and E_0 is the energy of the primary cosmic ray expressed in TeV [15, 17].

Figure 2 shows a real event taken from the EAS-UAP event display and Fig. 3 shows the measured particle densities and the fitted lateral distribution function for the same near-vertical shower. For this particular event, the number of electrons and positrons at the ground obtained by fitting the data to Eq. (1) were 174 600 and the reconstructed energy obtained by using Eq. (2) was 459 TeV.

Given that our array is not uniform and therefore the center-of-gravity method is not applicable to find the position of the shower core on the ground, we use the core position that provides the best fit (i.e., with the fit with a minimum for χ^2/dof) of the data data to the NKG formula given by Eq. (1). We have tested that this procedure and the energy reconstruction method work reasonably well for near-vertical MC showers by using Monte Carlo showers generated with the MC shower generator called Aires [18].

4 Muon/EM Separation

Finally, we discuss a number of different types of hybrid and composite arrays and their capabilities to separate the electromagnetic from the muonic component of EASs. As mentioned earlier, this separation constitutes the best handle to study the mass and identity of the primary cosmic ray. Note that this discussion represents work in progress and it is yet incomplete until the effect of photons, which are the dominant contribution among the particles in EAS, is taken into account.

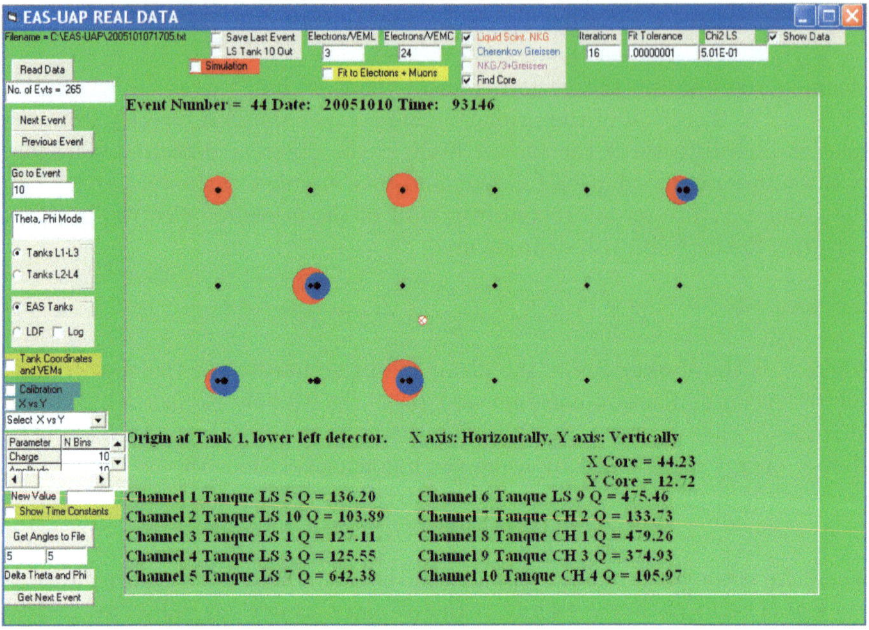

Fig. 2. Event display of a real shower event. The diameter of the circles is proportional to the signal strengths of water Cherenkov (blue) and liquid scintillator detectors (red). The small circle is the core location obtained as explained in the text.

4.1 Water Cherenkov-Liquid Scintillator Hybrid Array

In the case of hybrid arrays such as EAS-UAP composed of two types of detectors: water Cherenkov detectors with cross section A_C and liquid scintillator detectors of cross section A_L the set of equations for the muonic (ρ_{muon}) and electromagnetic ρ_{EM} particle densities is the following:

$$\frac{Q_{Cherenkov}}{VEM_C} = \rho_{muon} A_C + \frac{\rho_{EM} A_C}{24} \tag{3}$$

$$\frac{Q_{LiqScint}}{VEM_L} = \rho_{muon} A_L + \frac{\rho_{EM} A_L}{3} \tag{4}$$

where $Q_{Cherenkov}$ and $Q_{LiqScint}$ are the PMT charges collected in the Cherenkov and liquid scintillator detectors, respectively; VEM_C and VEM_L are the measured PMT charges that correspond to the detection of a vertical muon in the Cherenkov and liquid scintillator detectors, respectively and the numbers 24 and 3 correspond to our explicit measurement that on the average a penetrating muon deposits 24 times more Cherenkov signal than a 10 MeV electron in a 120 cm high water Cherenkov detector and 3 times more signal than a 10 MeV electron on a 13 cm high liquid scintillator detector [9].

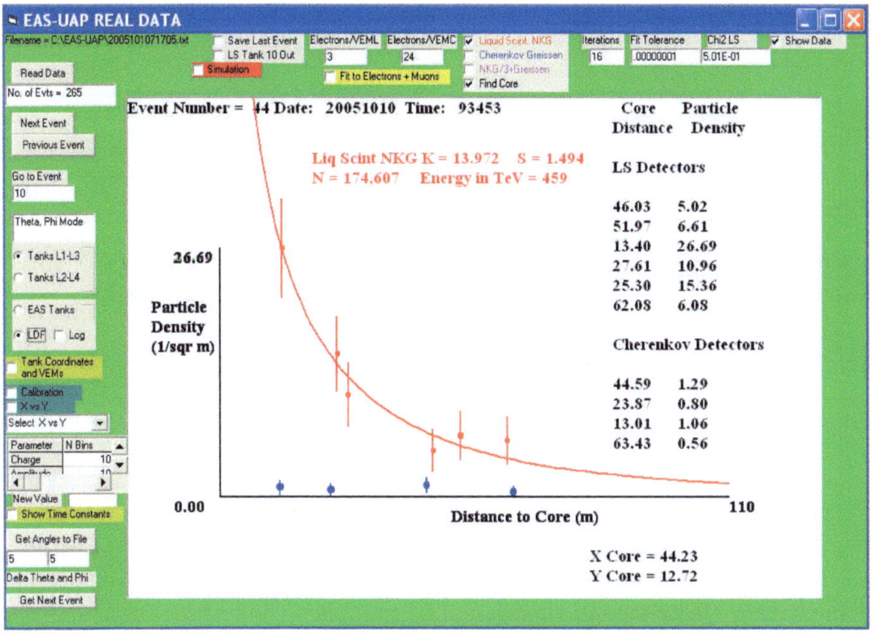

Fig. 3. Event display of the same event shown in Fig. 2. The solid curve is a fit of the NKG expression to the measured lateral distribution particle densities on the liquid scintillator detectors.

The solution to this set is

$$\rho_{EM} = \frac{24}{7}\left(\frac{Q_{LiqScint}}{A_L VEM_L} - \frac{Q_{Cherenkov}}{A_C VEM_C}\right) \quad (5)$$

$$\rho_{muon} = \frac{8}{7}\frac{Q_{Cherenkov}}{A_C VEM_C} - \frac{1}{7}\frac{Q_{LiqScint}}{A_L VEM_L}. \quad (6)$$

This scheme for independently measuring ρ_{muon} and ρ_{EM} is presently under tests both at the EAS-UAP and through MC simulations.

4.2 Composite Array with Two Types Water Cherenkov Detectors

Similarly, if we place water Cherenkov tanks with filled with half the volume of water, i.e., filled up to 0.60 m, side by side with fully filled tanks, i.e., filled up to 1.2 m, we have in this case the set of equations:

$$\frac{Q_{Cherenkov}}{VEM_C} = \rho_{muon} A_C + \frac{\rho_{EM} A_C}{24} \quad (7)$$

$$\frac{Q'_{Cherenkov}}{VEM'_C} = \rho_{muon} A_C + \frac{\rho_{EM} A_C}{12} \quad (8)$$

with solution given by

$$\rho_{EM} = \frac{24}{A_C}\left(\frac{Q'_{Cherenkov}}{VEM'_C} - \frac{Q_{Cherenkov}}{VEM_C}\right) \tag{9}$$

$$\rho_{muon} = \frac{1}{A_C}\left(\frac{2Q_{Cherenkov}}{VEM_C} - \frac{Q'_{Cherenkov}}{VEM'_C}\right) \tag{10}$$

where $Q_{Cherenkov}$ and $Q'_{Cherenkov}$ are the PMT charges collected in the water Cherenkov tank filled up to 1.20 m of water and 0.60 m, respectively.

4.3 Non-Hybrid Array of Water Cherenkov Detectors

The relevant equation in this case is

$$\frac{Q_{Cherenkov}}{VEM_C} = \rho_{muon}A_C + \frac{\rho_{EM}A_C}{24} \tag{11}$$

One possible solution to accomplish muon-EM separation in this case makes use of the fact that the lateral distribution functions for muons is steeper than that for electrons and therefore it is possible to do a three-parameter fit to a modified-NKG function to obtain the number of muons, the number of electrons and the shower age.

4.4 Use of Neural Networks

Another possibility is to use the different temporal structure of the PMT pulses in water Cherenkov detectors of bundles of EM particles (electrons, positrons and photons) with respect to muons. We have studied this possibility for a water Cherenkov detector with encouraging results [19].

5 Conclusions

We have discussed the importance of measuring the electromagnetic and muonic components of extensive air showers to study the mass and identity of the primary cosmic ray. We have also described the EAS-UAP array located in the campus of the University of Puebla, Mexico to study cosmic rays around the knee of the spectrum and some ideas in which this array can be used in a simple way to measure the relative contributions of muons in near-vertical extensive air showers.

Acknowledgements

We would like to thank University of Michoacan, University of Puebla and CONACyT for supporting this work.

References

1. S.P. Swordy et al., Astroparticle Physics, Volume 18 (2002) 129-150.
2. G.V. Kulikov and G.V. Khristiansen, Sov. Phys. JETP, 41 (1959) 8.
3. A.A. Petrukhin, Proc. XIth Rencontres de Blois "Frontiers of Matter" (The Gioi Publ., Vietnam, (2001) 401.
4. B. Alessandro, et al., Proc.27th ICRC, 1 (2001) 124-127.
5. KASCADE Collaboration (Klages H.O. et al.), Nucl. Phys. B (Proc. Suppl.), 52 (1997) 92.
6. National Instruments Catalog (2004).
7. J.F. Ziegler, Nucl. Instrum and Meths. 191 (1981) 419-424.
8. M. Alarcón et al., Nucl. Instrum and Meths. in Phys. Res. A 420 (1999) 39-47.
9. H. Salazar and L. Villasenor, Nucl. Instruments and Meths. in Phys. Res. A. Volume 553, Issues 1-2 (2005) 295-298.
10. M. Aglietta, et al., Nucl. Instr. and Meth. A, 277 (1989) 23-28.
11. H. Salazar, O. Martinez, E. Moreno, J. Cotzomi, L. Villasenor, O. Saavedra, Nuclear Physics B (Proc. Suppl.) 122 (2003) 251-254.
12. J. Cotzomi, O. Martinez, E. Moreno, H. Salazar and L. Villasenor, Revista Mexicana de Fisica, 51(1)(2005) 38-46.
13. J. Cotzomi, E. Moreno, T. Murrieta, B. Palma, E. Perez, H. Salazar and L. Villasenor, Nucl. Instrume. and Meths. in Phys. Res. A. Volume 553, Issues 1-2 (2005) 290-294.
14. J. Nishimura, Handbuch der Physik XLVI/2, (1967) 1.
15. EAS-TOP Collaboration and MACRO Collaboration, Phys Lett B, 337, (1994) 376-382.
16. M. Aglietta et al., Phys. Lett. B, 337 (1994) 376-382.
17. J. Knapp, D. Heck, Report FZKA KfK 5196B (1993) 8.
18. S.J. Sciuto, AIRES User's guide and reference manual Version 2.6.0, http://www.fisica.unlp.edu.ar/auger/aires/ (2002).
19. L. Villasenor, Y. Jeronimo, and H. Salazar, Proc. of the 28th International Cosmic Ray Conference, T. Kajita, Y. Asaoka, A. Kawachi, Y. Matsubara and M. Sasaki (eds.), Vol. 1, HE Sessions, , Universal Academy Press, Inc., Tokyo, Japan, (2003) 93-96.

Search for Gamma Ray Bursts at Sierra Negra, México

H. Salazar[1], L. Villaseñor[2], C. Alvarez[1], and O. Martínez[1]

[1] Facultad de Ciencias Físico-Matemáticas de la BUAP, Apdo Postal 1364, Puebla, 72000 México hsalazar@fcfm.buap.mx
[2] Instituto de Física y Matemáticas, Universidad Michoacana. Apdo Postal 282, Morelia, Mich. 58040 México villasen@ifm.umich.mx

We present results from a search for GRBs in the energy range from tens of GeVs to one TeV with an array of 6 water Cherenkov detectors located at 4500 m a.s.l. as part of the high mountain observatory of Sierra Negra (N18°59.1, W97° 18.76) near Puebla city in México. The detectors consist of light-tight cylindrical containers of 1 m^2 cross section filled with 750 l of purified water; they are spaced 25 m and have a 5" photomultiplier (EMI model 9030A) facing down along the cylindrical axis. We describe preliminary experimental results obtained by using a single-particle counting technique for a data taking period of 58 days.

1 Introduction

Discovered by military satellites in the 60's and more properly studied until 1991, when NASA launched the *Compton Gamma-Ray Observatory* (*CGRO*), Gamma Ray Bursts (GRBs) are probably the most energetic phenomena in the Universe. As its name indicates, GRBs are gamma ray explosions that can liberate up to 10^{53} ergs in about one second. In order to detect and study GRBs, *CGRO* carried onboard 4 instruments: *BATSE, EGRET, COMPTEL* and *OSSE* . These instruments were able to detect gamma-ray photons at different energy ranges; in particular, *BATSE* detected more than 2700 GRBs with photon energies in the range from 20 KeV to 1 MeV. On the other hand, 7 GRB events were observed with photon energies greater than 30 MeV by *EGRET*, with 6 of them with photon energies greater than 1 GeV. The event named GRB940217 had the highest energy photon, 18 GeV [8].

It is important to mention that so far *BATSE* and *EGRET* have not observed a cut-off in the GRB energy spectrum; this suggests that the spectrum may extend up to high energy components, with TeV photons, or even greater as some models predict [5,13]. GRBs have been very well studied in the range from KeV to MeV by the *CGRO* and *BEPPO-Sax* missions, however the high

A. Carramiñana et al. (eds.), Solar, Stellar and Galactic Connections between Particle Physics and Astrophysics, 253–261.

energy component from GeV to TeV is still unknown. The *GLAST* mission will be launched with this purpose next year.

In contrast, we are interested in detecting GRBs with energies in the 10 GeV to 1 TeV range using a ground-based detector array. This array is operating in a single-particle counting and coincidences modes. We describe this water Cherenkov detector array located at the high mountain observatory of Sierra Negra and we also describe the array's capabilities in comparison with other ground-based observatories.

Fig. 1. Sierra Negra Array site, showing 3 light-tight cylindrical water Cherenkov tanks.

2 Ground-Based Experiments

Since gamma rays coming from outside the Earth cannot penetrate the atmosphere, it is necessary to use detectors on balloons or satellites to detect them directly. In addition, as photon energies increase, the photon fluxes decrease as a power law. Therefore, in order to detect small fluxes of gamma radiation or high energy photons in the range of GeV to TeV is necessary to construct more sensitive detectors with larger areas. Satellite- borne detectors with large collecting areas become impractical due to their cost. However, with inexpensive ground-based experiments of large area, it is possible to detect the relativistic secondary particles produced by the interaction of GeV or TeV gamma-ray photons with the nuclei of the upper atmosphere.

Currently, there exists a handful of ground-based experiments around the world searching GRBs: Chacaltaya at 5200 m a.s.l. in Bolivia [6]; Argo at 4300 m a.s.l. in Tibet [4], China; Milagro at 2630 m a.s.l. in New Mexico [10], USA; the Pierre Auger Observatory at 1400 m a.s.l. in Malargüe, Argentina [1] and Sierra Negra at 4550 m a.s.l. in México. Of all these experiments only the prototype of Milagro called Milagrito has reported the possible detection of signals associated to a GRB, GRB 970417 [2]. Milagro is the largest area (60 m×80 m) water Cherenkov detector capable of continuously monitoring the sky at energies between 250 GeV and 50 TeV. Although the Pierre Auger Observatory was designed to study ultra high energy cosmic rays, it is also a competitive high energy GRB ground-based detector due to its large area and the good sensitivity to photons of its water Cherenkov detectors [1].

3 Sierra Negra Experiment

The high mountain array prototype of Sierra Negra is located near Puebla city, México, at 4550 m a.s.l. At present, the array consists of 3 cylindrical light-tight water Cherenkov detectors located at the vertices of a 25 m side pentagon and another detector will be shortly added 14 m away from an existing one to allow the possibility to detect secondary particles in coincidence between them. Each tank has a cross section of 1 m^2. The tanks are filled with 750 l of ultra-pure water (Fig. 1). The interior of each tank is covered with tyvek and all of them contain a PMT to collect the Cherenkov light produced in the water. The PMT signals are read out by a DAQ system that measures the rates of secondary particles each tenth of second.

4 Sensitivity of Sierra Negra to GRBs

GRBs can be detected with ground-based detectors if the secondary particles produced by their interactions with the atmosphere give rise to an excess in the counting rate significantly larger than the statistical fluctuations of the background rate. The method of counting every single particle that hits the tank is known as single particle technique [12]. It is important to mention that any observed counting excess due to GRBs should be temporally coincident with a detection by one of the satellite experiments that observe a common part of the sky, for example *SWIFT* [11]. In this way any other background processes that give rise to particle counting excesses are discarded. A GRBs can be detected with a statistical significance of n standard deviations if $N_s/\sigma_b > n$ [12], where N_s is the signal detected by the array. This signal is proportional to the area and to the flux of secondary particles; σ_b is the background noise and it is proportional to the square root of all the secondary particles produced by cosmic rays. In general, n is taken as 4.

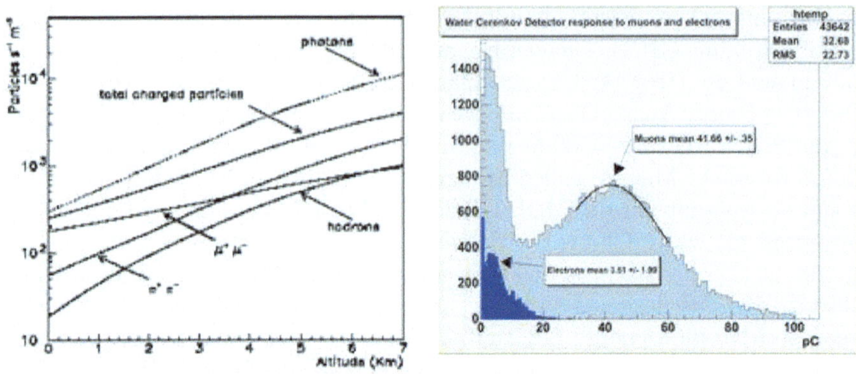

Fig. 2. (*Left*) Background rate due to secondary charge particles as a function of the altitude (Taken from Vernetto 1999, astro-ph/9904324). (*Right*) Water Cherenkov detector response to muons and electrons. The deposited charge ratio from muons and electrons is 41.7/3.5 =12, which is consistent with muon energies around 1 Gev crossing 70 cm of water and electrons with energies around 10 Mev.

The background consisting of all the secondary particles produced by cosmic rays entering into the terrestrial atmosphere varies with altitude as shown in Vernetto [12, Fig. 2 Left]. Then, for the altitude of Sierra Negra, the background rate of charge particles and photons is ≈1600 part m^{-2} s^{-1} and ≈4000 photons m^{-2} s^{-1} respectively. Knowing the background, we can calculate the minimum flux of particles detectable by an array of a given area, located at a given altitude. It is assumed that the shower is originated by a GRB with a total energy L = 10^{53} ergs and photons of E >1 GeV arriving vertically to the detector array during 1 second. For Sierra Negra, we expect a minimum detectable flux of secondary particles around 93 part m^{-2} s^{-1} for a detector area of 3 m^2. For Chacaltaya in Bolivia, which is the highest altitude array at 5.2 Km a.s.l, the minimum detectable flux is ≈26 particles m^{-2} s^{-1} considering an effective area of 48 m^2 and a background of ≈2100 particles m^{-2} s^{-1}.

The sensitivity increases strongly with the altitude of observation. Showers generated by primary photons of the same energy increase the size (number of secondary particles) with altitude. As an example, the mean number of particles generated by a photon of 16 GeV at 5200 m is 1 while at 2000 m (altitude of the EAS TOP experiment) is only 0.03, i.e., the sensitivity at Chacaltaya is better even though its detecting area (48 m^2) is considerable smaller than that of EAS TOP (350 m^2). The Sierra Negra array is sensible to GRBs of energies E < 200 GeV with the present detectors using the single-particle technique which is represents our first approach to detecting GRBs.

On the other hand, due to the geographic coordinates of Sierra Negra (N 18°59.1, W 97°18.76), we have the advantage of the almost zenithal transit of the Crab Nebula everyday. The Crab Nebula is a constant source of gamma rays that can be used as a standard candle for calibration [7]. Therefore, in Sierra Negra we have the possibility to detect it by using a simple method based on the rate of coincidences or showers detected by the array. First of all, we need to know the photon flux expected from the Crab Nebula when it is located at the zenith of Sierra Negra. Assuming that the source is observed during 4 hours everyday (±30° from the zenith) with 3 detectors in the vertices of a triangle that covers 300 m^2 of area. And knowing the Crab Nebula flux, f $= 2.68 \times 10^{-7} E^{-2.59}$ photons m^{-2} s^{-1} TeV^{-1} [2, 3], we obtain a flux of one photon per day. In other words, we expect to detect one shower per day coming from the Crab Nebula. On the other hand, from the coincidence data of the Sierra Negra array, we are presently detecting around 160 showers per minute (Fig. 3, Right). This means that if we are able to discriminate the 95% of the muonic component in the showers, and to detect 50% of the photons and reduce the uncertainty in the field of view to ±20 (corresponding to 16 minutes of Crab Nebula observation), we expect to detect the Crab Nebula with a significance of 5 in less than 6 months!

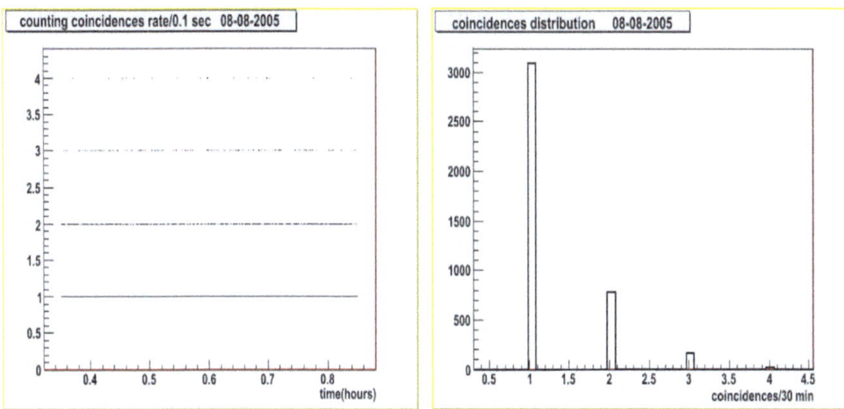

Fig. 3. (*Left*) .- Coincidence rate/0.1 sec for the three water Cherenkov detectors with a separation of 14 m. (*Right*).- In a period of 30 minutes we had about 160 showers per minute.

However, in order to achieve this goal, we need to be able to discriminate out the muonic component. The right plot of Fig. 2. shows that with a water Cherenkov detector and a fast digitization system we can indeed achieve a separation of the muon component from the electromagnetic one. Further steps planned will optimize the single-particle technique of detection with higher levels of coincidence triggers and a better shower reconstruction. It is worth mentioning that the Crab Nebula has already been detected at high energies by ground-based experiments such as Milagro, located in New Mexico, USA (2630 m a.s.l.) and ARGO, located in the Tibet, China (4300 m a.s.l.).

5 Data Analysis and Results

The data were taken from 3 water Cherenkov detectors operating in Sierra Negra, Puebla, México. The background rate measured in 0.1 s intervals versus time for each tank is not constant as shown in Fig. 4. The variations in the background rate are mainly due to two factors: the outdoor temperature and the atmospheric pressure, eventually solar activity may also be detectable [9]. Although, we have few atmospheric data at the site of Sierra Negra, we found that the background rates measured with each detector are correlated with temperature and pressure (Fig. 4). However, the time scale of these modulations is much larger than the typical time duration of GRBs and therefore they do not affect our GRB search. In addition, during night time the background rate of the 3 detectors is much more stable.

Figures 5 and 6 show the particle rate versus time and the particle rate distribution for each tank. The standard deviation and the mean rate of the fitted Gaussian for tank 1 is 20 part m^{-2}/0.1s and 206 part m^{-2} /0.1s. For tank 2 these parameters are 11 part m^{-2}/0.1s and 104 part m^{-2} /0.1s respectively. The coincidence rate/0.1s for the two closer detectors is shown in Fig. 3 Left. The measured background rate in tanks 1 and 3 is about a factor of 1.3 greater than that predicted by Vernetto [12] for the Sierra Negra altitude (about 1600 part m^{-2} s^{-1}). Out of the total background [12], approximately 41% corresponds to the electromagnetic component (electrons, positrons and photons), 33% corresponds to the muon component and all the rest is due to the hadronic component. It is also observed that at low altitude places, $<$3.5 Km, the muon component is dominant while at higher altitudes the electromagnetic component is dominant (Fig. 2 Left). According to our data, 4 standard deviations corresponds to 800 part m^{-2} s^{-1}, then a GRB will be detected if it shows a counting rate excess of at least this number of single particles. Notice that this flux of secondary particles is above the theoretical minimum flux detected by Sierra Negra array (93 part m^{-2} s^{-1}).

6 Conclusions

From the analysis of data and theoretical calculation, we expect that with a simple array at high altitude and a good method to discriminate the muonic component in the extended air showers we will be able to detect the Crab Nebula in less than 6 months. In addition, GRBs that produce a total energy above 10^{53} erg with photon energies $E > 1$ GeV are also within the detection reach capabilities of Sierra Negra.

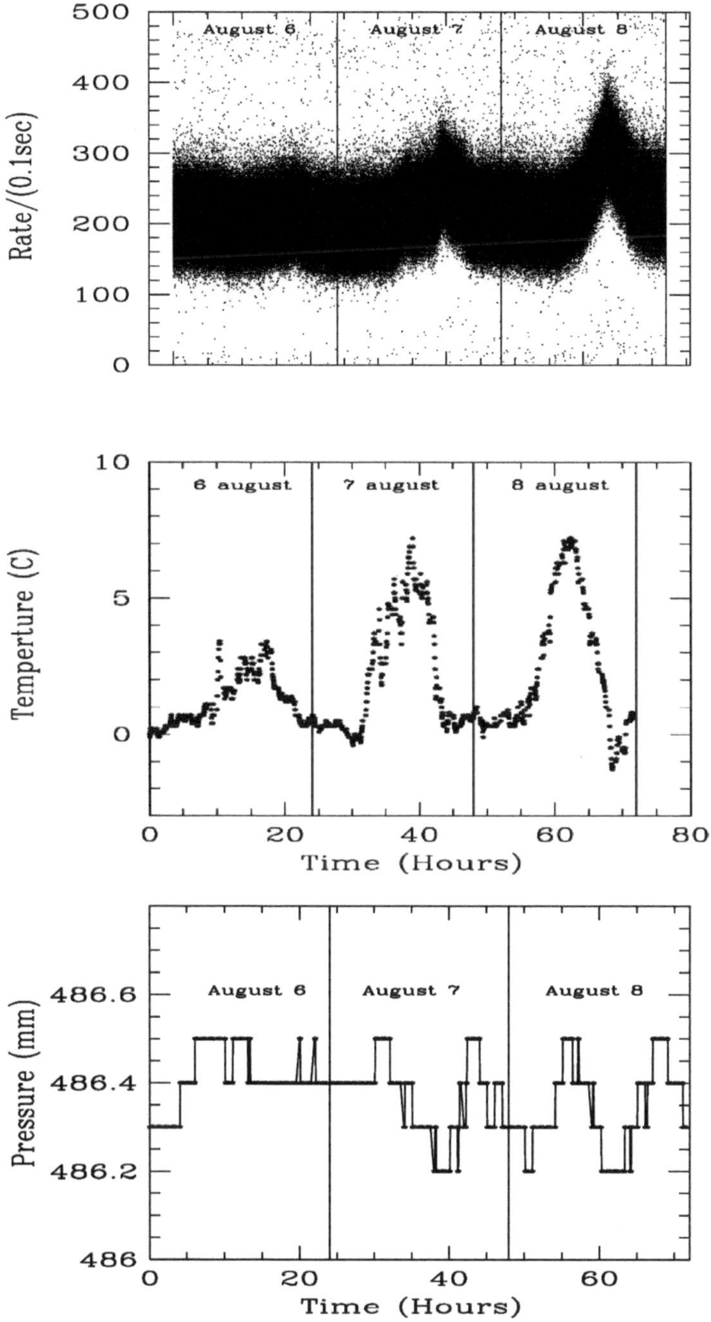

Fig. 4. Comparison or the background rate for one of the tanks in the array with the atmospheric conditions of temperature and pressure. It is observed that at night time the background rate is stable for all the detectors. The slight increase in the background rate of tank 3 is well correlated with measured changes in temperature and pressure.

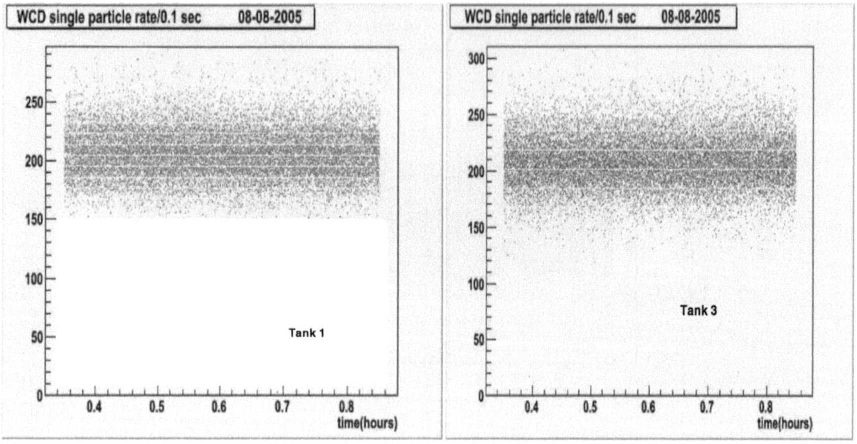

Fig. 5. Single particle rate/0.1sec for muons and high energy electrons for each tank in the array. Green and pink colors show data from tanks 1 and 3 respectively. The graphs show 30 minutes of data starting at 00 h 21 m 00 s during the night of August 8, 2005.

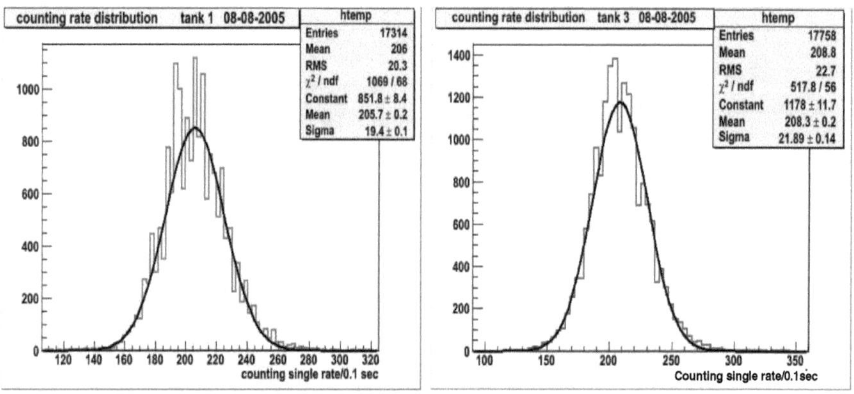

Fig. 6. (*Left*) Rate distribution observed in Sierra Negra for each tank during a period of 30 minutes, starting at 00 h 21 m 00 s during the night of August 8, 2005. It is observed that the mean value of particles flux for the tanks 1 and 3 is almost the same, about 207 part m^2/0.1s. while tank 2 shows a mean value of 104 part/m^2/0.1s. The typical dispersion rms/mean was lower than 9%.

7 Acknowledgements

We thank INAOE, LMT and especially Eduardo Mendoza for all the technical facilities at Sierra Negra that allow us to carry out this work. We thank also Tirso Murrieta, Saúl Aguilar and Ruben Conde for helping in the development of electronic devices and deployment of the detectors.

References

1. D. Allard et al. *Proceedings of ICRC* (2005)
2. R. Atkins et al. ApJ, **533**, L119, (2000)
3. R. Atkins et al. ApJ, **595**, 803, (2003)
4. C. Bacci et al. A&ASS, **138**, 597 (1999)
5. M. G. Baring, astro-ph/9711256
6. R. Cabrera et al. A&A **138**, 599, (1999)
7. A. M. Hillas et al. ApJ, **503**, 744, (1998)
8. K. Hurley et al. Nature **372**, 652-654, (1994)
9. A. Mahrous et al. *Proceedings of ICRC*, 3477 (2001)
10. P. M. Sas Pakirson, astro-ph/0505335
11. http://swift.gsfc.nasa.gov/docs/swift/swiftsc.html
12. S. Vernetto astro-ph/9904324
13. M. Vietri astro-ph/9705061

Are There Strangelets Trapped by the Geomagnetic Field?

J.E. Horvath[1], G.A. Medina Tanco[1,2] and L. Paulucci[3]

[1] Instituto de Astronomia, Geofísica e Ciências Atmosféricas IAG/USP, Rua do Matão, 1226, 05508-900 São Paulo SP, Brazil foton@astro.iag.usp.br
[2] Instituto de Ciencias Nucleares, UNAM, México
[3] Instituto de Física, Universidade de São Paulo, Rua do Matão, Travessa R, 187. CEP 05508-090 Cidade Universitária, São Paulo - Brazil paulucci@fma.if.usp.br

Basic aspects of strange quark matter (cold quark matter composed of roughly equal numbers of up, down and strange quarks) and the possible capture by the Earth's magnetosphere of the population of strangelets (hypothetical stable lumps of strange matter) are discussed to gauge the prospects for their detection.

1 Strange Matter and Strangelets

As an alternative to normal nuclei, strange matter (i.e. cold catalyzed quark matter composed of roughly equal numbers of u, d and s quarks) has been conjectured to be the ground state of hadronic matter (see [1, 2, 3]). The speculation on which this hypothesis has been based is that there is a gain in energy for converting some of the u and d quarks into s quarks in the soup by weak interactions which may be larger than the energy loss due to the finite s quark mass. Simple physical models [4] have shown that this is not unreasonable for a range of bagged QCD parameters. Since these initial works, a large number of papers have appeared devoted to the physics and astrophysics of strange matter (see the Proceedings of SQM Workshop for references and Madsen [5] for an update). Small chunks of strange matter having a low baryon number have been termed "strangelets". Roughly speaking, strangelets can be further divided into two broad classes depending on their baryon number. For $A \sim 100$ it has been shown that shell effects at the quark level are very important and mainly determine the properties of the states. When $A \geq 100$ the shell effects are less important and the states can be described by a liquid drop model in which a surface correction and so-called curvature terms are included in addition to the bulk volumetric terms in the free energy. Since for several parameter sets $E/A_{strangelet} \leq 930\,\mathrm{MeV}$, they

A. Carramiñana et al. (eds.), Solar, Stellar and Galactic Connections between Particle Physics and Astrophysics, 263–270.

are candidates for stable baryonic particles and may be produced in a variety of astrophysical environments.

Some cosmic rays (hereafter CR) events have been tentatively identified with primary strangelets in the past. However, no confirmation of these candidates has emerged and the identification of new ones has proved to be elusive. It is fair to say that strangelets are at most a rare component of CR. Of particular interest is the question of astrophysical mechanisms of production and the total mass present in the galaxy [6, 7].

If present among CR primaries, an alternative place to look for strangelets may be the Earth's magnetosphere, where analogously to the well-known trapped radiation belts [8], a geomagnetically trapped strangelet belt could be present and amenable to *direct* measurements. This hypothesis has been made several years ago [9] and recently revisited by us. Several key issues in strangelet phiscs and radiation belt physics must be addressed to evaluate the actual existence of a trapped strangelet component. Such calculations would also help to design and analyze the experiments to detect this component. It is the purpose of this work to discuss some of these questions in the following sections. Section 2 is devoted to discuss the state-of-the-art of strangelet physics. In Section 3 we address the expected trapped component features. Section 4 presents our first conclusions.

2 Strangelet Physics

Finite lumps of strange matter (strangelets) have been investigated by several authors using a spherical MIT bag model approach. For small $A \leq 100$ the mass (or energy per baryon) calculated by explicit mode-filling [4, 5, 10].

$$M = \sum_{f} \sum_{\kappa} N_{f\kappa} \left(m_f^2 + k_{f\kappa}^2 \right)^{1/2} + \frac{4}{3}\pi R^3 B \tag{1}$$

where the subindex f stands for the u, d and s flavors, k labels the order of the eigenfuntions corresponding to the Dirac solutions for a cavity of radius R and B is the MIT parametrization of the false vacuum energy. Due to finite mass of the s quark, the filling of the levels is quite cumbersome with increasing number of quarks and the strangelet "magic numbers" are not easily predicted. Moreover, the states are strongly degenerate around a minimum energy. Since many of a given set of those states are stable with respect to strong interactions and also long-lived because weak decays are Pauli-blocked, they may jointly contribute to laboratory experiments. However ,when dealing with catalyzed astrophysical stangelets, a single state for a given A will be selected. As a general feature, mode-filling calculations show that the charge Z of the most stable strangelet is $\ll A$ (see, for example [11]). Generally speaking, the presence of $A \leq 100$ "magic" strangelets is very dependent on the bulk limit energy per baryon of strange matter ϵ_b. If $\epsilon_b - m_n$ is small, the small strangelets are metastable at most, a result which has been explained in terms of the free energy expansion [12]. On the other hand, for $\epsilon_b - m_n \simeq$ tens of MeV, absolutely stable small strangelets exist and are interesting in CR research. It has been shown [12] that small strangelets are strongly disfavored energetically and, unless shell effects dominate [13] they should not be stable at all.

A further complication of this picture is the (very interesting) possible existence of attractive interaction among quarks generating a paired state instead of a pure, uncorrelated Fermi liquid. The best studied case is the so-called CFL (color-flavor locked), which was investigated in both the bulk limit [14] and strangelet [12] limits. In both cases the pairing interaction was found to enlarge substantially the parameter space for stability.

With these caveats in mind we adopt the charge-to-mass ratio for both ("normal" and "CFL" strangelets) found from a fitting of the exact calculations

$$Z = 0.1\,A\,, \quad Z = 0.3\,A^{2/3}\,, \tag{2}$$

in which we have fixed the mass of the s quark to a fiducial value $m_s = 150\,\mathrm{MeV}$.

Once $A \geq 100$ the strangelets are not critically dependent on shell filling and can be described by a free-energy expansion around the bulk limit including surface ($\propto R^2$) and curvature ($\propto R$) corrections. One unusual feature of strangelets physics is that the curvature term largely dominates the surface one, a result due to relativistic nature of adopted MIT model confined to a cavity, but believed to be more general than this. The curvature term is clearly less important as A (and therefore R) grows. For strangelets in the $A \geq 100$ regime (which may be important since several proposed candidates in CR [15] belong to it), a Thomas-Fermi analysis [16] renders a charge-to-mass relationship

$$Z = \frac{1}{\alpha}\left(\frac{\pi}{4}\right)^{1/3}\frac{m_s}{4\mu^2}\,A^{1/3} \tag{3}$$

for a strange quark mass $m_s = 80\,\mathrm{MeV}$ (this should be compared with the Farhi and Jaffe work [4], where Debye screening has not been taken into account). For even larger strangelets $A > 10^4$ the charge-to-mass ratio becomes $Z \propto A^{2/3}$, but since the flux of these massive species is expected reasonably to be a rapidly decreasing function of A in the cosmic flux, and in any case the particles would be very difficult to brake and trap, we need not worry about this regime for our purposes. The theoretical relations eqs. (2) will be the basis for our calculations.

3 Features of a Strangelet Belt

The features of the Earth's radiation belts have been reviewed by a number of authors [8]. The basic equation of charged particle motion was derived by Störmer at the beginning of the century when he solved the problem of determining the allowed and forbidden regions of the sky for a particle of momentum p coming from infinity to reach an observer at a given latitude λ, this condition reads

$$\sin\theta = \frac{2\gamma}{r\,\cos\lambda} - \frac{\cos\lambda}{r^2} \tag{4}$$

where θ is the angle between p and the meridian plane, $r = (qM/pc)^{1/2}$ is the Störmer variable with M the Earth's magnetic moment and q the charge of the particle, and 2γ is the impact parameter. Trapped (or closed) trajectories follow from the condition $\sin\theta \leq 1$. The actual relevance of Störmer's theory was recognized in the '60s after the discovery of radiation belts by Van Allen and collaborators. The discrimination between the inner (mainly protons) and outer (electrons) belts become clear after the Pioneer spacecraft flights and prompted several theoretical and experimental studies of these particle families.

Another important landmark in the study of magnetospheric particles was the discovery of trapped anomalous cosmic rays by Chan and Price [17] in the data taken onboard the Skylab and reviewed by Biswas et al. [18]. The origin and features of this trapped component was immediately addressed in several works and received definitive confirmation in a successive series of experiments (see [18] for a list of references). More recent data by SAMPEX (see [19] and references therein) has shown the consistency of several measured features with the suggested single-ionized, interplanetary ACR origin [20].

The important questions to be addressed about the hypothesis of strangelets being trapped analogously to ACR are the trapping conditions, their lifetime in the belt and the expected fluxes.

The simplest possibility is that the trapping mechanism discussed by Blake and Friesen [20] for anomalous cosmic ray nuclei (hereafter ACR) applies to the trapping of strangelets. According to these authors, the high mass-to-charge ratio of singly-ionized ACRs enables them to penetrate deeply into the magnetosphere. ACRs with trajectories near a low altitude mirror point interact with particles in the upper atmosphere, losing one or all their remaining electrons. Immediately after stripping, the particle gyroradius is reduced by a factor of $1/Z$, and the ion can become stably trapped. Since low A strangelets are expected to have already lost all their surrounding electrons due to interactions in the ISM, they will only be trapped if they already meet the Blake-Friesen conditions when fully ionized. The latter depend on the so-called "adiabaticity" parameter (ϵ), defined by Blake and Friesen as

$$\epsilon = 0.049\,(A/Z)\,L^2\,[\gamma^2 - 1]^{1/2} \tag{5}$$

where γ is the Lorentz factor of the particle. The parameter ϵ determines the maximum L-shell that allows stable trapping for a given particle, as characterized by the values of A, Z and momentum, or alternatively, the maximum A for a given L-shell and momentum, indicating, as already expected, that particles with high A and/or high p can not be stably trapped.

Direct observations of L-shell distribution [21] suggest that ions with $\epsilon < 1/10$ (and a suitable pitch angle) at the time of stripping would be stably trapped (fulfilling the requirements for *triply − adiabatic* motion).

For a particle to penetrate up to a certain region in the magnetosphere, its energy must be enough to overcome the local geomagnetic cutoff rigidity, a condition that can be written as

$$R_{particle} > \frac{59.6 \cos^4 \lambda}{L^2[1 + (1 - \cos\gamma \cos^3 \lambda)^{1/2}]^2} \frac{GV}{c} \tag{6}$$

where λ is the latitude and γ the arrival direction of the particle (East - West).

When considering charged particles trapped in a magnetic field their motion may be thought as the composition of three different motions: the bouncing motion of a guiding center along the magnetic field line; the rotational motion of the particle itself around that guiding center; and the longitudinal drift of the guiding center. In this way, an important condition intimately related to the validity of the guiding center approximation is that the magnetic field intensity must vary very slowly along a cyclotron orbit, imposing a *maximum* energy allowed for stable trapping in these conditions

$$\frac{p_\perp}{qB} \ll \frac{B}{\nabla_\perp B}. \tag{7}$$

Figure 1 shows graphically these constraints for CFL strangelets for a fixed value $L = 2$ in addition to the minimum baryon number which is required for strangelet stability [5]. The value adopted has been $A_{min} = 30$ and may be trivially altered for any other threshold.

The constraint (7) has been enforced to a 10% confidence level, and the assumption $E_\perp \sim E$, which means we are actually *underestimating* the number of particles that could be stably trapped in the geomagnetic field.

It is easy to understand that the conditions enforced by the Blake and Friesen model of trapping are much more restrictive for strangelets than for anomalous cosmic rays, mainly because their large inertia and small charge. In this way, if strangelets are a component of the ACR belt there should be a mechanism other than the Blake and Friesen one responsible for populating this region of the magnetosphere (the curves for normal strangelets, not shown in this work, lead to the same conclusion). Alternatively, the trapping may be still achieved, although not conserving the triple-adiabatic invariants. This case is much more complicated and the trajectories and other features must be found by numerical integration.

Additional considerations are relevant for the fate of a trapped population of strangelets. It is well-known that the solar wind has a strong influence on the ACR flux upon the Earth. The most abundant ACR heavy ion, oxygen, shows a strong intensity variation with the solar cycle, having its interstellar flux of 8-27 MeV/nucleon lowered up to two orders of magnitude during periods of solar maximum activity [18].

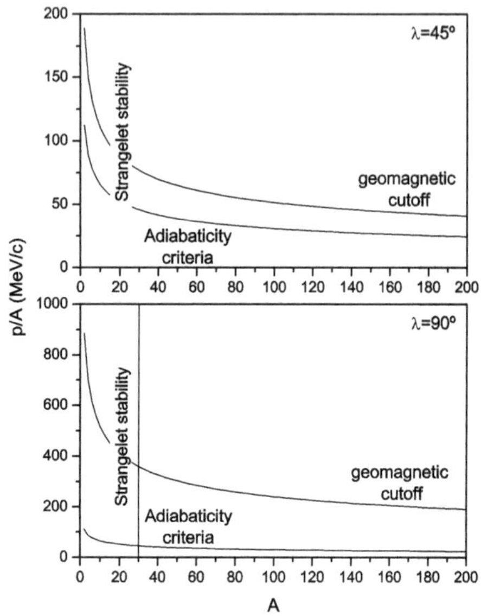

Fig. 1. Restriction curves 6 and 7 for $L = 2$ in the baryon number vs. momentum plane for CFL strangelets incident from the east ($\gamma = 0$) for vertical arrival (upper) and $\lambda = 45^o$ (lower),where field lines at $L = 2$ penetrate the earth surface in the dipole model.

Solar modulation would only act significantly on low-baryon number, low-energy strangelets. It could, however, have a measurable influence since the strangelet flux decreases towards higher values of baryon number and rigidity but probably not as important as the influence detected for ACR's. Therefore, the search for trapped strangelets in the geomagnetic field should be more successful if performed during the solar maximum activity whether they are an important component of the radiation belt or are to be measured penetrating the atmosphere towards to surface of the Earth due to reduced component of ACR.

4 Conclusions

We have analyzed here the possible trapping of non-relativistic strangelets with $A \lesssim 10^2$ already ionized by collisions with electrons in the ISM. A trapping mechanism similar to the one proposed by Blake and Freisen for ACRs can not work for the exotic strangelets, which display a very low charge-to-mass ratio. If strangelets are a component of the anomalous cosmic ray belt at $L \sim 2$, a more suitable mechanism must be found. These exotic baryons could in principle be detectable in the Earth magnetosphere whether stably trapped in a radiation belt (particularly during the period of maximum solar activity) or merely passing though the atmosphere, perhaps being captured and trapped in non-adiabatic conditions. We do not have yet firm estimates, which may require considerable numerical work. Independently of this proposal, ground-based methods and heavy ion collisions are still good sites for strangelet searches. The detection of those particles having low Z/A ratio (like the candidate presented in [22]) would be very important for determining the properties of cold, dense baryonic matter.

Acknowledgments

This work was supported by Fundação de Amparo à Pesquisa do Estado de São Paulo and CNPq Agencies (Brazil)

References

1. A. Bodmer, Phys. Rev. D **4**, 1601 (1971)
2. S. A. Chin and A. Kerman, Phys. Rev. Lett. **43**, 1292 (1979); H. Terazawa, Tokyo U. Report INS-336 (1979).
3. E. Witten, Phys. Rev. D **30**, 272 (1984)
4. E. Farhi and R. L. Jaffe, Phys. Rev. D **30**, 2379 (1984)
5. J. Madsen, astro-ph/9809032 (1998)
6. G. A. Medina Tanco and J. E. Horvath, Astrophys. J. **464**, 354 (1996)
7. J. Madsen, Phys. Rev. D**71**, 014026 (2005)
8. J. Roederer, *Dynamics of Geomagnetically Trapped Radiation* (Springer Verlag, 1970)
9. H. C. Rosu, hep-ph/9410028 (1994)
10. C. Greiner, D-H. Rischke, H. Stöcker and R. Koch, Phys. Rev. D **38**, 2797 (1988).
11. E. P. Gilson and R. L. Jaffe, Phys. Rev. Lett. **71**, 332 (1993).
12. J. Madsen, J.Phys. G **28**, 1737 (2002).
13. O. G. Benvenuto, J. E. Horvath, H. Vucetich, Phys. Rev. Lett. **64**, 713 (1990)
14. G. Lugones and J. E. Horvath, Phys. Rev.D **66**, 074017 (2002); see M. Alford, K. Rajagopal and F. Wilczek, Nucl. Phys. B **537**, **433** (1999); R. Rapp, T. Scháfer, E. V. Shuryak and M. Velkovsky, Ann.Phys. (N. Y.) **280**, 35 (2000); K. Rajagopal and F. Wilczeck, hep-ph/0011333 for an overview of the CFL state.
15. J. D. Bjorken e L.McLerran, Phys. Rev. D **20**, 2353 (1979); O. G. Benvenuto and J. E. Horvath, Mod. Phys. Lett. A **4**, 1085 (1989); M. Rybczynski, Z. Wlodarczyk e G. Wilk, hep-ph/0109225 (2001); T. Saito et al. Phys. Rev. Lett. **65**, 2094 (1990)
16. H. Heiselberg, Phys. Rev. D **48**, 1418 (1993)
17. J. H. Chan and P. B. Price: Composition and energy spectra of heavy nuclei of unknown origin detected on SKYLAB In: *14th Int. Cosmic Ray Conf., Conference Papers. Vol. 2. (A76-26851 11-93)* (Max-Planck-Institut für Extraterrestrische Physik, Munich, 1975), pp. 831-836.

18. S. Biswas, Space Science Rev. **75**, 423 (1996)
19. R. Selesnick et al., Jour. Geophys. Research **100**, 9503 (1995)
20. J. B. Blake and L. M. Friesen, Proc. 15th Int. Cosmic Ray Conf. **2**, 341 (1977)
21. A. J. Tylka, Proc. 23rd ICRC (Calgary) **3**, 436 (1993)
22. AMS home page: http://ams.cern.ch/ J. Madsen, hep-hp/0111417

Late Time Behavior of Non Spherical Collapse of Scalar Field Dark Matter

Argelia Bernal[1,2] and F. Siddhartha Guzmán[2]

[1] Departmento de Física, Centro De Investigación y De Estudios Avanzados Del IPN, AP 14-740,07000 México D.F., México <abernalresca@fis.cinvestav.mx>
[2] Instituto de Física y Matemáticas, Universidad Michoacana de San Nicolás de Hidalgo. Edificio C-3, Cd. Universitaria, C.P.58040 Morelia, Michoacán, México. <guzman@ifm.umich.mx>

We show for the first time the evolution of non-spherically symmetric balls of a self-gravitating scalar field in the Newtonian regime. In order to do so, we use a finite differencing approximation of the Shcrödinger-Poisson (SP) system of equations with axial symmetry in cylindrical coordinates. Our results indicate that spherically symmetric equilibrium configurations of the SP system are late-time attractors for non-spherically symmetric initial profiles of the scalar field, which is a generalization of such behavior for spherically symmetric initial profiles. Our system and the boundary conditions used, work as a model of scalar field dark matter collapse after the turnaround point. In such case, we have found that spherically symmetric halos are late time attractor solutions of possible axisymmetric initial scalar field overdensities.

1 Introduction

Recently, scalar fields have played the role in several scenarios related to astrophysical phenomena. The reason is that such fields are quite common in theoretical physics, specially branches related to theories beyond the standard model of particles, higher dimensional theories and brane world models of the universe. In the present research we deal with the scalar field dark matter model (SFDM), which assumes the dark matter to be a classical minimally coupled real scalar field determined by a cosh-like potential. Such potential provides the field with the necessary properties to mimic the behavior and successes of cold dark matter at cosmic scales. In fact in [1, 2, 3] it was shown that the mass parameter of the scalar field gets fixed by a desired cut-off of the power spectrum, which has two effects: i) the theory gets fixed and ii) there is no overabundance of substructure, which standard cold dark matter cannot achieve. One important consequence is that the boson has to be ultra-light with masses around $m \sim 10^{-21,-23}$eV. This is a substantially important

A. Carramiñana et al. (eds.), Solar, Stellar and Galactic Connections between Particle Physics and Astrophysics, 271–278.

bound, because in the standard dark matter models there are no such ultralight dark matter candidates. The benefit obtained however, is two fold: the scalar field can represent a Bose Condensate of such ultralight particles and the de Broglie wavelength forbids the scalar field to form cuspy structures.

After the fluctuation analysis about this candidate and its corresponding concordance with observations -the model mimics the properties of the ΛCDM at cosmic scales-, the next step has to be in the direction of the study of structure formation and the explanation of local phenomena, like rotation curves in galaxies. Fortunately there have been important advances in such direction [4, 5]. In [6] it was shown that relativistic self-gravitating scalar field configurations can be formed when they have galactic masses provided the mass of the boson is ultralight. Nevertheless, because the gravitational field in galaxies is weak, the race turned into the newtonian limit of the system of equations, which was developed in [7]. The price to be paid is that it is not possible to apply the approach at very early stages of the evolution of the universe, and the profit is that the scalar field in the non-relativistic regime provides a clear interpretation within the Bose Condensate formalism and classical quantum mechanics. In both cases, the strong gravity and the Newtonian regimes, a wide range of arbitrary spherically symmetric initial configurations collapse and form gravitationally bounded and virialized objects with a smooth density everywhere (except those that are related to unstable initial configurations that collapse into black holes in the strong field regime) [8, 9]. This property seems to be fundamental in order to form galactic halos, because several high resolution observations are consistent with regular galactic dark matter profiles in the center of the galaxies [10, 11, 12, 13].

It is clear then, that two approaches are in progress, both are complementary and are useful to explore the SFDM hypothesis. Turning back to the astrophysical scenario, at stages after the turnaround point where weak field applies, the question is whether dark matter halos are gravitationally bounded objects of scalar field which have been formed through a gravitational collapse of initial scalar field overdensities. The newtonian version of the Einstein-Klein-Gordon system of equations is provided by the Schrödinger-Poisson equations (SP), which are the ones that lead the gravitational collapse of the system. This approximation should work for the evolution of an initial density profile after the epoch when the overdensity fluctuation starts to evolve independently of the cosmological expansion.

In the recent past, it has been found that in spherical symmetry the SP system has stationary solutions, from which only the nodeless one is stable [8, 14]; even further, it has been found that such configurations behave as late-time attractors for initially quite arbitrary spherically symmetric density profiles [7, 15]. The goal of the present manuscript is to show that such attractor behavior extends beyond spherical symmetry, still within axially symmetric initial density profiles.

In the next section we present the code to solve the axially symmetric SP system and enumerate the physical quantities that characterize the axial configurations. In order to provide testbeds of our code, in section 2 we show the evolution of spherically symmetric equilibrium configurations, which behavior is well known; latter in the same section we show the evolution of non equilibrium axial symmetric configurations and the attractor behavior of the spherically symmetric ground states. Finally in section 3 we draw some comments and conclusions.

2 Numerical Evolution of Axially Symmetric Initial Profiles

In order to solve the SP system with axial symmetry we developed a code that uses cylindrical coordinates (x, z). The scalar field Φ is of the form $\Phi = \Phi(x, z)$ and the SP system in such coordinates reads

$$i\frac{\partial \Phi}{\partial t} = -\frac{1}{2}\left(\frac{\partial^2 \Phi}{\partial x^2} + \frac{1}{x}\frac{\partial \Phi}{\partial x} + \frac{\partial^2 \Phi}{\partial z^2}\right) + U\Phi \tag{1}$$

$$\frac{\partial^2 U}{\partial x^2} + \frac{1}{x}\frac{\partial U}{\partial x} + \frac{\partial^2 U}{\partial z^2} = \Phi^* \Phi$$

where U is the gravitational potential due to the presence of Φ. Notice that we are using units in which $c = \hbar = 1$ and the associated mass of the scalar field has been normalized to one. The way we solve the whole system is as follows: given an initial scalar field profile, the evolution process consist in solving Poisson's equation using centered finite differencing, then, once U is known the Schrödinger equation is solved using a second order accurate explicit time integrator, then we use such new wave function to solve Poisson's equation and repeat.

The term containing the $1/x$ factor deserves special attention; it happens that when $r \rightarrow 0$ such term does not converge (when it does not in fact diverge) with second order. In order achieve second order convergence at such point we have staggered the grid in the x direction and used the discretized version $2\frac{d\Phi}{d(x^2)}$ instead of $\frac{1}{x}\frac{\partial \Phi}{\partial x}$ and $2\frac{dU}{d(x^2)}$ instead of $\frac{1}{x}\frac{\partial U}{\partial x}$, where $\frac{d}{d(x^2)}$ is a derivative with respect to x^2.

Another important issue has to do with the boundary conditions. We assume the system is physically open, however Schrödinger equation is far from being the wave equation and no sommerfeld boundary conditions can be applied as far as we are aware. However we emulate open boundaries by applying fully reflecting boundary conditions and implementing a sponge region in the outermost grid points of the domain. The sponge consists in adding in such a region an imaginary potential to the Schrödinger equation, whose effects are those of a sink of density of probability. Thus the particles near the boundaries are trapped by the sponge. Details about the implementation of the sponge for spherical symmetry can be found in [8], and we just implemented the same idea in a two dimensional domain.

Because we are dealing with a classical quantum mechanical system, it is possible to calculate expectation values of physical operators. Among the important quantities we want to monitor are: the density of probability $\rho = \Phi\Phi^*$, the mass of the configuration $M = \int \rho d^3x$, the expectation value of the kinetic $K = -(1/2)\int \Phi^* \nabla^2(\Phi) d^3x$, potential $W = (1/2)\int \rho U d^3x$ and total $E = K + W$ energies. The number of particles allows us to calculate the material content in a region of space. The expression $2K + W$ determines whether the system is dynamically virialized or not; we say that the system is nearly virialized when $2K + W \sim 0$; it is fair to say that because the systems we deal with are intrinsically perturbed due to discretization errors (discrete approximation of the equations), the relation will never be strictly equal to zero and we only demand that it converges to zero with second order in the continuum limit.

2.1 Tests of the Code

As a testbed for the code, it is necessary to verify whether in these non-spherical coordinates the code is able to reproduce the expected results for an equilibrium configuration, that is, a spherically symmetric stationary solution. In spherical symmetry, in order to construct stationary configurations the scalar field is assumed to have the form $\Phi = e^{-i\sigma t}\phi(x)$; provided the boundary conditions $\phi = 0$ at infinity and demanding the gravitational potential to be smooth at the origin, the SP system becomes an eigenvalue problem. The eigenvalues σ related to the eigenfunctions Φ are non degenerate and take a different values for different central field $\phi(0)$. Because the time dependence of Φ is harmonic, $\rho = \Phi\Phi^*$ and therefore the gravitational potential U should be time independent. Another property of the equilibrium configurations constructed in this way is the number of nodes of the wave function; the solution with a nodeless wave function is said to be in the ground state and it has been shown (see [8]) that only the ground state solutions are stable. In fact a linear perturbation analysis has revealed that such ground state systems oscillate with a specific angular frequency when slightly perturbed, a property that is quite useful when dealing with the discretized version of the SP equations. Another important property of the system (1) is the following scale invariance:

$$\{t, x, U, \Phi\} \to \{\lambda^{-2}\hat{t}, \lambda^{-1}\hat{x}, \lambda^2\hat{U}, \lambda^2\hat{\Phi}\} \tag{2}$$

$$\{\rho, M, K, W\} \to \{\lambda^4\hat{\rho}, \lambda\hat{M}, \lambda^3\hat{K}, \lambda^3\hat{W}\} \tag{3}$$

where λ is a scaling parameter. Property (2-3) implies that if a solution is found for a given central field value $\hat{\phi}(0) = \hat{\phi}_0$ (e.g. $\hat{\phi}(0) = 1$) it is possible to build the whole branch of equilibrium configurations. For instance, if the plot M *vs* $\phi(0)$ is to be constructed, we know from [8] that for $\hat{\Phi}(0) = 1$ we have $\hat{M} = 2.0622$; using the relations (2-3) for Φ and M we find $\lambda = \sqrt{\Phi/\hat{\Phi}} =$

$M/\hat{M}$, which implies that the desired diagram is given by the function $M = 2.0622\sqrt{\Phi(0)}$ for all central values of the scalar field. This function is used later on when showing the attractor behavior of these equilibrium configurations.

Therefore, without lost of generality we tested this code by evolving the particular ground state equilibrium configuration with $\phi(0) = 1.0$; in practice we construct initial data for this particular central field value in spherical coordinates and interpolate the value of Φ and U into our $xz-$plane. We know from the construction of these solutions that the density and the gravitational potential are time independent. However, as happens to the virial theorem, both the density and the potential are slightly time-dependent due to the discretization error that is present at all times during the evolution of the system. The effect of such an error is a perturbation that can be analyzed through linear perturbation theory; in fact it was found in [8] that for the present system the central density of the system should oscillate with a frequency $\gamma \sim 0.046$. Then, a test consists in demanding that the central density actually oscillates with this frequency within our axisymmetric code, and that the central density converges to the correct value in the continuum limit, that is $\rho(t,0,0) = |\Phi(t,0,0)|^2 = 1$. In Figure 1 we show both results. The virial relation is also convergent to zero in the continuum limit. These results indicate that we are obtaining the expected results for an equilibrium configuration.

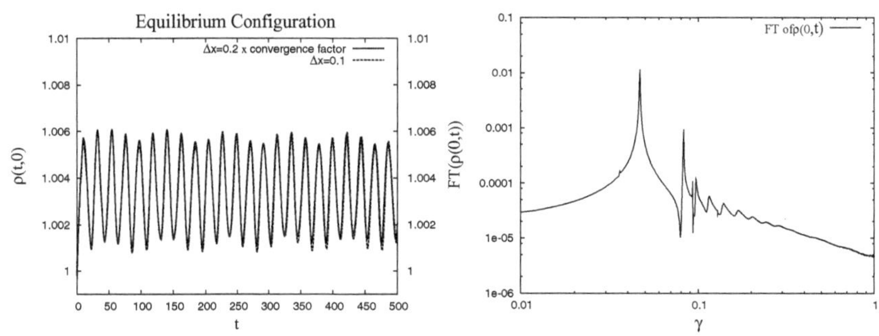

Fig. 1. Test of the axisymmetric code. Evolution of the ground state for the SP system in spherical symmetry. Left: The central value ρ in time using different resolutions $\Delta xz = 0.2$ and $\Delta xz = 0.1$, $\rho_{0.2}$ and $\rho_{0.1}$ respectively; $\rho_{0.2}$ has been scaled by $(\rho_{0.2} - 1)/4 + 1$ and the fact that it lies upon $\rho_{0.1}$ indicates the second order convergence of the value 1. Right: The Fourier Transform of the central density shows its main peak -which is associated to the fundamental mode frequency of the system- at $\gamma \sim 0.046$ (see text for details). These plots, together with the convergence of the virial relation indicate that our code works properly.

2.2 Evolution of Non Spherical Initial Profiles

In this section we show that spherically symmetric ground states, are late time attractors for initial configurations which are not spherically symmetric.

A first step would be to show that equilibrium configurations are stable against axisymmetric perturbations, and a second step would be to show how quite arbitrary axisymmetric initial data evolve toward a spherical equilibrium configuration. We decided to skip the first step and go straight to the second one. We carried out a series of simulations with initial data consisting of the superposition of an equilibrium configuration with $\phi(0) = 1$ and a gaussian-like profile given by the expression $\delta\psi = Ae^{[-x^2/\sigma_x^2 - z^2/\sigma_z^2]}$ that we call invading profile. The parameters of the invading profile for the cases shown here correspond to $\sigma_x = 1$ and $\sigma_z = 1.5$, with amplitudes $A = 0.1, 0.2$, which implies an addition of 5% and 9% of the total mass of the original equilibrium configuration. The reason to proceed in such way is that beyond the perturbation of the system we wonder whether the invading particles attach to the spherical configuration and together they evolve toward a rescaled equilibrium configuration.

The result is that the whole system evolves toward a stationary configuration through the gravitational cooling process [8, 15], which is powered by the ejection of a small amount of scalar field mass of 0.05% and 0.12% respectively. In the left panel of Figure 2, we show the evolution of the mass M versus the central density $\rho_c = \rho(x = 0, z = 0)$ for both initial profiles. The solid line is the branch of all the equilibrium configurations constructed as described above. What can be observed in the plot is that the initial configurations oscillate around the branch of equilibrium configurations and at late times it tends to converge to one of such solutions. This attractor behavior has been shown for spherically symmetric configurations in [15] and is shown here for the first time for the case of non-spherical profiles. However, the information related to the convergence to an equilibrium configuration is not enough, and in the right panel of Figure 2 we show the ellipticity of the system, which we define as the integrated difference between $\rho_z = \rho(0, z)$ and $\rho_x = \rho(x, 0)$ measured from the center of mass of the configuration; we can see that after a transient with initial ellipticity the system relaxes and becomes spherical in the continuum limit.

3 Conclusions

We have presented a new code that solves the SP system of equations with axial symmetry using finite differencing and an explicit time integrator. We tested our code with the evolution of an equilibrium spherically symmetric ground state configuration, and obtained the correct values of the fundamental frequency of perturbations due to the discretization errors; in fact we showed that such error is spherical and no axisymmetric modes are excited through the discretization of the SP equations in cylindrical coordinates. Moreover we verified that the virial state of such system behaves in the expected way in the continuum limit.

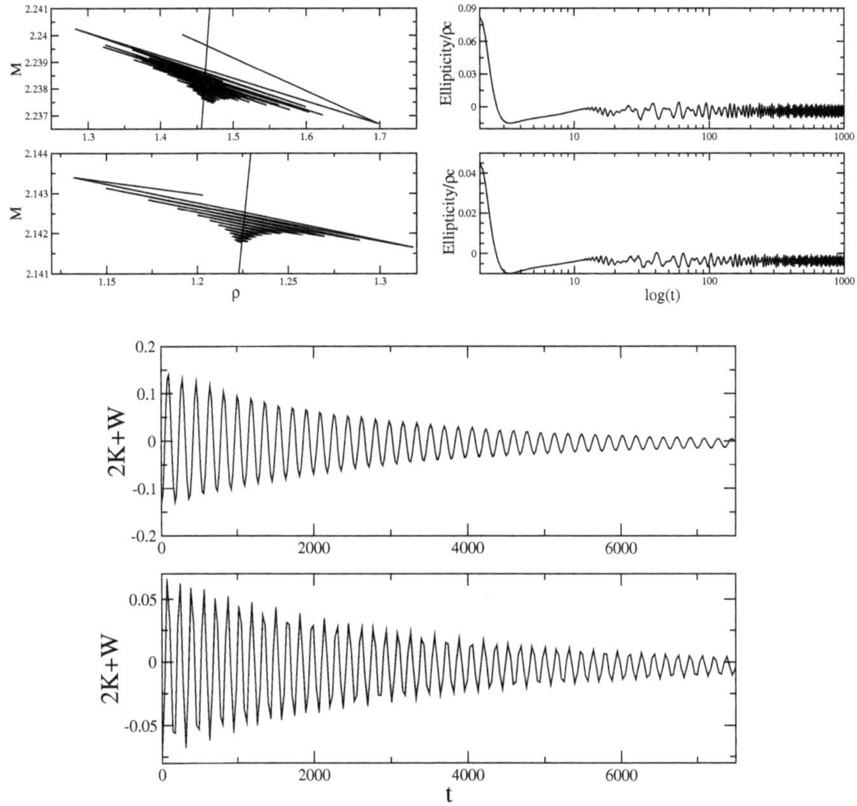

Fig. 2. Evolution of two axially symmetric initial data made of an equilibrium configuration plus a non-spherical gaussian like profile. Left: we show that the initial axially symmetric configurations evolve toward spherical equilibrium configurations (points in the solid line) through the emission of scalar field. Right: the ellipticity is shown for both simulations. Bottom: we show the value of the expression $2K + W$; as it oscillates around zero with a decreasing amplitude we conclude that the system tends to a virialized state.

Using such code we evolved non-spherically symmetric configurations made with the superposition of an equilibrium configuration and a gaussian-like axially symmetric profile. The gaussian-like profile was more than just a perturbation, because the amount of matter added to the equilibrium system was of the same order of its mass; therefore we actually tracked the evolution of an axially symmetric initial profile. We showed that at late times, the configurations tend toward an equilibrium spherical ground state solution. This result suggests a late time attractor behavior of spherical ground states for axisymmetric initial scalar field density profiles, which would be a generalization of such behavior for spherically symmetric profiles.

In the context of the scalar field dark matter model we have quite a new result: the collapse of overdensities tolerates an initial non-sphericity of the profiles, and moreover, initially axisymmetric profiles tend toward a spherical ground state.

Acknowledgments

This research is partly supported by grants PROMEP UMICH-PTC-121 and CIC-UMSNH-4.9. The runs were carried out in the Ek-bek cluster of the "Laboratorio de Supercómputo Astrofísico (LASUMA)" at CINVESTAV-IPN. A. B. acknowledges a scholarship from CONACyT.

References

1. V. Sahni, and L. M. Wang, *Phys. Rev.* **62**, 103517 (2000).
2. T. Matos, and L. A. Ureña-López, *Class. Quantum Grav.* **17**, L75 (2000).
3. T. Matos, and L. A. Ureña-López, *Phys, Rev. D* **63**, 063506 (2000).
4. A. Arbey, J. Lesgourgues, and P. Salati, *Phys, Rev. D* **64**, 123528 (2001). *Ibid.* **65**, 083514 (2002). *Ibid.* **68**, 023511 (2003).
5. J. P. Mbelek *A & A* **424**, 761-764 (2004).
6. M. Alcubierre, F. S. Guzmán, T. Matos, D. Núñez, L. A. Ureña, and P. Wiederhold. *Class. Quantum Grav.* **19**, 5017 (2002).
7. F. S. Guzmán, and L. A. Ureña, *Phys, Rev. D* **68**, 024023 (2003).
8. F. S. Guzmán, and L. A. Ureña, *Phys, Rev. D* **69**, 124033 (2004).
9. M. Alcubierre, R. Becerril, F. S. Guzmán, T. Matos, D. Núñez, and L. A. Ureña, *Class. Quantum Grav.* **20**, 2883 (2003).
10. S. S. McGaugh, V. C. Rubin, and E. de Block, *Astron. J.* **122**, 2831 (2001).
11. W. J. G. de Blok, S. S. McGaugh, and V. C. Rubin, *Astron. J.* **122**, 2396 (2001).
12. P. A. S. Blais-Ouellette, and C. Carignan, *Astron. J.* **121**, 1952 (2001).
13. A. D. Bolato, J. D. Simon, A. Leroy, and L. Blotz, *Astrophys. J.* **565**, 238 (2002).
14. R. Harrison, I. Moroz, and K. P. Tod, *math-ph/0208045*.
15. F. S. Guzmán, and L. A. Ureña, *ApJ*, in press; arXiv: astro-ph/0603613.

Inhomogeneous Dark Matter in Non-trivial Interaction with Dark Energy

Roberto A. Sussman[1], Israel Quiros[2] and Osmel Martín González[2]

[1] Instituto de Ciencias Nucleares, Apartado Postal 70543, UNAM, México DF, 04510, México <sussman@nuclecu.unam.mx>.
[2] Departamento de Física, Universidad Central de las Villas, Santa Clara, Cuba <israel@uclv.edu.cu; osmel@uclv.edu.cu>.

We study interacting dark energy (DE) and cold dark matter (DM) in the context of an inhomogeneous and anisotropic spacetime. DM and DE are modeled as an interactive mixture of inhomogeneous dust (DM) and a generic homogeneous dark energy (DE) fluid. By choosing an "equation of state" linking the energy density (μ) and pressure (p) of the DE fluid, as well as a free function governing the radial dependence, the models become fully determinate and can be applied to known specific DE sources, such as quintessense scalar fields or tachyonic fluids. For the case of the simple equation of state $p = (\gamma - 1)\,\mu$ with $0 \leq \gamma < 2/3$, the free parameters and boundary conditions can be selected for an adequate description of a local DM overdensity evolving in a suitable cosmic background that accurately fits current observational data. If the particular case when DE fluid corresponds to a quintessense scalar field, the interaction term can be associated with a well motivated non–minimal coupling to the DM component. The effects of inhomogeneity and anisotropy yield different local behavior and evolution rates for observational parameters in the local overdense region.

1 Introduction

Observational data on Type Ia supernovae strongly suggests that the universe is expanding at an accelerated rate [1, 2]. This effect has lead to the widespread assumption that the inventory of cosmic matter–energy could contain, besides baryons, photons, neutrinos and cold dark matter (DM)[3], an extra contribution generically known as "dark energy" (DE), whose kinematic effect could be equivalent to that of a fluid with negative pressure. While the large scale dynamics of the main cosmic sources (DE and DM) is more or

[3] We shall assume henceforth that DM is of the "cold" variety, *i.e.* CDM.

A. Carramiñana et al. (eds.), Solar, Stellar and Galactic Connections between Particle Physics and Astrophysics, 279–294.

less understood, their fundamental physical nature is still a matter for debate, thus various physical explanations have been suggested. Cold DM is usually conceived as a collisionless gas of supersymmetric particles (neutralinos), while DE can be modeled as a "cosmological constant", quintessense scalar fields, tachyonic fluids, generalized forces, etc [3, 4]. The standard approach is mostly to consider a Friedman-Lemaître-Robertson-Walker (FLRW) metric, with linear perturbations, making also the simplest assumption that DE only interacts gravitationally with DM. However, there are still some unresolved issues, such as the so–called "coincidence problem", concerning the odd apparent fact that the critical densities of DM and DE approximately coincide in our cosmic era [5, 6]. Aiming at a solution to this problem and bearing in mind our ignorance on the fundamental physics of DM and DE, various models have been proposed recently which include assorted forms of interaction between these sources [7, 8, 9, 10, 11].

It is customarily assumed that DE dominates large scale cosmic dynamics, so that DM inhomogeneities in galactic clusters and superclusters can be considered a local effect or can be treated by means of linear perturbations in a FLRW background. Thus, a reasonable generalization of existing models could be to assume inhomogeneous DM interacting with homogeneous DE, so that large scale dynamics is governed by the latter. We propose in this paper a class of analytic models which provide a reasonable description of inhomogeneous DM interacting with a generic homogeneous DE source. The models are based on the decomposition of a perfect fluid tensor as a mixture of an inhomogeneous dust component (DM) plus a homogeneous perfect fluid with negative pressure (generic DE), as the matter source of the spherically symmetric subcase of the Szafron–Szekeres exact solutions of Einstein's field equations [12]. However, the underlying geometry of the models we present can be easily generalized to include non–spherical symmetries or even the case without any isometry, since Szafron–Szekeres solutions do not usually admit Killing vectors.

The decomposition of a perfect fluid tensor as a mixture of an inhomogeneous dust component plus a homogeneous fluid has been considered previously [13, 14] but in the context of mixtures of baryons and radiation. We consider in this paper only the type of models examined in [14], by assuming the homogeneous fluid to describe a generic DE source, while the dust component corresponds to inhomogeneous DM, all of which is a reasonable assumption since the dynamical effects of quintessence mostly become dominant in very large scales, larger than the "homogeneity scale" (100–300 Mpc), while DM (galactic clusters and superclusters) is very inhomogeneous at scales of this magnitude and smaller.

In order to determine the time evolution of the sources, we need to assume a physical model, or "equation of state" for the generic DE source (the homogeneous fluid). Thus, we assume in section VIII a simple "gamma law" equation of state of the form $p = (\gamma - 1)\,\mu$, where p, μ are the pressure and matter–energy density of the DE source. Such an equation of state leads to a

DE homogeneous fluid evolving like a FLRW fluid with flat spacelike sections with a scaling law of the form $\mu \propto t^{-2}$, which is compatible with a scalar field with an exponential potential [15]. Although this is a very simple type of DE source, it yields analytic forms for the DM density, observational parameters and all other relevant physical and geometric quantities.

The assumption of a gamma law equation of state fully determines the time dependence of all relevant quantities, while the spacial dependence is determined once we select an arbitrary function whose form depends on the choice of suitable boundary conditions associated with a description of a local DM overdense region in a DE dominated cosmic background that accurately complies with observational constraints on observational parameters: Ω for DM and for DE and the deceleration parameter q. We provide a full graphical illustration of the interplay between "local" and "cosmic background" effects on these observational parameters: for example, anisotropy emerges in the local dependence of these quantities on the "off-center observation angle" ψ (Figure 2), while inhomogeneity leads to local conditions in the overdense region (DM dominates over DE and q is positive) that are different from those of the cosmic background: DE dominates and $q < 0$, as required by an "accelerated" universe whose large scale dynamics is dominated by a repulsive force associated with DE (Figures 3).

The issue of the interaction between DE and DM is dealt with in section VI. We show that the individual momentum–energy tensors for DM and DE are not independently conserved, thus the models are incompatible with these components interacting only gravitationally. However, if we assume the DE fluid to be a scalar field quintessense type of source, then the models can accommodate various prescriptions for a DE–DM interaction, like those proposed in the literature [7, 8, 9, 10, 11]. Finally, in section VII we present a discussion and summary of our results. The results presented in here have been further generalized and expanded in [16].

2 A Mixture of Dark Matter and Dark Energy

Consider the spherically symmetric inhomogeneous line element

$$ds^2 = -c^2\,dt^2 + R_0^2\,a^2\left[\frac{W^2}{V^{2/3}}\,dr^2 + r^2 V^{4/3}\,d\,\Omega^2\right], \quad d\,\Omega^2 \equiv d\,\theta^2 + \sin^2\theta\,d\,\varphi^2, \tag{1}$$

where R_0 is a constant with length units, $a = a(t)$ is dimensionless and V, W are:

$$V \;=\; 1 + f(r)\,T(t), \quad W \;=\; 1 + \left[f(r) + \frac{2}{3} r f'(r)\right]\,T(t), \tag{2}$$

with $f(r)$ arbitrary and $f' = df/dr$. For an interacting mixture of inhomogeneous dust-like DM and homogeneous generic DE we consider the momentum–energy tensor $T^{ab} = T^{ab}_{\text{DM}} + T^{ab}_{\text{DE}}$, with

$$T^{ab}_{\rm DM} = \rho(t,r)\, c^2\, u^a\, u^b, \quad T^{ab}_{\rm DE} = [\,\mu(t) + p(t)]\, u^a\, u^b + p(t)\, g^{ab}, \tag{3}$$

where the comoving 4–velocity is $u^a = \delta^a_t$. Because of their construction, the two energy–momentum tensors are not separately conserved: $T^{ab}_{\rm DM;b} = -T^{ab}_{\rm DM;b} = Q \neq 0$, hence we must have a non–trivial coupling between DM and DE characterized by the interaction term Q. We will address this interaction in section VI.

Einstein's field equations for (1)–(3) yield:

$$\dot H = \frac{\kappa}{2}\,(\mu + p), \tag{4}$$

$$\kappa\,\mu = 3H^2, \tag{5}$$

$$\dot T = \frac{c_0\, H_0}{a^3}, \tag{6}$$

where $\kappa = 8\pi G/c^4$, c_0 is a dimensionless constant, $\dot H = dH/dt$ and

$$H = \frac{\dot a}{a}. \tag{7}$$

The matter–energy density of the DM component is given by:

$$\kappa\,\rho\, c^2 = \frac{4\,(3\,T\,H + \dot T)\, f\, F + 6\, H\,(f + F)}{3\,(1 + f\,T)\,(1 + F\,T)}\,\dot T, \quad \text{where} \quad F = f + \frac{2}{3}\, r\, f'. \tag{8}$$

So far, the homogeneous DE fluid with μ, p is a generic form of DE. Once we choose an "equation of state" $p = p(\mu)$, corresponding to a specific DE model for this fluid (for example, a scalar field), we can find H by integrating (4) and (5), and then a and T by integrating (6) and (7). Once H and T have been determined, we can obtain the DM density ρ for a given choice of the arbitrary function $f(r)$. This free function represents the freedom to choose an initial DM density profile along a given surface of constant t.

Other important quantities are the expansion kinematic scalar $\Theta = u^a{}_{;a}$ and the traceless symmetric shear tensor $\sigma_{ab} = u_{[a;b]} - (\Theta/3)\, h_{ab}$

$$\Theta = 3\,H + \mathcal{Z}\,\dot T, \quad \sigma^a{}_b = \textbf{diag}\,[0, -2\Sigma, \Sigma, \Sigma], \tag{9}$$

where

$$\mathcal{Z} = \frac{f + F + 2\, f\, F\, T}{(1 + f\,T)\,(1 + F\,T)}, \quad \Sigma = \frac{(F - f)\,\dot T}{(1 + f\,T)\,(1 + F\,T)}. \tag{10}$$

The metric (1) looks like a FLRW line element modified by the terms containing T, F and f. In fact, all r–dependent variables derived above reduce to their FLRW forms: $\rho = \sigma = 0$, $\Theta/3 = H$, if these "perturbations" vanish, i.e. if either $T = 0$ or $f = f' = 0$. This homogeneous subcase is a FLRW spacetime whose source is the DE perfect fluid with matter–energy density and pressure given by μ and p. In a sense, if $f\,T \ll 1$ and $F\,T \ll 1$ the models would correspond formally to specific exact perturbations of FLRW cosmologies.

Notice that the form of the two equations (4) and (5) is identical to the field equations of a FLRW spacetime with flat space sections and matter source with μ, p. This suggest that we identify μ, p, a and H with variables somehow associated with a FLRW background. In fact, the function f can be suitably selected so that such a background can be identified with conditions $r \to \infty$ along surfaces of constant t. We must stress, though, that the correct interpretation of physical and observational quantities must be given in terms of tensorial quantities like (9-10) characteristic of the inhomogeneous spacetime (1-8).

3 Observational Parameters

The quantity H in (7) is the Hubble expansion factor associated with a FLRW geometry, for the inhomogeneous metric (1) the proper generalization of this parameter is given by [17, 18]

$$\mathcal{H} = \frac{\Theta}{3} + \sigma_{ab}\, n^a n^b, \tag{11}$$

where the vector n^a complies with $n_a n^a = 1$, $u_a n^a = 0$. For a spherically symmetric spacetime, it is necessary to evaluate n^a for general comoving observers located in an "off–center" position in the spherical coordinates (r, θ, ϕ) centered at $r = 0$. For the metric (1) equation (11) becomes in general

$$\mathcal{H} = H + \mathcal{F}\dot{T}, \quad \mathcal{F} = \frac{2\left[f\,(1+F\,T) + \frac{3}{2}\,(F-f)\,\cos^2\psi\right]}{3\,(1+f\,T)\,(1+F\,T)}. \tag{12}$$

where ψ is "observation angle" between the direction of a light ray and the "radial" direction for a fundamental observer located in (r, θ, φ) [18]. Therefore, the exact local values of the observational parameters Ω for DE and DM are

$$\Omega_{\rm DE} = \frac{\kappa\,\mu}{3\,\mathcal{H}^2} = \frac{H^2}{\left[H + \mathcal{F}\dot{T}\right]^2}, \quad \Omega_{\rm DM} = \frac{\kappa\,\rho\,c^2}{3\,\mathcal{H}^2} = \frac{\kappa\,\rho\,c^2}{3\,\left[H + \mathcal{F}\dot{T}\right]^2} = \frac{\rho\,c^2}{\mu}\,\Omega_{\rm DE}, \tag{13}$$

while the acceleration parameter is [17]

$$q = \frac{6\Sigma^2}{\mathcal{H}^2} + \frac{\Omega_{\rm DE} + \Omega_{\rm DM}}{2}\left[1 + \frac{3\,p/\mu}{1 + \rho c^2/\mu}\right], \tag{14}$$

where Σ is given by (10).

If we consider the flow of cosmic DM with density ρ at the length scale of the observable universe ($\sim 3\,h$ Gpc), the present day values of shear and

DM density gradients in comparable scales are severely restricted by the CMB near isotropy [19].

$$\left[\frac{|\sigma_{ab}\,\sigma^{ab}|^{1/2}}{\Theta}\right]_0 = \left[\frac{\sqrt{6}\,|\Sigma|}{\Theta}\right]_0 \lesssim 10^{-5}, \qquad \left[\frac{h_a^b\,\rho_{,b}}{\rho}\right]_0 = \left[\frac{\rho'}{\rho}\right]_0 \lesssim 10^{-5}. \quad (15)$$

Hence the free function f must be suitably chosen so that the large scale spatial dependence of the observational parameters (11), (13), (13) and (14) fits these bounds. However, these restrictions can be strongly relaxed, at a local level, if we examine the spatial variation of local values of DM density and observational parameters in scales smaller than the homogeneity scale ~100–300 Mpc.

A convenient choice for the asymptotic behavior of f as $r \to \infty$ is to choose it monotonously decreasing in r so that

$$f \to f^*, \quad F \to f^*, \qquad \text{as} \quad r \to \infty, \quad (16)$$

where f^* is a positive constant. This yields $\Sigma_\infty = 0$ so that $[\sigma^a{}_b]_\infty = 0$ and $(1/3)\Theta_\infty = \mathcal{H}_\infty$, while the remaining space dependent quantities have the asymptotic forms

$$\kappa\,\rho_\infty\,c^2 = \frac{4}{3}\,f^*\,\frac{3\,H\,[1+f^*\,T] + f^*\,\dot{T}}{[1+f^*\,T]^2}\,\dot{T}, \quad \mathcal{H}_\infty = H + \frac{2\,f^*\,\dot{T}}{3\,[1+f^*\,T]}, \quad (17)$$

$$\Omega_{\text{DE}}|_\infty = \frac{\kappa\mu}{3\,\mathcal{H}_\infty^2} = \frac{H^2}{\mathcal{H}_\infty^2}, \qquad \Omega_{\text{DM}}|_\infty = \frac{\kappa\rho_\infty}{3\,\mathcal{H}_\infty^2}, \quad (18)$$

$$q_\infty = \frac{\Omega_{\text{DE}}|_\infty + \Omega_{\text{DM}}|_\infty}{2}\left[1 + \frac{3\,p/\mu}{1+\rho_\infty c^2/\mu}\right].$$

The choice (27) is well suited to examine a large scale (supercluster scale or larger) spherical inhomogeneity whose evolution requires that we somehow "plug in" the effects of a cosmological background, while the asymptotic behavior $f \to 0$ as $r \to \infty$ (case $f^* = 0$) may be preferable for a relatively small scale and/or large density contrast description of an homogeneity (cluster of galaxies) that ignores cosmic effects.

4 A Simple Example: The "Gamma Law"

In order to illustrate how to work out the expressions we have derived, we consider now the simple case of a homogeneous DE fluid satisfying a simple equation of state known as the "gamma–law"

$$p = (\gamma - 1)\,\mu, \quad (19)$$

where γ is a constant. The dust plus homogeneous fluid mixtures that we are studying were examined previously [13], assuming (among other choices) this equation of state, but placing especial emphasis in a dust and radiation ($\gamma = 4/3$) mixture. Since our emphasis is now on modeling DE sources, we will assume $0 < \gamma < 2/3$, so that $-1 < p/\mu < -1/3$. In this case we have from (4), (5), (6) and (7)

$$a \equiv \frac{R}{R_0} = \tau^{2/3\gamma}, \qquad \tau = \frac{3}{2}\gamma\, h\, H_0\, t, \tag{20}$$

$$\kappa\,\mu = 3\,H^2 = \frac{3\,(hH_0)^2}{\tau^2}, \tag{21}$$

$$T = T^* - \frac{\gamma_1}{\tau^{1/\gamma_1}}, \qquad \gamma_1 = \frac{\gamma}{2-\gamma} \tag{22}$$

where $H_0 = 100\,\mathrm{km/(sec\,Mpc)}, h = 0.7$ and T^* is a dimensionless constant denoting the asymptotic value of T (we have then in (6) the choice $c_0 = \frac{3}{2}\gamma\, h$).

Notice that the assumptions (20-22) have been obtained from the FLRW equations (4) and (5) and yield a power law form for the function a (equivalent to the FLRW scale factor). Therefore, following [15], this form of the homogeneous DE fluid is equivalent to a scalar field with an exponential potential following the so–called “scaling law”.

The density of the dust component and the generalized Hubble factor are found by inserting (20-22) into (8) and (12)

$$\kappa\rho c^2 = \frac{3\,\gamma^2\,H_0^2}{\tau^2}\,\frac{[2\,T^*\,f\,F + (1/\gamma)\,(f+F)]\,\tau^{1/\gamma_1} + \gamma_2\,f\,F}{\left[(1+f\,T^*)\,\tau^{1/\gamma_1} - \gamma_1\,f\right]\,\left[(1+F\,T^*)\,\tau^{1/\gamma_1} - \gamma_1\,F\right]}, \tag{23}$$

$$\gamma_2 = \frac{2-3\,\gamma}{2-\gamma}, \tag{24}$$

$$\mathcal{H} = \frac{h\,H_0}{\tau}\left\{1+\gamma\,\frac{\left[(1+F\,T^*)\,\tau^{1/\gamma_1} - \gamma_1\,F\right]f + \frac{3}{2}\,(F-f)\,\tau^{1/\gamma_1}\,\cos^2\psi}{\left[(1+f\,T^*)\,\tau^{1/\gamma_1} - \gamma_1\,f\right]\,\left[(1+F\,T^*)\,\tau^{1/\gamma_1} - \gamma_1\,F\right]}\right\}. \tag{25}$$

From (20-22), we see that R scales as t^k, so that $k > 1/3$ for $\gamma < 2$, hence using (21) and (24) and assuming an arbitrary but finite f and F, we obtain for $\tau \gg 1$

$$\frac{\rho\,c^2}{\mu} \to \frac{2\,T^*\,f\,F + (1/\gamma)\,(f+F)}{(1+f\,T^*)\,(1+F\,T^*)}\,\frac{3\,\gamma^2}{\tau^{1/\gamma_1}} \to 0, \quad \mathcal{H} \to \frac{h\,H_0}{\tau} = H\,, \tag{26}$$

indicating that for all cosmic observers the mixture homogenizes and isotropizes as the homogeneous DE fluid dominates asymptotically over the cold DM component.

5 Numerical Exploration

Having found T, H and $a = R/R_0$ for the particular case of a gamma law (19), we only need to select the function $f = f(r)$ in order to render the models fully determinate. A convenient form for f is

$$f \;=\; f^* + \frac{\delta}{1+r^2}, \tag{27}$$

where $\delta > 0$ so that $f(0) = f^* + \delta > f^*$. Notice also that asymptotically as $r \to \infty$, we have: $\mathcal{M} \to r$ if $f^* = 0$ and $\mathcal{M} \to r^3$ if $f^* > 0$.

The regularity of models requires ρ, $\sqrt{g_{rr}}$, $\sqrt{g_{\theta\theta}}$ to be non–negative, but $\sqrt{g_{rr}} = 0$ must not occur in all the range in which $\sqrt{g_{\theta\theta}} > 0$. The regular evolution range for the models is the coordinate range (τ, r) for which this condition holds with a, H, T given by (19-22) and f, F given by (8) and (27). Thus, the coordinate surface $1 + fT = 0$ marks a non–simultaneous initial central singularity or "big–bang", we can take then the big bang time as the value $\tau = \tau_{\rm bb}$ at this surface corresponding to $r \to \infty$, that is:

$$\tau_{\rm bb}^{1/\gamma_1} = \frac{\gamma_1\,(f^* + \delta)}{1 + (f^* + \delta)\,T^*} \tag{28}$$

Thus, considering the "age of the universe" roughly as $\Delta t_0 \approx 14$ Gys and $h \approx 0.7$, the present cosmic era corresponds to

$$\tau_0 = \tau_{\rm bb} + \frac{3}{2}\gamma\, h\, H_0\, \Delta t_0 \approx \tau_{\rm bb} + 3.17\,\gamma \tag{29}$$

Since the big bang surface $\sqrt{g_{\theta\theta}} = 0$ is not simultaneous, for any hypersurface $\tau =$ constant, the regions near the center at $r = 0$ will be "younger" than those asymptotically far at large values of r.

The parameter $\delta = f_{(c)} - f^* > 0$ in (27) provides a measure of inhomogeneity contrast, or "spatial" variation of all quantities along the rest frames ($t =$ constant) between the symmetry center $r = 0$ and $r \to \infty$. Thus, a sufficiently large/small value of δ makes the values at $r = 0$ and $r \to \infty$ sufficiently close/far to each other, thus indicating small/large "contrast" or degree of inhomogeneity.

The appropriate numerical value for the asymptotic constant, f^*, can be found by demanding that the cosmological observational parameters $\Omega_{\rm DE}$, $\Omega_{\rm DM}$, q, evaluated in the cosmic background ($r \to \infty$) at the present era ($\tau = \tau_0$), take (for a given γ) reasonably close values to those currently accepted from observational data. Since (25) with $T^* = 0$ and f given by (27), evaluated at $\tau = \tau_0$ and $r \to \infty$, is independent of δ and ψ, we can plot $\Omega_{\rm DE}$, $\Omega_{\rm DM}$, q as functions of f^* and γ. As Figures 1a, 1b and 1c illustrate, the desired value of f^* for any given γ can be selected so that the forms (13), (13) and (14) at $\tau = \tau_0$ and $r \to \infty$ yield:

$$0.6 < \Omega_{\rm DE} < 0.7, \quad 0.3 < \Omega_{\rm DM} < 0.4, \quad -0.5 < q < -0.4\,. \tag{30}$$

In particular, if we select $\gamma = 0.15$, an appropriate value is $f^* \approx 100$, leading to $\Omega_{\rm DE} \approx 0.64$, $\Omega_{\rm DM} \approx 0.35$ and $q \approx -0.5$. Notice that $\Omega_{\rm DE} + \Omega_{\rm DM} \approx 1$, but the present Omega for DM would be slightly higher than the currently accepted value $\Omega_{\rm DM} \approx 0.3$. Thus, we could argue that these parameter values would

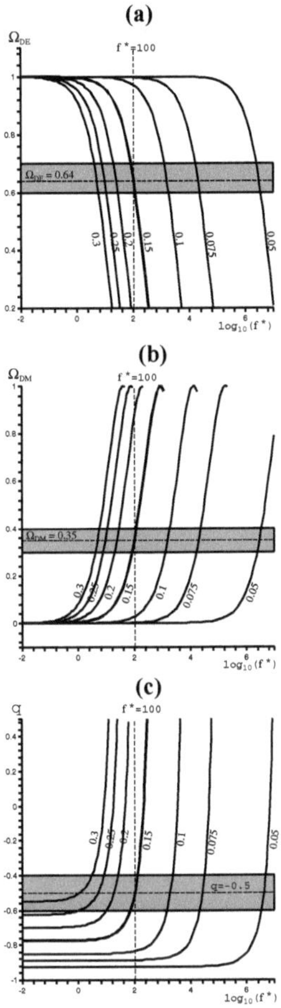

Fig. 1. The asymptotic value of f is given by f^*. Assuming $T^* = 0$ and an arbitrary δ, we can find a suitable value for f^*, for any given γ, by demanding that the observational parameters Ω_{DE}, Ω_{DM}, q (panels (a),(b) and (c)) have appropriate "present" cosmological values as $r \to \infty$ for $\tau = \tau_0$ (gray stripes). Each curve is marked by a given value of γ. Notice that for $\gamma = 0.15$, the choice $f^* \approx 100$ yields $\Omega_{\mathrm{DE}} = 0.64$, $\Omega_{\mathrm{DM}} = 0.35$, $q = -1/2$, which are reasonably close to currently accepted observational data. For γ closer to 0 (cosmological constant), we would have to select larger values for f^*, while larger γ close to 0.3 correspond to $f^* \sim 1$.

become a very accurate approximation to actually inferred cosmological parameters if we would consider ρ as the compound density of DM and baryonic matter.

The effects of anisotropy emerge in the dependence of $\mathcal{H}$, as given by (25), on the off–center "observation angle" ψ. This implies dependence on ψ for Ω_{DE}, Ω_{DM}, q. Considering the free parameter values

$$T^* = 0, \quad f^* = 100, \quad \delta = 200, \quad \gamma = 0.15, \tag{31}$$

complying with the cosmic background ranges (30-30), Figure 2 displays Ω_{DE} and Ω_{DM}, evaluated at $\tau = \tau_0$, as a function of $\log_{10} r$ for assorted fixed values of ψ. The same profiles of Ω_{DE} and Ω_{DM} occur for $\psi = 0$ and $\psi = \pi$ (thick black lines) and, in general for any two values of ψ that differ by a phase of π, with the highest "peaks" corresponding to gray curves with $\psi = \pi/2$ and $\psi = (3/2)\pi$. This singles out two "preferential" distinctive directions:

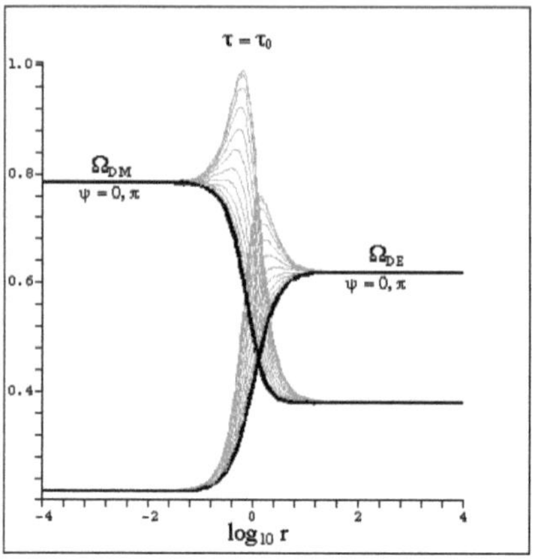

Fig. 2. The effects of anisotropy through the dependence of the profiles of $\Omega_{\rm DE}$ and $\Omega_{\rm DM}$, along $\tau = \tau_0$, on the off–center "observation" angle ψ. The thick black curves correspond to $\psi = 0, \pi$, while the gray curves provide the profiles of $\Omega_{\rm DE}$ and $\Omega_{\rm DM}$ for various other values of ψ. In particular, the larger "peaks" of these parameters occur for $\psi = \pi/2, 3\pi/2$ and around $r \sim 1$. Notice how these effects of anisotropy are negligible in the overdensity region around $r = 0$ and in the cosmic background for large r. For the deceleration parameter q these effects are negligible for all r.

one along the axis $\psi = 0, \pi$ and the other along $\psi = \pi/2, (3/2)\pi$. This is a clear representation of a quadrupole pattern, as expected for a geodesic but shearing 4-velocity [17, 18]. Also, as revealed by Figure 2, the curves for the various ψ differ from each other only in the transition region near $r \sim 1$, thus the effects of this quadrupole anisotropy are negligible near the center of the local overdensity and in the cosmic background asymptotic region.

The effects of inhomogeneity are illustrated by Figures 3a, 3b and 3c, displaying $\Omega_{\rm DE}$, $\Omega_{\rm DM}$ and q as functions of τ and $\log_{10} r$ for $\psi = 0, \pi$ and using the free parameters (31). It is particularly interesting to remark how as τ grows we have: $\Omega_{\rm DE} \to 1, \Omega_{\rm DM} \to 0$ and $q \to -1/2$ for all r, as expected for a DE dominated asymptotically future scenario and associated with an ever accelerating universe that follows a "repulsive" dynamics. However, in the present cosmic era (thick black curve) this repulsive accelerated dynamics on which $\Omega_{\rm DE}$ dominates and $q < 0$ only happens in the cosmic background region, with $\Omega_{\rm DM} > \Omega_{\rm DE}$ and $q > 0$ (*i.e.* "attractive" dynamics) in the local overdensity region with a relatively large DM density contrast $\delta = 200$.

Conditions (15) and (15) place stringent limits on large scale deviations from homogeneity and anisotropy, but these bounds do not apply to local values of these quantities. If we plot these quantities (see [16]) for the cases depicted in the previous figures, all characterized by (19), (20-22), (27) and (31), and even considering the relatively large value $\delta = 200$, condition (15) holds throughout most of the coordinate range (τ, r) including the far range "cosmological background" region of large r and the local overdensity region near the symmetry center $r = 0$, so that for the present cosmic time $\tau = \tau_0$ it only excludes the relatively small scale local region around $r \sim 1$ that marks the "transition" from the local overdensity to the cosmic background. However, a choice like $\delta = 0.01$ would yield similar level curves, but with values three orders of magnitude smaller, thus denoting a state of almost global homogeneity, since (15) would hold in almost all local scales in $\tau = \tau_0$. A graph that is qualitatively very similar emerges for condition (15).

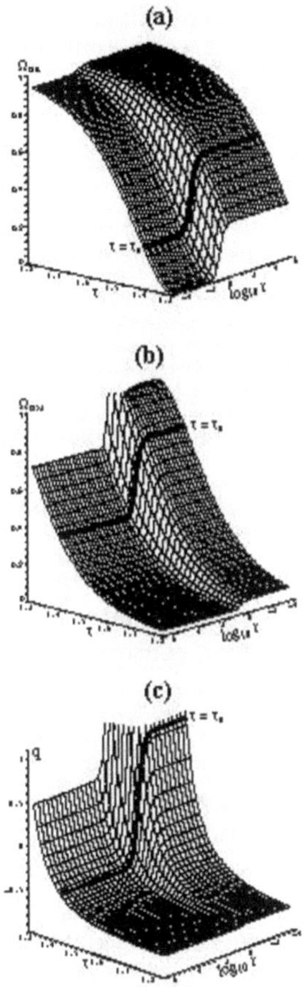

Fig. 3. The full dependence of $\Omega_{\rm DE}$, $\Omega_{\rm DM}$ and q on τ and r (panels (a), (b) and (c)) for $\psi = 0$. The thick black curves denote the hypersurface of present cosmic time $\tau = \tau_0$. The figures clearly show the differences among the overdense region around $r = 0$, the cosmic background for large r and the transition zone between them around $r \sim 1$. As in the previous figure, it is evident that DE dominates over DM in the cosmic background, but DM dominates over DE in the center, with both converging as $\tau \to \infty$ to asymptotically homogeneous states with values $\Omega_{\rm DE} \to 1$, $\Omega_{\rm DM} \to 0$. Notice in (c) how the deceleration parameter, q, is negative in the cosmic background (accelerating universe), but is positive in the overdense region where the dynamics of local gravity should not be repulsive.

This difference between the dynamics of local inhomogeneities and that of the cosmic background cannot be appreciated in such a striking and spectacular way if one examines DE and DM sources by means of the usual FLRW models and their linear perturbations.

6 Interaction Between the Mixture Components

As we mentioned before, the two mixture components: DM (inhomogeneous dust) plus DE (homogeneous fluid) are not separately conserved. Considering (9) and (10), the energy balance for the total energy–momentum tensor, $\dot{e} + (e+p)\,\Theta = 0$, can be written in terms of the DM and DE components as

$$\left[\dot{\rho} + \rho\left(3\,H + \mathcal{Z}\,\dot{T}\right)\right] + \left[\dot{\mu} + (\mu + p)\left(3\,H + \mathcal{Z}\,\dot{T}\right)\right] = 0, \qquad (32)$$

Since each term in square brackets in the left hand side of (32) corresponds to the energy balance of each mixture component alone, a self–consistent form for describing the interaction between the latter is given by

$$\dot{\mu} + (\mu + p)\left[3\,H + \mathcal{Z}\,\dot{T}\right] = -Q, \quad \dot{\rho} + \rho\left[3\,H + \mathcal{Z}\,\dot{T}\right] = Q, \tag{33}$$

where $Q = Q(t,r)$ is the interaction term. Since the physics behind DM and DE remains so far unknown, we cannot rule out the existence of such interaction. Notice that once a given model has been determined by specifying an "equation of state" $p = p(\mu)$ and a form for $f = f(r)$, as we have done in the previous sections, this interaction term would also be fully determined. In general, if the interaction term in (33-33) is a negative valued function, then DM transfers energy into the DE and viceversa. Considering the free parameters given by (19), (27) and (31), we plot in Figure 4 the interaction term Q in (33-33), as a function of τ and $\log_{10} r$. Notice that Q is initially positive and remains so today (DE transfers energy to DM at $\tau = \tau_0$) but will change sign in a future time (DM transfers energy to DE), tending to zero asymptotically as $\tau \to \infty$. The time–asymptotic state is that of only gravitational interaction between DE and DM (*i.e.* separate conservation of each component).

However, the relevant question is not so much the explicit functional form of Q, but its interpretation in terms of a self–consistent physical theory that would be regulating the interaction between DM and DE. In fact, one of the challenges of modern cosmology is to propose such a self–consistent theoretical model of this interaction, while agreeing at the same time with the experimental and observational data. In this context, the interaction between DM and DE has been considered, using homogeneous FLRW cosmologies, in trying to understand the so–called "coincidence problem", that is, the suspiciously coincidental fact the DE and DM energy densities are of the same order of magnitude in our present cosmic era [5, 6, 7, 8, 9, 10, 11].

If the homogeneous DE fluid corresponds to a quintessense scalar field, $\phi = \phi(t)$, with self-interaction potential $V(\phi)$, we have instead of (19):

$$\mu = \frac{\dot{\phi}^2}{2} + V(\phi), \qquad p = \frac{\dot{\phi}^2}{2} - V(\phi). \tag{34}$$

In this case, the interaction term in (33-33) can be associated with a well motivated non–minimal coupling to the DM component. Consider a scalar-tensor theory of gravity, where the matter degrees of freedom and the scalar field are coupled in the action through the scalar-tensor metric $\chi(\phi)^{-1}g_{ab}$[21]:

$$S_{\mathrm{ST}} = \int d^4x\sqrt{|g|}\{\frac{R}{2} - \frac{1}{2}(\nabla\phi)^2 + \chi(\phi)^{-2}\mathcal{L}_m(\nu, \nabla\nu, \chi^{-1}g_{ab})\}, \tag{35}$$

where $\chi(\phi)^{-2}$ is the coupling function, $\mathcal{L}_m$ is the matter Lagrangian and ν is the collective name for the matter degrees of freedom. Equations (33-33) become

$$\ddot{\phi}+\dot{\phi}\left[3H+\mathcal{Z}\dot{T}\right] = -\frac{\mathrm{d}V}{\mathrm{d}\phi}+\frac{\rho}{2\dot{\phi}}\frac{\dot{\chi}}{\chi},\quad \dot{\rho}+\rho\left[3H+\mathcal{Z}\dot{T}\right] = -\frac{\rho}{2}\frac{\dot{\chi}}{\chi}. \tag{36}$$

so that, the coupling function $\chi(\phi)$ and the interaction term Q are related by

$$Q = -\frac{\rho}{2}\frac{\dot{\chi}}{\chi}. \tag{37}$$

Therefore, once we determine a given model, so that Q can be explicitly computed, we can use (37) to find the coupling function χ that allows us to relate the underlying interaction with the theoretical framework associated with the action (35).

For the models under consideration, we can integrate (in general) the constraint (37) with the help of (1), (4-6), (8), (9), (36) and using $\Theta = (d/dt)\left[\ln(Y^2\, Y')\right]$. This yields

$$\chi^{-1/2} = \xi(r)\,\rho Y^2 Y',= \frac{2c_0\,H_0\,r^2\,\xi(r)}{3\,a^3}\frac{d}{dt}\left[(2\,f\,F\,T+f+F)\,R^3\right], \tag{38}$$

where $\xi(r)$ is an arbitrary function that emerges as a constant of integration. Notice that the models require $\phi = \phi(t)$, so that the assumption $\chi = \chi(\phi)$ implies $\chi = \chi(t)$. However, from (38), we have in general $\chi = \chi(t,r)$, with the case $\chi = \chi(t)$ occurring for the following particular cases, associated with very special forms of f and ξ

$$F = 0, \qquad f \propto r^{-3/2},\quad \xi \propto r^{-1/2}, \Rightarrow \chi^{-1/2} \propto H, \tag{39}$$

$$f+F = 0, \qquad f \propto r^{-3},\quad \xi \propto r^{4}, \Rightarrow \chi^{-1/2} \propto \dot{T}+3TH \tag{40}$$

or, if $F \neq 0$ and $f+F \neq 0$, then f and ξ must be obtained from the constraints:

$$\frac{3}{2}\frac{f''}{f}+r\frac{f'^2}{f^2}+\frac{3f'}{f}+\frac{6}{r} = 0, \qquad \frac{\xi'}{\xi} = -\frac{6f+6rf'+r^2 f''}{r(r\,f'+3f)}. \tag{41}$$

However, since (39-40) and (41) yield very special forms of f, ξ and of χ, we prefer to apply (35) under the most general assumption that the coupling function χ should be a function of ϕ and of position, *i.e.* $\chi = \chi(\phi(t), r)$, as given by (38) for suitable forms of the free functions f, H and T (thus, ϕ and $V(\phi)$), hence ξ can be considered a wholly arbitrary free function.

However, we should point out that relating the interaction term, Q, to the formalism represented by (35) is strictly based on the formal similitude between the field equations derived from the action (35), on the one hand, and equations (33-33) and (34), on the other. Also, the interaction between DM and DE in the context of (35) is severely constrained by experimental tests in the solar system [20]. A more detailed and careful examination of the relation between χ and ϕ that incorporates properly these points, as well as the application of (35) to the models presented here, will be undertaken in future papers (see also [16]).

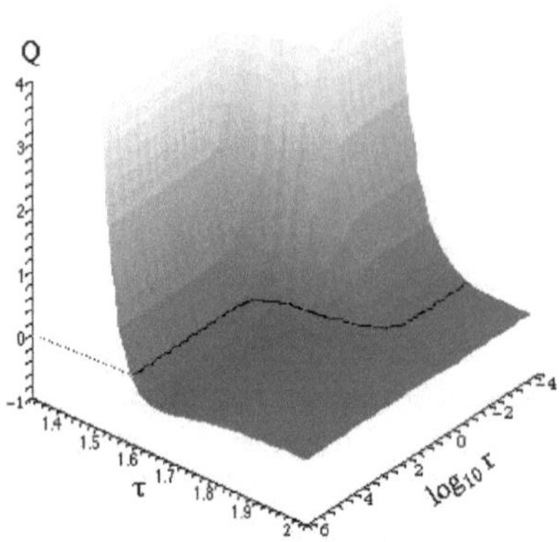

Fig. 4. The interaction function $Q(\tau, r)$. The level curve $Q = 0$ is the thick black curve. Notice how $Q > 0$ in the present cosmic era (DE transfers energy to DM), then changes sign (DM transfers energy to DE), with $Q \to 0$ for large τ, thus indicating an asymptotic state in which DE and DM only interact gravitationally.

7 Conclusion

We have presented a class of inhomogeneous cosmological models whose source is an interacting mixture of DM (dust) and a generic DE fluid. The relevance of the present paper emerges when we realize that there are surprisingly few studies in which DE and DM are the sources of inhomogeneous and anisotropic spacetimes (see [22]), as practically all study of the dynamical evolution of these components is carried on in the context of homogeneous and isotropic FLRW cosmologies or linear perturbations on a FLRW background. There are also very few papers that examine the possibility of non–gravitational interaction between DE and DM.

Once we assume or prescribe an "equation of state" (*i.e.* a relation between pressure, p, and matter–energy density, μ, of the DE fluid), we have a specific DE model (quintessense scalar fields, tachyonic fluid, etc) and all the time–dependent parameters can be determined by solving differential equations reminiscent of FLRW fluids. Since the spatial dependence of all quantities is governed by the function $f = f(r)$, once the latter is selected the models become fully determined. In order to work out this process we chose the simple "gamma law" equation of state: $p = (\gamma - 1)\mu$ (equation (19)), leading to analytic forms for all relevant quantities, including the main observational parameters, $\Omega_{\rm DE}$, $\Omega_{\rm DM}$ and q. Our choice for a DE fluid complying with (19) is equivalent to a scalar field with exponential potential, satisfying a scaling power law dependence on t [15]. Although this is a very idealized quintessense model, our aim has been to use it as a guideline to illustrate how more sophisticated DE scenarios can be incorporated in future work involving the models.

As shown in Figures 3, the models homogenize and isotropize asymptotically in cosmic time for all fundamental observers and/or assumptions on the DE fluid, thus they are well suited for studying the interaction between DE and DM in the context of the evolution of large scale inhomogeneities (of the order of the scale of homogeneity $\sim 100 - 300$ Mpc). By selecting appropriate boundary conditions (form of f), we can examine inhomogeneities at various scales and/or asymptotic conditions (see Figures 2 and 3). In particular, we have explored the case of a local DM overdense region, whose scale can be arbitrarily fixed and with an asymptotic behavior that accurately converges to a cosmic background characterized by observational parameters that fit currently accepted observational constraints: $0.6 < \Omega_{\rm DE} < 0.8$, $0.2 < \Omega_{\rm DM} < 0.4$ and $q \approx -0.5$ (see Figure 4). As illustrated by the various graphical examples that we have presented, this interplay between a local overdensity, a cosmic background and a transition region between them, shows in a spectacular manner how inhomogeneity and anisotropy lead to interesting and important information that cannot be appreciated in models based on FLRW metrics and/or linear perturbations. For example, as revealed by Figure 2, the effect of anisotropy emerges as a dependence of observational parameters on local observation angles in "off center positions", an effect which is only significant in the transition between the overdensity and the cosmic background. On the other hand, as shown by Figures 3, inhomogeneity allows for radically different ratios between DM and DE in the overdensity and the cosmic background, so that DM dominates over DE, locally, in the overdense region, as a contrast with DE dominating DM, asymptotically, in the cosmic background (as expected). Also, while q is negative in the cosmic (DE dominated) background, thus denoting the expected "repulsive" accelerated expansion at large scales, we have $q > 0$ along smaller scales in the local overdensity. For all parameters there is a smooth convergence between local and asymptotic values in the transition region.

We have also examined the non–gravitational interaction between DE and DM. Plotting the interaction term, Q, (Figure 4) shows that energy flows from DE to DM at the present cosmic era, with the flow reversing direction in the future and evolving towards an asymptotic future state characterized by pure gravitational interaction: $Q \to 0$. If we take the DE fluid to be a quintessense scalar field, the DE vs DM interaction can be incorporated to the theoretical framework of an action like (35), associated with a non–minimal coupling of scalar fields and DM. Since DM is inhomogeneous while the scalar field is homogeneous, only for some particular forms of spacial dependence (*i.e.* the function f) we obtain a coupling function expressible as $\chi = \chi(\phi)$. In general, we have to allow for the possibility that $\chi = \chi(\phi(t), r)$. However, we have examined this interaction just in qualitative terms, with the purpose of illustrating the methodology to follow in future applications. As guidelines for future work, we have the application of the models to more sophisticated and better motivated DE formalisms, perhaps in the context of the "coincidence" problem [5, 6, 7, 8, 9, 10, 11]. See [16] for a more comprehensive version of this article.

Acknowledgements

RAS acknowledge financial support from grant PAPIIT–DGAPA–IN117803.

References

1. R.R. Caldwell, R. Dave and P.J. Steinhardt, *Phys. Rev. Lett.*, **80**, 1582, (1995); M.S. Turner and M. White, *Phys. Rev.* D **56**, R4439, (1997); Bahcall *et al*, *Science*, **284**, 1481, (1999); A.G. Riess, *et al.*, Astron. J. **116**, 1009-1038 (1998); S. Perlmutter, *et al.*, Astrophys. J. **517**, 565-586 (1999).
2. A.G. Riess, *et al.*, *Astrophys. J.*, **536**, 62 (2000); A.G. Riess, *et al.*, *Astrophys. J.*, **560**, 49-71 (2001); J.L. Tonry, *et al.*, *Astrophys. J.*, **594**, 1-24 (2003); A.G. Riess, *et al.*, *Astrophys. J.*, **607**, 665-687 (2004); M. Tegmark *et. al.* astro-ph/0310723; A. Upadhye, M. Ishak and P.J. Steinhardt, astro-ph/0411803.
3. T. Padmanabhan, *Phys Rept*, **380**, 235–320, (2003).
4. J.A.S. Lima, *Braz.J.Phys.*, **34**, 194–200, (2004).
5. L. Amendola, Phys. Rev. D**62** (2000) 043511; 063508 (astro-ph/0005070).
6. L.P. Chimento, A.S. Jakubi and D. Pavón, *Phys. Rev. D*, **62**, 063508, (2000).
7. W. Zimdahl, D. Pavón and L.P. Chimento, *Phys Lett B*, **521**, 133, (2001).
8. L.P. Chimento *et. al.*, *Phys Rev D*, **67**, 083513, (2003).
9. L.P. Chimento and A.S. Jakubi, *Phys Rev D*, **67**, 087302, (2003); L.P. Chimento, A.S. Jakubi and D. Pavon, *Phys. Rev. D*, **67** (2003) 087302.
10. W. Zimdahl and D. Pavón, *Gen Rel Gravit*, **35**, 413, (2003).
11. L.P. Chimento and A.S. Jakubi, *Phys Rev D*, **69**, 083511, (2004).
12. A. Krasinski, *Inhomogeneous Cosmological Models*, Cambridge Univ. Press 1997.
13. J.A.S. Lima and J. Tiomno, *Gen Rel Gravit*, **20**, 1019, (1988).
14. R.A. Sussman, *Classical and Quantum Gravity*, **9**, 1881-1915, (1992).
15. J. J. Halliwell, *Phys. Lett. B*, **185**, 341 (1987); C. Wetterich, *Nucl. Phys. B*, **302**, 668 (1988); D. Wands, E. J. Copeland, and A. R. Liddle, *Ann. N.Y. Acad. Sci.*, **688**, 647 (1993). P. G. Ferreira and M. Joyce, *Phys. Rev. D*, **58**, 023503 (1998); L.P. Chimento and A.S. Jakubi, *Int J Mod Phys D*, **5**, 313, (1996).
16. R.A. Sussman, I Quiros, OM González, *Gen. Rel. Grav.* **37** (2005) 2117-2143.
17. G.F.R Ellis and H. van Elst, Cosmological Models, Cargèse Lectures 1998, gr-qc/9812046.
18. N. Humphreys, R. Maartens and D. Matravers, *Ap J*, **477**, 47, (1997).
19. R. Maartens, G.F.R. Ellis and W.R. Stoeger, *Phys Rev D*, **51**, 1525-1535, (1995); J. Barrow and R. Maartens, *Phys Rev D*, **59**, 043502, (1999).
20. C.M. Will, *Theory and Experiment in Gravitational Physics* (Cambridge University Press, 1993); gr-qc/0103036.
21. N. Kaloper and K.A. Olive, *Phys Rev D* **57** (1998) 811-822 (hep-th/9708008); L. Amendola, *Phys. Rev. Lett.* **93** (2004) 181102 (hep-th/0409224).
22. L.P. Chimento, A.S. Jakubi and D. Pavón, *Phys Rev D*, **60**, 103501, (1999).

Mini-review on Scalar Field Dark Matter

L. Arturo Ureña–López[1]

Instituto de Física de la Universidad de Guanajuato, A.P. E-143, C.P. 37150, León, Guanajuato, México.<lurena@fisica.ugto.mx>

Our aim in this review is to briefly describe the main results in the topic of scalar field dark matter, in which a scalar field is the dark matter particle. We will start with a revision of the concept of scalar fields at the classical and quantum levels. We will then continue with some of the relevant solutions for self-gravitating scalar fields in a cosmological setting. At the end, we will discuss the constraints cosmological observations impose upon the free parameters of a scalar field dark matter model.

1 What is a Scalar Field?

The properties of any system are given by its Lagrangian, and for a *classical* scalar field $\phi(t,\mathbf{x})$ minimally coupled to gravity is given by [1, 2, 3]

$$\mathcal{L}_\phi = -\sqrt{-g}\left[\frac{1}{2}\partial^\mu\phi\partial_\mu\phi + V(\phi)\right] , \tag{1}$$

where $V(\phi)$ is called the *scalar potential*, and encodes in itself all of the scalar field self-interactions. We shall assume that the scalar field lives in a curved spacetime which is described by the metric $g_{\mu\nu}$ with signature $(-1,1,1,1)$ (our units are such that $c=1$), and $g = det(g_{\mu\nu})$.

The equation of motion for the scalar field, also known as the Klein-Gordon (KG) equation, arises from the Euler-Lagrange equations, and then is of the form

$$\Box\phi = -\frac{dV(\phi)}{d\phi} , \tag{2}$$

where we identify the (covariant) d'Alembertian operator $\Box \equiv (1/\sqrt{g})\partial_\mu (\sqrt{-g}\partial^\mu)$.

The scalar potential $V(\phi)$ depends on the kind of scalar system we would like to describe. The simplest choice is the *free scalar field* potential [1], $V(\phi) =$

[1] The KG equation can be seen as the relativistic version of the Schrödinger equation, in the following manner. If the free particle version of the latter appears from

A. Carramiñana et al. (eds.), Solar, Stellar and Galactic Connections between Particle Physics and Astrophysics, 295–302.

$(1/2)m^2\phi^2$, where m is called the mass of the scalar field[2]. For simplicity, we will work out the free case only, as it will help us to illustrate the properties of a scalar field in a cosmological setting.

1.1 Quantization of a Free Scalar Field

The accepted interpretation of scalar fields is that their excitations, once we *quantize* them, represent *particle states* so that the single-particle interpretation of the wave function ϕ should be abandoned. In this section, we briefly describe the quantization of scalar fields, (for details see [1, 2, 3]).

To begin with, we consider the (flat) Minkowski spacetime in which $g_{\mu\nu} = \eta_{\mu\nu} \equiv diag(-1,1,1,1)$. Thus, the KG equation is just the wave equation for a massive particle, and then the general solution is given as a superposition of (properly normalized) plane waves of the form $\phi_p(t,\mathbf{x}) \propto e^{i\mathbf{p}\cdot\mathbf{x}-iEt}$, with momentum $\mathbf{p}$ and energy $E^2 = p^2 + m^2$ ($p = |\mathbf{p}|$). This solution reinforces the interpretation of parameter m as the *mass* of the scalar field.

At the *quantum* level, the scalar field ϕ is taken as a *quantum operator*, so that its expansion is of the form

$$\hat{\phi}(t,\mathbf{x}) = \sum_p \left[\hat{a}_p \phi_p(t,\mathbf{x}) + \hat{a}_p^\dagger \phi_p^*(t,\mathbf{x})\right] , \tag{3}$$

where we recognize the annihilation $\hat{a}_p$ and creation $\hat{a}_p^\dagger$ operators which obey the usual commutation relations. We should think of functions $\phi_p(t,\mathbf{x})$ as a complete basis of functions formed with the solutions of the KG equation (2). That is, the quantum scalar field is a *time-dependent quantum operator* (within the so-called Heisenberg picture), and the coefficients in (3) are solutions of the *classical* KG equation (2).

Defining the number operator $\hat{N}_i = \hat{a}_i^\dagger \hat{a}_i$, the state vectors on which the quantum operators act are such that

$$\hat{N}_i |N_1, N_2, ...\rangle = N_i |N_1, N_2, ...\rangle , \tag{4}$$

where N_i will represent the number of particles with momentum $\mathbf{p}_i$. The aforementioned states can be constructed if we assume that there is a vacuum state $|0\rangle = |0,0,...\rangle$, and then we obtain the other states according to the known formula for creation and annihilation operators

$$\hat{a}_i^\dagger |N_1, N_2, .., N_i, ..\rangle = \sqrt{N_i + 1} |N_1, N_2, .., N_i + 1, ..\rangle , \tag{5}$$

$$\hat{a}_i |N_1, N_2, .., N_i - 1, ..\rangle = \sqrt{N_i} |N_1, N_2, .., N_i - 1, ..\rangle . \tag{6}$$

The states constructed in this manner are orthonormal, and represent particle states, each composed of many scalar particles of mass m.

---

substituting $E \to i\partial_t$ and $\mathbf{p} \to i\nabla$ in the *classical* energy conservation equation $E = p^2/(2m)$, then the KG equation appears from using instead the *relativistic* equation $E^2 = p^2 + m^2$.

[2] In general, we define the mass of the scalar field as $\partial_\phi^2 V(\phi_c) = m^2$, where ϕ_c denotes the scalar field at the minimum of the potential.

1.2 Self-gravitating Scalar Fields

All of the above formalism can be generalized to the case of an arbitrary curved spacetime with metric $g_{\mu\nu}$. We would like to stress in here that expansion (3) can be used in any spacetime, curved or not. Hence, functions ϕ_p are (appropriately) orthonormal functions that form a complete basis of solutions of the KG equation in a curved spacetime. The rest of the formalism is preserved: assuming the existence of a vacuum state, a set of quantum states can be constructed using the creation and annihilation operators[3].

The gravitational interaction for classical scalar fields is taken into account by the Einstein equations, which are generically written in the form $G_{\mu\nu} = 8\pi G T_{\mu\nu}$. The so called Einstein tensor $G_{\mu\nu}$ is constructed from the geometrical properties of the spacetime, while $T_{\mu\nu}$ is the energy-momentum tensor of matter. For a classical scalar field, the energy-momentum tensor can be obtained from the Lagrangian (1), and then

$$T_{\mu\nu} = \phi_{;\mu}\phi_{;\nu} - \frac{1}{2} g_{\mu\nu} \left[\phi^{;\sigma}\phi_{;\sigma} + 2V(\phi)\right] . \tag{7}$$

However, if we are taking scalar fields as quantum operators, so is the energy-momentum tensor, and then the *expectation value* of the latter is what appears on the r.h.s. of the Einstein equations; this is known as *semiclassical approximation* that have revealed so many surprises.

With this in mind, we can define the expectation values of the energy density and momentum density flux, respectively, as $\rho_\phi = -\langle|\hat{T}^0_0|\rangle$ and $P^i = \langle|\hat{T}^{i0}|\rangle$, where the bra-ket operation is done over all of the quantum number states, that is,

$$\langle| \; |\rangle \equiv \sum_l \langle N^{(l)}_1, N^{(l)}_2, ..| \; |N^{(l)}_1, N^{(l)}_2, ..\rangle_l \, , \tag{8}$$

where the l-state has $N^{(l)}_i$ particles with momentum p_i.

When working with scalar fields, whether classical or quantum, we require to solve the classical EKG equations in a curved spacetime. Hence, in the following sections, we will review (typical) classical solutions of the EKG system in Cosmology.

2 Cosmological Solutions

For a homogeneous and isotropic universe, the spacetime is described by the Friedmann-Robertson-Walker metric. For simplicity, we shall only consider the spatially flat case. In such a space time, the scalar field is homogeneous and

[3] The quantum theory in a curved spacetime is far richer than the brief description I have given here; for many more details see the classical text [1].

isotropic too, and then the (basis) functions depend only on time, $\phi = \phi(t)$. The KG equation we have to solve is

$$\ddot{\phi} + 3H\dot{\phi} + m^2\phi = 0 \,, \tag{9}$$

where a dot means derivative with respect to time.

Eq. (9) can be solved easily if there is a regime in which $H < m$. In this case, the scalar field ϕ oscillates with a time scale (given by m^{-1}) shorter than the expansion time scale (given by H^{-1}), so that the universe only feels the energy density averaged on a large number of oscillations. The averaged energy density goes as $\langle \rho_\phi \rangle \sim a^{-3}$, and then the scalar field behaves as *cold dark matter* [4, 5, 6]. This is the main result of this section.

The discussion in the above paragraph is quite general, and applies to any *massive* scalar field, that is, to any scalar field potential with a minimum around which it can be written quadratically in first approximation.

3 (Real) Scalar Field Stars: Oscillatons

We are also interested in studying the kind of gravitationally-bounded objects scalar fields can form. The motivation for this is to investigate whether a scalar field, which in some circumstances can behave as cold dark matter, can also form stable cosmological structure.

It is known that there are long-lived solutions of the EKG equations in a spherically symmetric spacetime, which are generically called *oscillatons* [7]. We will describe below these objects in both the relativistic and Newtonian regimes.

3.1 Relativistic Oscillatons

The most general spherically-symmetric spacetime in the polar-areal slicing reads

$$ds^2 = \alpha^2(t,r)dt^2 + a(t,r)dr^2 + r^2 d\Omega^2 \,, \tag{10}$$

where $d\Omega$ is the usual solid angle element. The EKG equations in this spacetime are simple, but have to be solved numerically. It was a bit surprising that the most general solution can be given as a (convergent) Fourier series of the form [7, 8]

$$\phi(t,r) = \sum_{j=0} \phi_j(r) \cos(j\omega t) \,, \tag{11}$$

and similar expansions are used for the metric functions α and a, too.

Parameter ω is called the *fundamental frequency*, and its values are determined from an eigenvalue problem, which arises once one imposes the boundary conditions of regularity at the center ($\partial_r \phi_j(0) = 0$), and of asymptotic flatness at large distances ($\phi_j(r \to \infty) = 0$). The total mass of each configuration, i.e., for each eigenvalue of ω, is shown in Fig. 1 as a function of a representative radius $R_{\max}$. Notice that there is a maximum point, indicating the existence of a critical configuration. As a matter of fact, the configurations on the left of the critical point are intrinsically unstable, whereas those on the right are long-lived[4].

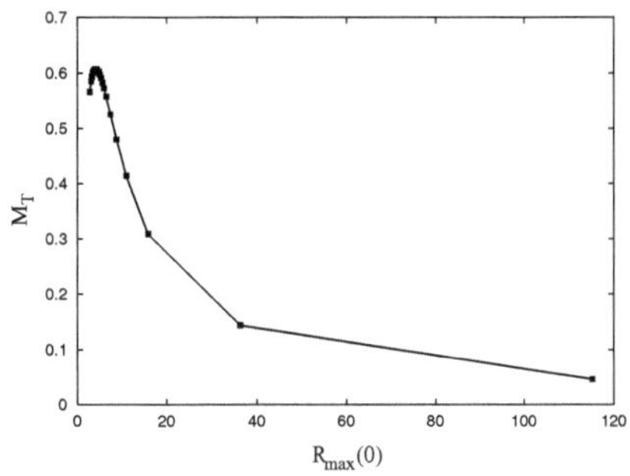

Fig. 1. Each point in this figure represents an oscillaton of total mass M_T and (typical) radius $R_{\max}$. The distance units is given in terms of $m^{-1} = \lambda_C$, where λ_C is the Compton length of the scalar field. The mass units are given by $m_{\rm Pl}^2/m$, where $m_{\rm Pl}$ is the Planck mass. This figure was taken from [8].

3.2 Newtonian Oscillatons

The relativistic configurations described above are in the strong-field regime, and can be very compact objects. However, it is also our interest to study the weak-field limit, in which gravity interactions and velocities ($v/c \ll 1$) are small [9, 10].

One first step is to realize that higher order modes in expansion (11) contribute less and less as the scalar field becomes weak, more precisely, as $\sqrt{8\pi G}\phi \sim \mathcal{O}(v^2/c^2) \ll 1$. Hence, in the weak field limit one expects the contribution of only the first term. It is then convenient to write a weak-field expansion as

$$\sqrt{8\pi G}\phi(t,r) = \psi(t,r)e^{imt} + \text{C.C.}, \tag{12}$$

where ψ is a complex function in general. Notice that the weak-field limit translates into the constraint $|\psi| \sim \mathcal{O}(v^2/c^2) \ll 1$.

Likewise, we have to expand the metric coefficients in order to separate out its weak-field limit contributions. We will not go into the details, but it is enough to expand the g_{tt} metric coefficient in the form

[4] We are being careful of not saying *stable*, since the numerical runs we arranged to test stability can only prove it for finite (though certainly long) times.

$$\alpha^2(t,r) = 1 + 2U(t,r)\,, \tag{13}$$

where function $U(t,r)$ can be identified with the usual Newtonian potential, which is also of order $U(t,r) \sim \mathcal{O}(v^2/c^2) \ll 1$.

Preserving only the terms of order $\mathcal{O}(v^2/c^2) \ll 1$ in the EKG equations, we find the so-called Schrödinger-Newton (SN) equations that govern the dynamics of the scalar field ϕ in the weak-field limit,

$$i\partial_\tau \psi = -\frac{1}{2}\nabla^2\psi + U\psi\,, \tag{14}$$

$$\nabla^2 U = |\psi|^2\,, \tag{15}$$

where we have made use of dimensionless time ($\tau = mt$) and distance variables ($x = mr$), which are normalized by the scalar field mass m.

To better visualize the properties of (weak-field) Newtonian oscillatons, we define a weakness parameter as $\lambda^2 = \sqrt{8\pi G}\phi$, so that $\lambda \sim \mathcal{O}(v/c)$. Thus, it can be verified that the SN system possesses (i.e., is invariant under) the following scaling symmetry

$$\tau,\, r,\, \psi,\, U \rightarrow \lambda^{-2}\tau,\, \lambda^{-1}r,\, \lambda^2\psi,\, \lambda^2 U\,. \tag{16}$$

Such an invariance implies some expected properties of the SN system. To begin with, it confirms that the complex field ψ and the Newtonian potential U are of order $\sim \mathcal{O}(v^2/c^2)$.

But it also tells us about the time and distance scales typical of Newtonian oscillatons. As in the relativistic case, these scales are determined by the mass of the scalar field. In the Newtonian case, though, time and distance scales are larger as the system becomes weaker ($\lambda \rightarrow 0$). In other words, Newtonian oscillatons can be much larger and live much longer that their relativistic counterparts.

Equilibrium configurations can be also found for Newtonian oscillatons, in which the complex field is of the form $\psi = \varphi(r)e^{-i\gamma t}$. Under the same assumptions of regularity at the origin and asymptotic flatness, the solution of the SN system becomes an eigenvalue problem from which we can determine the (eigen) frequency γ. Notice that in order to preserve the scaling symmetry of the SN system, the frequency should scale as $\gamma \rightarrow \lambda^2\gamma$.

Due to the scaling symmetry, all of the possible Newtonian oscillatons can be obtained through a scale transformation applied to a properly chosen equilibrium configuration. For simplicity, one usually chooses the one in which $\varphi(0) = 1$.

We show in Fig. 2 a plot M (mass) vs $R_{\rm max}$ for Newtonian oscillatons.

4 Final Discussion

We have left for this last section the discussion about the parameters of a scalar field model for dark matter. Up to this point, we have described the

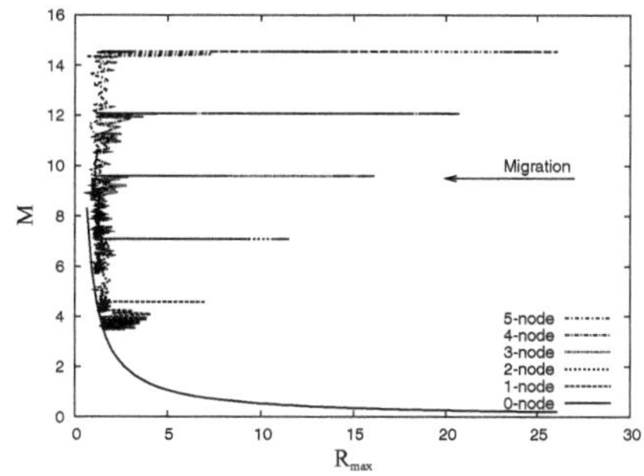

Fig. 2. M vs radius plot for Newtonian oscillatons, for which we have taken a weakness parameter $\lambda = 1$; for other values of it is just necessary to apply the scaling transformation (16). Also shown are the migration paths of other configurations that end up at the line representing 0-node (ground) equilibrium configurations. This shows that (0-node) Newtonian oscillatons are in general attractor solutions. The mass units are given by $m_{\rm Pl}^2/m$, where $m_{\rm Pl}$ is the Planck mass. Figure from [9].

properties of self-gravitating scalar fields endowed with a quadratic scalar potential. We argued that such a potential is a good approximation to any scalar potential with a minimum, and then the relevant free parameter is the scalar field mass m.

We have just to find out which values of the scalar field mass m could have physically interesting consequences. The constraints we have to take into account are the following.

- The standard cosmological model requires the existence of a matter dominated era in order to allow the formation of structure. This in turn implies that the scalar mass cannot be too small, and that $m \geq H_{\rm eq}$, where $H_{\rm eq}$ denotes the value of the Hubble parameter at the time of equivalence between radiation and matter [4, 6].
- On the other hand, real galaxies are Newtonian objects, for which $v/c \sim 10^{-3}$, so that Newtonian oscillatons should be the appropriate scalar objects to describe them, if we think of a scalar field as a good dark matter candidate. Hence, as the size of a Newtonian oscillaton would have to be of the order $(\lambda m)^{-1} > 10\,{\rm kpc}$, this implies that $m \leq 0.1{\rm pc}^{-1} \sim 10^{-22}{\rm eV}$ [10].

Resuming in, we see that the more massive a scalar field is, the earlier that it behaves as cold dark matter. On the other hand, the lighter a scalar field is, the larger a Newtonian oscillaton is, and then the more likely it can fit galactic observations.

Therefore, the two constraints above are complementary: cosmological evolution imposes a lower bound on the scalar mass, whilst galactic evolution imposes on it an upper bound. That is, we cannot expect any scalar field model to fit observations at all scales, but it must be a very particular model.

Our group has determined, through a more detailed study, that a massive scalar field should have a mass of about $m \sim 10^{-23}$eV in order to accomplish all of cosmological observations in a decent manner [4, 5].

However, there may be properties we are not taken into account properly, as extra terms in the scalar potential may add desirable properties and then alleviate the constraints imposed upon the free scalar field. We expect to report on them in a future communication.

Acknowledgments

This work was partially supported by grants from CONACYT (32138-E, 34407-E, 42748), CONCYTEG (05-16-K117-032), and PROMEP UGTO-CA-3.

References

1. N. D. Birrell, P. C. W. Davies: *QUANTUM FIELDS IN CURVED SPACE*, Cambridge, Uk: Univ. Pr. (1982).
2. L. H. Ryder: *QUANTUM FIELD THEORY*, Cambridge, Uk: Univ. Pr. (1985) 443p.
3. David Lyth and Andrew Liddle: *Cosmological Inflation and Large Scale Structure*, Cambridge University Press (2000).
4. T. Matos and L. A. Ureña-Lopez: Int. J. Mod. Phys. **D13**, 2287-2292 (2004).
5. T. Matos et al.: Lect. Notes Phys. **646**, 401-420 (2004).
6. V. Sahni and L-M Wang: Phys. Rev. **D62**, 103517 (2000).
7. E. Seidel and W-M. Suen: Phys. Rev. Lett. **66**, 1659-1662 (1991).
8. M Alcubierre et al.: Class. Quant. Grav. **20**, 2883-2904 (2003).
9. F. S. Guzmán and L. A. Ureña-López: Phys. Rev. **D69**, 124033(2004).
10. F. S. Guzmán and L. A. Ureña-López: Phys. Rev. **68**, 024023 (2003).

Index

Baryons
inventory, 123
Blazars, 227
Bremsstrahlung, 221

Compton scattering, 217–222
inverse Compton scattering, 221
Cosmic Microwave Background, 126–160

Dark Matter, 120–160
non–baryonic, 131

Galaxies, 116–126
anatomy, 119
environment, 123
evolution, 124
interstellar medium, 118
Lyman α, 126
Lyman break, 125
properties, 116
stellar populations, 117
sub-millimeter, 125
taxonomy, 118
Gamma-ray astronomy, 215–228
Cerenkov telescopes, 215, 219
Compton telescopes, 217
EGRET sources, 223
pair production telescopes, 218
water Cerenkov, 220
Gamma-ray bursts, 227

Hubble, Edwin, 115, 118

Interstellar Medium, 118

Large Scale Structure
formation, 126–138

Matter
baryonic, 131

Neutral pions, 222
Neutron stars, 43–75
manifestations, 50
milestones, 44

Pair annihilation, 220
Pair production, 218–220
Photoelectric effect, 216
Pulsars, 43–75
classes, 52
distances, 56
gamma-ray emission, 225
gravity probes, 68
kicks, 56
magnetospheres, 55
surveys, 64
velocities, 56

Stellar evolution
end states, 46
Stellar populations, 117
Supernova remnants, 226
Synchrotron radiation, 222
curvature radiation, 222

Astrophysics and Space Science Proceedings

Diffuse Matter from Star Forming Regions to Active Galaxies, edited by T.W. Hartquist, J.M. Pittard, S.A.E.G. Falle
Hardbound ISBN 978 1-4020-5424-2, December 2006

Solar, Stellar and Galactic Connections between Particle Physics and Astrophysics, edited by A. Carramiñana, F.S. Guzmán, and T. Matos
Hardbound ISBN 978 1-4020-5574-4, December 2006

For further information about this book series we refer you to the following web site: www.springer.com

MIX
Papier aus verantwortungsvollen Quellen
Paper from responsible sources
FSC® C105338

If you have any concerns about our products,
you can contact us on
ProductSafety@springernature.com

In case Publisher is established outside the EU,
the EU authorized representative is:
Springer Nature Customer Service Center GmbH
Europaplatz 3, 69115 Heidelberg, Germany

Printed by Libri Plureos GmbH
in Hamburg, Germany